GREG COWIE

D1575079

Understanding
THE EARTH

Understanding
THE EARTH
A Reader in the Earth Sciences

edited by

I. G. Gass, *Professor of Earth Sciences*
Peter J. Smith, *Senior Lecturer in Earth Sciences*
R. C. L. Wilson, *Senior Lecturer in Earth Sciences*

at the Open University

2ND EDITION

PUBLISHED FOR
THE OPEN UNIVERSITY PRESS
BY
THE ARTEMIS PRESS
SUSSEX   MCMLXXII

Copyright © 1971, 1972, The Open University

All rights reserved. No part of this publication
may be reproduced or transmitted, in any form
or by any means, without the written permission
of the Publisher, except by a reviewer who wishes
to quote brief passages in connection with a review.

FIRST PUBLISHED 1971

REPRINTED TWICE 1971

SECOND EDITION 1972

REPRINTED March, 1973

REPRINTED January, 1974

REPRINTED October, 1975

REPRINTED April, 1977

SBN 85141 308 0 (paperback edition)

SBN 85141 309 9 (library edition)

*The paperback edition of this book is sold subject to the condition
that it shall not, by way of trade or otherwise, be lent, re-sold,
hired out, or otherwise circulated without the publisher's prior
consent, in any form of binding or cover other than that in which
it is published and without a similar condition including this
condition being imposed on the subsequent purchaser.*

Set in 11pt Bembo and printed on Evensyde White Offset

Printed and bound in Great Britain by
Balding & Mansell Ltd., Wisbech, Cambridgeshire.

CONTENTS

Acknowledgements:

We are indebted to Messrs. Branch/Horan Ltd., of Crawley, Sussex, for their intelligent and careful rendering of many of the diagrams in this book; to *The New Scientist* (I.P.C. Magazines Ltd.) for permission to reproduce the article by Lloyd V. Berkner and Lauriston C. Marshall, and that on *The Primitive Earth* by Preston Cloud Jr.; to *Scientific American* for permission to reproduce the article on *The Oxygen Cycle* by Preston Cloud Jr. and Aharon Gibor; and to our printers, Messrs. Balding & Mansell for the detailed attention they gave this book in the press.

A number of illustrations and some statistics were directly based on, or loosely inspired by, material first published by the following authors and organizations, to whom we hereby make grateful acknowledgement on behalf of our contributors and of ourselves:

D. W. Anderson; M. Barazangi and J. Dorman; G. Barrow; W. Bascom; B. J. Bluck and G. Kelling; G. M. Brown; S. W. Carey; C. O. Dunbar and J. Rodgers; F. J. V. Fisher; M. F. Glaessner; N. J. Guest; R. N. Hamilton; J. W. Harbaugh; J. R. Heirtzler *et al.*; A. Holmes; P. M. Hurley; B. Isacks; V. E. Khain and M. V. Muratov; M. A. Khan; W. C. Krumbein; M. Ksiazkiewicz; A. H. Lachenbruch; L. Mansinha and D. E. Smylie; H. Martin; R. G. Mason; A. E. Maxwell *et al.*; P. McCurry; P. M. Melchior; W. J. Morgan; R. O. Muir; W. H. Munk; B. Nagy; D. Nichols; B. J. O'Brien; J. Oliver; N. D. Opdyke; S. W. Richardson; P. C. Rickwood and M. Mathias; D. Rossignol-Strick; R. F. Roy, D. D. Blackwell and E. R. Decker; J. W. Schopf; P. C. Scruton; J. S. Shelton; A. E. Smith; E. F. C. Somerscales and D. Dropkin; M. N. Toksöz; J. W. Minear and B. R. Julian; K. E. Torrance; J. H. F. Umbgrove; L. R. Wager; C. Wang; L. J. Wills; J. T. Wilson; P. A. Wilson; and A. O. Woodford.

The Smithsonian Institution; N. A. S. A. and the United States Embassy, London; The British Museum (Natural History); Scripps Oceanographic Institute; Aerofilms Ltd.; Associated Press Ltd.; The Ecologist (Ecosystems Ltd.); U. S. Bureau of Mines; U. S. National Academy of Science.

ARTEMIS PRESS

The Editors:

Professor I. G. Gass
Dept. of Earth Sciences
The Open University

Dr. Peter J. Smith
Dept. of Earth Sciences
The Open University

Dr. R. C. L. Wilson
Dept. of Earth Sciences
The Open University

The Contributors:

Dr. D. I. Black
I. C. I. Petrochemicals Division
Billingham, Teesside

Professor Sir Edward Bullard, F.R.S.
Dept. of Geodesy and Geophysics
University of Cambridge

Professor Michael Chinnery
Dept. of Geological Sciences
Brown University, Rhode Island

Professor T. N. Clifford
Dept. of Geology
University of the Witwatersrand

Professor Preston Cloud Jr.
Dept. of Geology
University of California

Dr. Keith Cox
Dept. of Geology and Mineralogy
University of Oxford

Dr. D. Davies
Lincoln Laboratory
M. I. T., Massachusetts

Dr. Stephen Drury
Dept. of Earth Sciences
The Open University

Professor Aharon Gibor
Dept. of Biological Sciences
University of California

Dr. Roland Goldring
Dept. of Geology
University of Reading

Daniel S. Greenberg
'Science'
Washington, D.C.

Professor Peter Harris
Dept. of Earth Sciences
University of Leeds

Professor M. R. House
Dept. of Geology
University of Hull

Professor Zdeněk Kopal
Dept. of Astronomy
University of Manchester

Professor Robert L. Kovach
Dept. of Geophysics
Stanford University

Professor Lauriston C. Marshall
Southwest Center for Advanced
Studies, Dallas, Texas

Dr. Brian Mason
Dept. of Mineral Sciences
Smithsonian Institution

Dr. Stephen Moorbath
Dept. of Geology and Mineralogy
University of Oxford

Dr. E. R. Oxburgh
Dept. of Geology and Mineralogy
University of Oxford

Dr. John H. Sass
U.S. Geological Survey
Menlo Park, California

Dr. Alan Gilbert Smith
Dept. of Geology
University of Cambridge

Dr. Peter J. Smith
Dept. of Earth Sciences
The Open University

Professor John Sutton, F.R.S.
Dept. of Geology
Imperial College, London

Professor P. C. Sylvester-Bradley
Dept. of Geology
University of Leicester

Dr. F. J. Vine
Princeton University and
University of East Anglia

Professor E. K. Walton
Dept. of Geology
University of St. Andrews

J. B. Wright
Dept. of Earth Sciences
The Open University

Editor for the Publishers: M. T. Bizony

PREFACE

DURING THE LAST DECADE, there has been a revolution in the Earth sciences, largely due to the fruitful researches of the oceanographic geophysicists, which has led to the wide acceptance that continents drift about the face of the Earth and that sea floor spreads, continually being created and destroyed. Finally, in the last two to three years, it has culminated in an all-embracing theory known as 'plate tectonics'. The success of the plate tectonics theory is not only that it explains the geophysical evidence, but that it also presents a framework within which geological data, painstakingly accumulated by land-bound geologists over the past two centuries, can be fitted. Furthermore, it has taken the Earth sciences to the stage where they can not only explain what has happened in the past and is happening at the present time, but can also predict what will happen in the future.

While the evolution of the plate tectonics theory was rapid and spectacular, the 1960's also saw remarkable advances in other branches of the Earth sciences. Notable among these were the accurate dating of rocks by radiometric methods, and a deeper understanding of the Earth's internal composition and of our planet's relation to its satellite, the Moon, the Sun and the other planets in the solar system. Sophisticated analytical techniques and advances in biochemistry threw new light on the origin of life on Earth, and detailed studies showed how it had evolved through geological time. Finally, there were the contributions made by the Earth scientists to man's environment—researches aimed at conserving natural resources and at predicting, and possibly modifying, some of the Earth's natural hazards.

This was the situation earlier this year when we, as the Earth scientists on the Open University's team preparing the Foundation Course in Science, were searching for a text to which we could refer our 7000 students. What were our needs? We needed an inexpensive text which covered the same subject area as the Earth Sciences component of the Open University's Science Foundation Course but which, while dealing with these subjects in more detail than our time allowed, did not contain too much irrelevant material. The book would have to be at the right level—no higher mathematics or advanced physics—but with plenty of ideas that were mentally stimulating and conceptually demanding. Perhaps more than anything else, we needed a book which would support our course in emphasizing that the Earth sciences are alive and vital; that present-day Earth scientists are very far removed from the all too prevalent mental picture of elderly, bewhiskered gentlemen diligently brushing the dust from museum collections of rocks, minerals and fossils.

A tall order? Too tall—we could find no such book. Much of the material we wanted, and not just that dealing with plate tectonics, was incorporated in research papers written by specialists and published during the last few years in a wide variety of scientific journals. So we decided to commission our own book, and asked twenty-two scientists of international reputation to write contributions, at first-year university level, on their own speciality. In most cases, our authors have written on the same subject at research level, but you will also find herein three articles taken direct from scientific journals. This book is, then, a Reader. You cannot, even by reading it from cover to cover, get a complete survey, even at the shallowest level, of *all* the aspects of the Earth sciences. We primarily intend that it should be used by someone taking a university or school course in the Earth sciences to get an up-to-date and well-informed synopsis of particular facets. Indeed, some research results make their first appearance in this book even before they have been presented in scientific journals!

What, then, does the book contain and how should it be used? We start with a chapter on minerals and rocks, followed by one on radiometric dating—dating geological time. As well as providing a wealth of information in their own right, these contributions provide the technical vocabulary required by later authors. Then follow four chapters dealing with the Earth as a planet. In Chapters 9 to 11 we delve into the Earth's remote past. How did life start on Earth, how did the atmosphere evolve? Questions such as these are here answered to the best of our present ability. The oldest rocks on Earth are about 3 500 million years old (you will find a range from 3 400 to 3 600 amongst our contributions!) but how has the surface of the Earth, and the life on it, changed during this time? Chapters 12 to 14 précis the latest information.

Interest then concentrates on the Earth's surface features—the titles *Continental Drift* and *Sea-floor Spreading* indicate, well enough, the contents of the next two chapters. We then take a short diversion enquiring into the possible causes and consequences of the change in polarity of the Earth's magnetic field, before dealing with plate tectonics, orogeny (mountain building) and the part which volcanic activity has played in the evolution of the Earth's crust. In the next three contributions examples are given of the rôle of the Earth sciences in society before, in the final chapter, we enter the realm of geopolitics with the cautionary, yet illuminating, story of the Mohole project.

The sequence, we think, is logical, if you wish to read straight through. However, we have endeavoured to see that each contribution is as self-contained as possible, enhancing, rather than being dependent upon, others in the volume.

This book was conceived in March, 1970, and published in December of the same year— an uncomfortably short gestation period! It is truly remarkable that so much has been done in so short a time, and our sincere thanks are due to all concerned. First our contributors—twenty-two were invited to contribute in April, and *all* manuscripts, text figures and photographs were in our hands by the end of August. We thank them not only for their promptness, their good-natured forbearance with numerous queries and editorial comments, but also for the dispatch with which they corrected proofs. We are grateful to Mark Gummer for his early liaison work with our publisher, and for this signal service we are greatly in his debt. For, without the wholehearted and superbly efficient collaboration of Michael Bizony of Artemis Press, this volume would not exist. We accept full responsibility for the contents of this book, but the format and the excellence of presentation are entirely his and reflect, we feel, the enthusiasm, equalling ours, that he feels about this volume. Collaboration between publisher and editors has been close and cordial, remarkable perhaps when one is a mathematician and the others Earth scientists with strange, unpredictable foibles. Finally, the Open University has been good to us; the Director of Publishing, Leslie Lonsdale-Cooper, has given invaluable advice and counsel and watched over us throughout. Mary Gass, appointed as our Editorial Assistant, organized us and our contributors, corrected proofs, saw that manuscripts went to the publisher in good order and acted as literary dogsbody (her term) in improving English, unsplitting infinitives and anglicising some of our American authors' spelling. Our sincere thanks are also due to Mary Smith for undertaking the onerous task of preparing the index, to Tina Brocklebank and Merynne Barratt, secretaries in the Department of Earth Sciences, for prompt and efficient clerical service, and to our academic colleagues for their forbearance throughout this short but somewhat traumatic period.

Department of Earth Sciences,   I. G. GASS

The Open University,   PETER J. SMITH

Walton Hall, Walton,   R. C. L. WILSON

Bletchley, Buckinghamshire.

November, 1970.

PREFACE TO THE SECOND EDITION

EIGHTEEN MONTHS HAVE PASSED since the first edition of this book was published. During that time we have become increasingly conscious that it owed its continuing success largely to the fact that its contributors spoke with the authority of internationally known scientists, and that it was completely up-to-date. This second edition retains these advantages: apart from there being two additions, the authors are the same, and each has been asked to up-date his contribution. In some cases this was only a matter of altering a few numbers, but in others it has meant a complete revision in the light of new knowledge now to hand. In particular, Chapters 7 and 24 on the Earth-Moon System and on Nuclear Explosions and Earthquakes have been greatly modified. But even a change in a single number can reflect a major advance; for instance, the date for the oldest known rocks on Earth (Chapter 2) was some 3 400 million years—it is now 3 800 million years. This discovery, made by Dr. Moorbath and his colleagues, has lengthened the geological time scale by 12%.

Chapter 12 has been completely rewritten by Dr. Goldring, and a paper on *The Oxygen Cycle* by Professors Preston Cloud and Aharon Gibor (originally published in *Scientific American*) has been added as Chapter 26. In the first edition we were apparently successful in conveying the enthusiasm, stimulation and impact that the 'new global tectonics' had on the Earth Sciences. Another impact—perhaps the most crucial to the Earth Sciences and society—is the growing realisation that the Earth's natural physical resources are finite and in many cases rapidly being exhausted. In Chapter 27, Dr. Stephen Drury of the Open University examines this problem and its effect on the environment. Finally, a glossary has been added, and the index has been expanded.

October, 1972   I. G. G.,   P. J. S.,   R. C. L. W.

A knowledge of the nature and origin of rocks is fundamental to an understanding of the Earth, and so the first chapter in our book is devoted to this subject. It is written by Dr. K. G. Cox of the Department of Geology and Mineralogy at the University of Oxford, and he introduces it as follows:

This chapter is concerned with the branches of geology called mineralogy and petrology. The meaning of the first term is obvious enough: simply the study of minerals, of substances like calcite and quartz, for example, which are crystalline, have relatively fixed compositions, and which when aggregated together form rocks. Petrology, however, requires a little explanation. It means the study of rocks as rocks, that is to say, what they are made of and how they originated. It is not primarily concerned with how old they are, or what fossils they contain, or where they are, or with any of the multitude of other questions one might care to ask. These studies belong to other branches of geology, though any or all of them might be put to petrological use.

Petrology is a relatively young science and may be said to have its origins in the mid-19th century, when H. C. Sorby devised a method of preparing slices of rock so thin that they would transmit light and hence could be made the subject of microscopic examination. This advance gave rise to the field of study known as petrography, the descriptive side of petrology. The genetic study of rocks has flourished more recently, mainly as a result of a great increase in the application of analytical chemistry, including isotope studies, and the experimental investigation of rock and mineral compositions at elevated pressures and temperatures.

The rise of understanding of rocks, and particularly of petrogenic processes, was initially highly dependent on the way in which microscopy assisted in the study of the constituent minerals; these often occur in aggregates so finely grained that they defied the efforts of even those numerous 19th century geologists who were virtuosi of the magnifying glass. However, even before Sorby's day descriptive mineralogy was well advanced, and the chemical composition of a large range of minerals was known, the rise of mineralogy having been in its turn substantially dependent upon the great advances in chemistry which the 18th century had seen. We shall follow a similar path, starting with minerals and leading on to rocks, with the chemical theme constantly recurring.

1 Minerals and Rocks

by Keith Cox

Minerals are naturally occurring, mainly crystalline, chemical compounds which, in their aggregates, form the three great classes of rocks: *igneous* (solidified from the molten state), *sedimentary* (formed by the erosion of existing rocks followed by re-deposition) and *metamorphic* (formed by the action of heat and pressure on existing rocks).

To DEFINE PRECISELY what is meant by minerals is by no means easy, but to think of them as naturally occurring, usually crystalline substances, with chemical compositions which are either fixed or can only vary between prescribed limits is adequate for nearly all purposes. From this one can see that the terms 'mineral' and 'rock' are respectively closely comparable to the terms 'compound' and 'mixture' used by chemists. Minerals show a great diversity of composition, as indicated in Table 1. It is often useful to distinguish that relatively small group which is important in the formation of rocks as the 'rock-forming minerals', and most of this chapter will be concerned with them. The silicates (in which can be included the very common mineral quartz, SiO_2) is the most important single group, followed by the carbonates calcite and dolomite. The other minerals, for example the oxides and sulphides, do not as a rule form large masses of rock but occur more sporadically, for example in veins. They include many of the economically valuable ores.

Nearly all minerals are crystalline, that is to say they have a regularity of internal structure on the atomic scale, which is often expressed as an external regularity of form (see Figs. 1. 1 and 1. 2). When minerals occur as finely divided aggregates, their crystallinity may not be apparent to the naked eye, but it can usually be detected by polarisation microscopy or by the use of X-ray diffraction techniques.

Crystals can be studied in two complementary ways, either by a consideration of their external form, or by using X-ray methods to determine their internal structure. Although most people make their first acquaintance with minerals as hand specimens and are therefore immediately confronted with external morphological problems, it is probably an aid to understanding if some consideration is given to crystal structure first.

1. 1 Crystals of quartz (SiO_2).

1. 2 Crystals of the magnesium-iron aluminosilicate, garnet.

NATIVE ELEMENTS

Gold	Au
Silver	Ag
Diamond	C
Graphite	C
Sulphur	S

NOTE: the chemical compositions of many of the minerals, notably amongst the silicates, are given in a simplified form.

HALIDES

Halite	NaCl
Fluorite	CaF_2

SULPHIDES

Galena	PbS
Chalcopyrite	$CuFeS_2$
Pyrite	FeS_2

OXIDES

Haematite	Fe_2O_3
Magnetite	Fe_3O_4
Ilmenite	$FeTiO_3$

CARBONATES

Calcite	$CaCO_3$
Dolomite	$CaMg(CO_3)_2$

SULPHATES

Gypsum	$CaSO_4.2H_2O$
Anhydrite	$CaSO_4$
Barites	$BaSO_4$

PHOSPHATES

Apatite	$Ca_5(F,OH)P_3O_{12}$

SILICATES

Quartz	SiO_2
Feldspathoid Group	
Leucite	$KAlSi_2O_6$
Nepheline	$NaAlSiO_4$
Feldspar Group	
Orthoclase	$KAlSi_3O_8$
Plagioclase series	
Albite	$NaAlSi_3O_8$
Anorthite	$CaAl_2Si_2O_8$
Clay Minerals	
Kaolinite	$Al_4(OH)_8Si_4O_{10}$
Micas	
Biotite	$K_2Mg_6(Mg, Fe, Al)_6(Si, Al)_8O_{20}$
Muscovite	$KAl_2(OH)_2Si_3AlO_{10}$
Chlorite	$(Mg, Fe, Al)_6(OH)_8(Si, Al)_4O_{10}$
Pyroxene Group	
Augite	$Ca(Mg, Fe)Si_2O_6$ (clinopyroxene)
Enstatite	$(Mg, Fe)SiO_3$ (orthopyroxene)
Amphibole Group	
Hornblende	$Ca_2(Mg, Fe)_5(OH)_2Si_8O_{22}$
Garnet Group	
Pyrope	$Mg_3Al_2Si_3O_{12}$
Almandine	$Fe_3Al_2Si_3O_{12}$
Cordierite	$(Mg, Fe)_2Al_3Si_5AlO_{18}$
Zircon	$ZrSiO_4$
Olivine	$(Mg, Fe)_2SiO_4$
Sillimanite	
Kyanite	Polymorphs of
Andalusite (Chiastolite)	Al_2SiO_5
Staurolite	$HFe_2Al_9O_8Si_4O_{16}$

The Internal Structure of Minerals

The internal structure of crystals is characterised by an arrangement of atoms in space such that the same configuration is repeated at regular intervals in all three dimensions, the structure being loosely termed a lattice. As a simple example consider common salt, NaCl, which in its natural occurrence is the mineral *halite*. The halite lattice is built on a simple pattern in which sodium particles and chlorine particles occupy alternate corners of a continuously repeated set of cubes (see Fig. 1. 3). The structure of halite, the first crystal structure to be determined by X-rays, illustrates the important general point that the atoms are not grouped into discrete molecules. In halite each sodium is surrounded by six equidistant chlorines (and vice versa). Thus there are no molecules of sodium chloride present as such, and if we use the symbol NaCl to indicate the composition, this is simply a way of saying that the ratio of sodium atoms to chlorine atoms in this mineral is 1:1.

Another relevant experimental observation is that fused NaCl conducts electricity. This, together with the structural evidence given above, prompted the suggestion that in halite the sodium particles are positive ions (Na^+) and the chlorine are negative (Cl^-). The structure illustrated in Figure 1. 3 therefore arises because the most stable arrangements are achieved by juxtaposing oppositely charged *ions*.

X-ray investigations of many solids disclose other structures with similar properties and extend the scope of this *ionic model*. It is interesting that by assuming that the ions are 'hard spheres' the radii of ions can be deduced.

As can be seen in Figure 1. 3, the distance measured along the cube edge from the centre of one sodium ion to the next is about 5.6 Å (1 Ångstrom $= 10^{-8}$ cm). This distance is obtained by X-ray methods, and when taken in conjunction with measurements of other compounds of sodium and chlorine gives the Na^+ ion a radius of 0.97 Å, and the Cl^- ion a radius of 1.81 Å.

The lattice thus illustrates two important features, firstly that the equal numbers of positive and negative ions (usually termed cations and anions respectively) ensure that the whole structure is electrically neutral. Secondly, that the ions are packed closely together, each one being surrounded by large numbers of others of opposite charges. This latter feature confers stability on the structure because it ensures that the inter-ionic forces are at

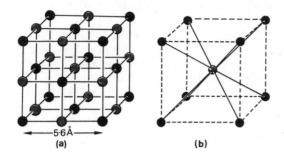

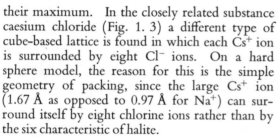

1. 3 **Representation of the crystal structures of (a) sodium chloride, (b) caesium chloride. Symbols show the positions of the two types of ion in each structure, but are not drawn to scale.**

1. 4 **The SiO_4 tetrahedron. Solid lines represent bonds linking the central silicon atom to four oxygen atoms. The broken line indicates the form of the tetrahedron, a solid figure bounded by four equal equilateral triangles.**

their maximum. In the closely related substance caesium chloride (Fig. 1. 3) a different type of cube-based lattice is found in which each Cs^+ ion is surrounded by eight Cl^- ions. On a hard sphere model, the reason for this is the simple geometry of packing, since the large Cs^+ ion (1.67 Å as opposed to 0.97 Å for Na^+) can surround itself by eight chlorine ions rather than by the six characteristic of halite.

So far we have been considering ionic crystals, and for these, relatively simple rules concerning the balancing of charges and the geometry of packing are remarkably successful in predicting crystal structures. However, in many minerals some or all of the atoms are held together by a rather different type of bond termed *covalent*, as opposed to the *ionic* bonds discussed above. A detailed account of bonding is beyond the scope of this work, and here it is sufficient to note that covalent bonds involve the sharing of electrons between adjacent atoms, and they are therefore specific for the atoms concerned.

The Silicates

The silicates are the most important group of rock-forming minerals. In the inorganic world these compounds are somewhat analogous to the compounds of carbon in the organic world, because of the astonishing variety of structures that they display. The keystone of this versatile group is the SiO_4 grouping in which each silicon is linked to four oxygens situated at the corners of a tetrahedron (see Fig. 1. 4). This structure is very

stable and the strong bonds between silicon and oxygen are usually regarded as covalent. However, in discussing the silicate minerals it is very convenient to regard the silicon as an Si^{4+} ion, especially when describing its substitution by Al, Mg, Fe etc. It should be emphasised that this is done as an aid to understanding the substitution of these metals for silicon, and that it does *not* imply that the silicon in silicates is an Si^{4+} ion in the sense that sodium in sodium chloride is an Na^+ ion.

The SiO_4 tetrahedra in some cases act as independent units within the structure of silicate minerals. The common mineral *olivine* (Mg, Fe)$_2$$SiO_4$ has a structure of this type and is an example of the structural group known as *nesosilicates* (independent tetrahedral units). More commonly, SiO_4 tetrahedra are linked in various ways by the sharing of two or more of their oxygen atoms between adjacent units. *Quartz* (SiO_2) is an extreme case in which every oxygen atom is shared by two tetrahedra in a three-dimensional frame-work. The Si: O ratio is thus 1:2, and all charge requirements are satisfied without the necessity for other ions to be present. Related to quartz in so far as each O is shared by two tetrahedral groups are the *feldspars* (aluminosilicates of K, Na and Ca), amongst the most important of the rock-forming minerals. In the alkali feldspars *orthoclase* ($KAlSi_3O_8$) and *albite* ($NaAlSi_3O_8$), one silicon in four is replaced by one aluminium. The resultant deficiency of one positive unit of charge per four tetrahedral groups is balanced by a single K^+ or Na^+ ion, as the case may be, occupying lattice sites formed by 'holes' within the con-

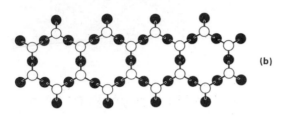

1. 5   The structure of SiO_4 chains in (a) pyroxenes, (b) amphiboles.   Only oxygen atoms are shown. Each white atom with its three adjacent atoms represents a single tetrahedral group.   The white atoms lie above the others (i.e. nearer the reader) and conceal the silicon atoms.   Bonds are indicated by short lines.

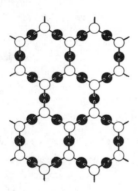

1. 6   The arrangement of SiO_4 groups in the sheet structure of the phyllosilicates.   Symbols as in Figure 1. 5.

tinuous framework of $(Si, Al)O_4$ units.   It is instructive to work out how this structure results in a molecular formula such as $KAlSi_3O_8$.   As we have seen for quartz, the Si:O ratio when each oxygen is shared by two tetrahedra is 1 : 2.   When one silicon in four is replaced by aluminium the Al:Si:O ratio becomes 1:3:8.   Consideration of charges however $(Al^{3+}, Si^{4+}, O^{2-})$, shows that such a structure is not electrically neutral, and has a single negative charge for every group of $AlSi_3O_8$. Thus a single potassium ion (K^+) can be added to

confer electrical neutrality, the resulting structure having a K:Al:Si:O ratio of 1:1:3:8, which we write as the molecular formula given above.

In the calcium feldspar *anorthite* $(CaAl_2Si_2O_8)$ half the silicon is replaced by aluminium, the charge difference being balanced by the inclusion of one Ca^{2+} ion per four tetrahedral groups.   The feldspars and quartz are examples of *tectosilicates* (three-dimensional framework structures) and, perhaps not surprisingly, such minerals tend to be rather hard and they do not cleave easily (cleavage in minerals is the tendency to split along definite crystallographic planes).

In contrast to the tectosilicates there are several important mineral groups in which the SiO_4 tetrahedra are linked into *chains* (see Fig. 1. 5). In the *pyroxenes* (silicates of Ca, Mg and Fe^{2+}) the chains are single so that in any one SiO_4 group two oxygen atoms are shared, each with an adjacent tetrahedral group.   The chains are of indefinite length so that there is a silicon:oxygen ratio of 1:3. The resultant double negative charge per tetrahedral group is balanced by a single divalent ion such as Mg^{2+}, Ca^{2+} or Fe^{2+}.   Closely related to the pyroxenes are the minerals known as *amphiboles*, in the lattices of which the tetrahedra are linked into double chains with a silicon:oxygen ratio of 4:11.   The structure is electrostatically balanced by cations as in the pyroxenes but there are OH^- groups in addition.   Like feldspars and quartz, the amphiboles and pyroxenes are rather hard, but they have a very prominent cleavage direction parallel to the chain structure.

The remaining important structural group is known as the *phyllosilicates* in which the tetrahedra are linked into sheets of indefinite lateral extent with a silicon oxygen ratio of 2:5.   A hexagonal pattern within the sheet structures (see Fig. 1. 6) is reflected in the external form of many phyllosilicate minerals which include the *clay minerals* and the *micas*, both of great importance as rock-forming minerals.   In phyllosilicates the cations (and water) are accommodated between the layers of linked tetrahedra.   As a group they show a very prominent cleavage in one direction, parallel to the sheets.

Solid Solutions

In a definition of the term 'mineral' at an earlier stage I noted that minerals had a constant composition or compositions which varied between certain fixed limits.   In fact it is the latter which is typical and minerals of more or less fixed composition are comparatively uncommon (quartz is the most familiar example).

The *olivines*, common minerals in many igneous rocks (rocks which solidify from a molten state), provide a simple example of the way in which composition varies between fixed limits, in this case between Mg_2SiO_4 and Fe_2SiO_4. These 'molecules' are spoken of as end-members in a solid solution series, and almost all natural olivines have compositions intermediate between the end-members. To speak of this as solution is perhaps slightly misleading, since there is no direct analogy with the more common phenomenon of solution in liquids. What we are seeing in fact is the effect of the free substitution of either Fe^{2+} or Mg^{2+} in a single lattice site. These ions have almost the same radius and have the same charge so that the lattice can accommodate either with almost equal facility. The composition of the olivine formed by crystallization from a silicate melt is of course a function of the availability of the two elements. The replacement of Mg^{2+} by Fe^{2+} is quite a general phenomenon and is also typical of pyroxenes and amphiboles, though these groups also exhibit additional solid solution effects.

The *plagioclase* feldspars show a slightly more complex type of ionic replacement and form the series $NaAlSi_3O_8$ (albite)-$CaAl_2Si_2O_8$ (anorthite). Like the olivines the solid solution between the end-members is complete, but in this case the univalent ion Na^+ (ionic radius 0.97 Å) is replaced by the similarly-sized divalent ion Ca^{2+} (ionic radius 0.99 Å). To maintain electrical balance there is a simultaneous replacement of Si^{4+} (0.42 Å) by Al^{3+} (0.51 Å) in the $AlSi_3O_8$ framework. The complexity of feldspar solid-solution effects does not end here, because most plagioclases contain small amounts of potassium which can be expressed in terms of the end-member $KAlSi_3O_8$ (orthoclase). The substitution here is simply K^+ for Na^+, but since there is a considerable difference in ionic radius (1.33 Å and 0.97 Å respectively) only a limited number of Na^+ ions can be replaced by K^+ unless the feldspar is formed at a high temperature. In this case the solid solution between $NaAlSi_3O_8$ and $KAlSi_3O_8$ becomes complete, though the solid solution between $KAlSi_3O_8$ and the more sodium-poor members of the plagioclase series remains only partial. Figure 1.7 shows approximately the range of solid solutions of the feldspars at high and low temperatures. A high temperature feldspar, for example having a composition about half way between albite and ortho-

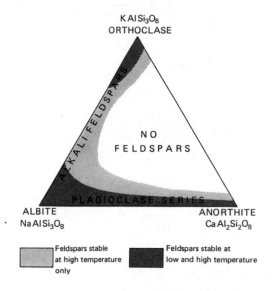

1.7 **Approximate limits of solid solution in the feldspars at two arbitrary temperatures.**

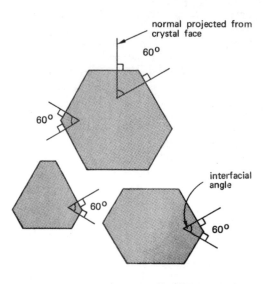

1.8 **Cross-sections of quartz crystals illustrating the law of constancy of interfacial angles. The angles are measured between normals projected from the faces.**

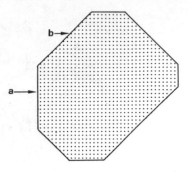

1. 9 Illustration of the relationship of faces to a lattice. Faces such as 'a' have a parallel relationship to the lattice points while faces such as 'b' show a diagonal relationship. The size of the faces is variable, but angular relationships are constant.

clase, frequently 'unmixes' if cooled slowly to form a crystal composed of discrete blebs or laminae of two compositionally different feldspars, one more sodic and the other more potassic.

Crystal Morphology

Good specimens of crystals (see Figs. 1. 1 and 1. 2) are bounded by planar surfaces termed *faces*, which group themselves into different types. For example, the quartz crystals illustrated in Figure 1. 1 have faces of one type forming the pointed terminations of the crystals and of another type bounding the elongated sides. This introduces the concept of the *form*, a term used to mean a group of all the faces of a crystal which are of one type, and which share a common geometrical relationship to the crystal lattice. In a regularly developed crystal all the faces belonging to one form will have the same size and shape. However reference to the quartz crystals in Figure 1. 1 will show that in this particular case there is some irregularity of development, and indeed this is typical to some degree of practically all natural mineral specimens. For example, in Figure 1. 8 cross-sections of quartz crystals are shown, and it will be noticed that there is no particularly rational pattern in the development of the faces as regards their *size*. However what we do see is that the angle between adjacent faces is always constant at 60°. This illustrates one of the fundamental laws of crystallography, *the law of constancy of interfacial angles*. This means that for any given mineral the angle between corresponding faces is constant.

This is of course a useful diagnostic property of a particular mineral. The regularity of morphology revealed is a reflection of the regularity of internal structure. In Figure 1. 9 a very simple two-dimensional analogy is given to illustrate the way in which faces have a rational relation to the lattice and how this is shown externally by constancy of interfacial angles, but not necessarily by constancy of shape.

To describe crystals morphologically it is necessary to measure interfacial angles, which can be carried out with high accuracy using an optical goniometer, a device which utilises a narrow beam of light reflected from the faces of the crystal. A projection method (usually the stereographic projection) is used to record the data in diagrammatic form. Figure 1. 10 illustrates the principle of the stereographic projection, while some examples of the resultant stereograms are shown in Figures 1. 11 and 1. 12. The most important feature of such projection methods is that each face of the crystal is represented by a point termed a *face pole* which is the direction of the normal (pole) to the face. In this way all the angular properties of the crystal faces are retained, and irregularities of growth in terms of face-sizes are lost.

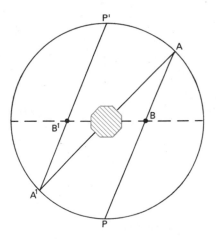

1. 10 Principle of the stereographic projection. The crystal is imagined to be at the centre of a sphere. A face pole is projected to intersect the sphere (point A). Point A is projected to the opposite pole of the sphere (P) and generates point B where it cuts the equatorial plane of the sphere. The equatorial plane becomes the stereogram as illustrated in Figures 1. 11 and 1. 12, the crystal faces being represented by points such as B.

The most striking feature of the resulting projections is the *symmetry* which they show, the most important symmetry elements being *rotation axes* and *reflection planes*. Rotation axes are referred to as two-fold, three-fold, four-fold or six-fold. To illustrate this, consider the centre of the stereogram of quartz shown in Figure 1. 11. This is the point of emergence of a three-fold rotation axis, which means that if we imagine the stereogram to be rotated about this point (rotated in the plane of the paper) it will fall into a position of congruence, that is to say it will look exactly the same as it did before rotation, three times in one complete revolution. The stereogram of zircon shown in Figure 1. 12 in contrast shows a four-fold rotation axis in the centre and also shows a number of reflection planes, all of which are vertical (lying normal to the paper) and whose positions are shown by the dotted lines. The halves of the stereogram separated by a dotted line are mirror images of each other.

The study of the symmetry of crystals and crystal lattices enables crystalline substances to be divided into seven groups known as *crystal systems*, the essential symmetry requirements of which are shown below :-

Crystal System	*Symmetry*
cubic	four three-fold axes
tetragonal	one four-fold axis
hexagonal	one six-fold axis
trigonal	one three-fold axis
orthorhombic	three two-fold axes
monoclinic	one two-fold axis
triclinic	no axes

Of the morphological examples we have met so far, quartz belongs to the trigonal system (though the crystals often have a pseudohexagonal appearance), zircon belongs to the tetragonal system and garnet (Fig. 1. 2) is cubic. Recognition of the crystal system is of course a very considerable aid to identification of a mineral, particularly since it may be employed in microscopy as well as in the examination of hand specimens.

IGNEOUS ROCKS

The recognition of the true nature of igneous rocks was due to James Hutton of Edinburgh, whose celebrated *Theory of the Earth*, published in 1795, laid the foundations of modern geology. Hutton's work was based essentially on field

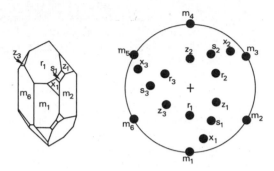

1. 11 **Quartz crystal and stereogram. The stereogram can be envisaged as a map of a globe viewed from the north pole, the long (vertical) axis of the crystal coinciding with the axis of the globe. Normals to faces 'm' plot round the equator; the other faces shown are those forming the pointed upper termination of the crystal.**

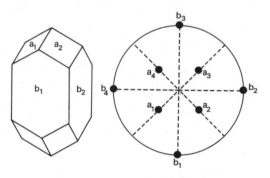

1. 12 **Zircon crystal and stereogram. Reflection planes shown by broken lines.**

studies which demonstrated that certain masses of rock (which we now call intrusive igneous bodies) could only have gained their position within the rocks which surround them if they had been emplaced as mobile liquids (*magmas*) which subsequently solidified. His friend and collaborator, Sir James Hall, demonstrated the plausibility of this hypothesis by taking a piece of 'whinstone' (dolerite) and melting it, sealed inside a cannon barrel.

The igneous origin of present-day lavas emitted from volcanoes is of course self-evident. From field characters it is also usually possible to recognise older lavas fairly easily. When the science of petrography (the description of rocks) became centred round the microscope in the latter part of the 19th century, it rapidly became clear that igneous, sedimentary and metamorphic rocks were highly distinctive when examined in thin sections. We can therefore generally assign a specimen of rock to its appropriate group without any knowledge of its field character.

Sedimentary rocks are usually distinguished by their fragmental texture. Young sedimentary rocks frequently have considerable porosity, and even when this is reduced or removed by compaction or by the deposition of secondary minerals within the voids the original texture is still often recognizable. Most of the igneous and metamorphic rocks are in contrast compact, with all the mineral grains in contact with their neighbours. These two groups are sometimes more difficult to distinguish, but most metamorphic rocks show some sort of banding or foliation because of the deformation they have undergone, and this feature is rare in igneous rocks. In many cases the mineralogy of the two groups is also substantially different, particularly if we compare igneous rocks with the very common metamorphic rocks derived from sedimentary parentage. Figures 1.15, 1.21, and 1.28 show thin sections of typical rocks from each group, illustrating most of the differences I have mentioned.

Chemical and Mineralogical Variation in Igneous Rocks

Virtually all the igneous rocks are largely composed of silicate minerals, the only important non-silicate constituents being oxides of iron and titanium such as magnetite and ilmenite (see Tables 1 and 2). The rocks known as *carbonatites* provide the only notable examples almost devoid of silicate materials, and are by contrast made up of carbonates of calcium, magnesium and iron.

We can think about variation in igneous rocks either in terms of chemistry or in terms of mineralogy, and it is worth while to spend a moment considering how these two different facets are related. Consider for example the rock-type granite. Granite is a very common intrusive rock (i.e. one injected in a fluid state into a pre-existing rock) consisting of about 10–30 per cent of quartz (SiO_2) accompanied by a large amount of feldspar of a more or less alkalic variety, and by relatively minor amounts of ferromagnesian minerals such as hornblende or biotite. Granites when analysed are found to contain about 70 per cent of SiO_2, which is more than in any other igneous rock-type. The quartz content is also higher than that of other igneous rocks. Clearly there is some sort of rational relationship here if we find that the rock which contains most SiO_2 when analysed in bulk also has a high content of the most SiO_2-rich mineral. Is it the mineral which creates, as it were, the chemical features we find on analysis, or is it the chemistry which creates the mineralogy? This question may well seem analogous to the one about the chicken and the egg, but in fact the answer is easily found and illuminates some of the fundamental characters of igneous rocks.

Chemical composition can be said to create mineralogy in rocks which crystallize from a more or less totally liquid magma. The magma has a given composition; it crystallizes and gives rise to a perfectly definite assemblage of minerals. A different magma would crystallize to a mineral assemblage which differed either in the actual minerals present or in their proportions. As I shall try to show later, many volcanic rocks are essentially of this type. In contrast, suppose we imagine a body of magma which is crystallizing in a subterranean chamber. Crystals being denser than magma (in general) sink to the floor of the chamber and, as the weight of overlying crystals increases, the interstitial liquid is more or less effectively squeezed out. The rocks so formed are called *cumulates*, and of course it is the supply of mineral grains which controls the bulk composition they eventually have. Here then we have a clear case of bulk chemistry generated by mineralogy. These two totally different mechanisms of igneous rock formation, the crystallization of liquids on the one hand, and the formation of cumulates on the other, are, as it were, two end-members in a gradational series of mechanisms. In fact we can find numerous good examples of each type, as well as examples of rocks which are somewhere in between, the latter having been liquids enriched by the addition of crystals from elsewhere.

Let us look now at the range of variation, both mineralogical and chemical, which is encountered in the igneous rocks. Some average analyses are given in Table 3. The rock types *granite*, *diorite* and *gabbro* are all coarse-grained and usually found in intrusions. They are matched compositionally

TABLE 2: APPROXIMATE CHEMICAL COMPOSITIONS OF COMMON MINERALS OF IGNEOUS ROCKS (% WEIGHT)

	SiO_2	Al_2O_3	$MgO + FeO$	CaO	$Na_2O + K_2O$
FERROMAGNESIAN MINERALS					
olivine[a]	40	—	60	—	—
augite[b]	50	3	23	20	—
hornblende[b]	40	10	30	12	trace
biotite[b]	36	15	30	1	10
PLAGIOCLASE SERIES					
calcic plagioclase[c]	54	29	—	12	5
sodic plagioclase (albite)	68	20	—	—	12
ALKALI FELDSPARS					
potash feldspar (orthoclase, microcline etc.)	65	18	—	—	17
FELDSPATHOIDS					
nepheline	42	36	—	—	22
leucite	55	23	—	—	22
SILICA MINERALS					
quartz etc.	100	—	—	—	—

[a] Composition of the common olivine of basic igneous rocks, containing about 25% of Fe_2SiO_4 and 75% of Mg_2SiO_4.
[b] In addition to the constituents shown, augite, hornblende, and biotite contain significant amounts of TiO_2 and Fe_2O_3. Hornblende and biotite also usually contain about 2% of water.
[c] Composition of the common plagioclase of basic igneous rocks, containing about 60% of $CaAl_2Si_2O_8$ and 40% of $NaAlSi_3O_8$.

TABLE 3: APPROXIMATE AVERAGE COMPOSITIONS OF SOME COMMON IGNEOUS ROCKS (AFTER DALY)

Rock type	Granite (Rhyolite)	Diorite (Andesite)	Gabbro (Basalt)	Syenodiorite (Trachyandesite)	Syenite (Trachyte)	Nepheline Syenite (Phonolite)	Ijolite (Nephelinite)	Syenogabbro (Trachybasalt)
SiO_2	70.8	57.6	49.0	56.7	62.5	55.4	43.2	49.3
TiO_2	0.4	0.9	1.0	1.1	0.6	0.9	1.6	1.9
Al_2O_3	14.6	16.9	18.2	17.1	17.6	20.2	19.1	18.2
Fe_2O_3	1.6	3.2	3.2	3.0	2.1	3.4	3.9	4.4
FeO	1.8	4.5	6.0	4.1	2.7	2.2	4.9	5.7
MgO	0.9	4.2	7.6	3.3	0.9	0.9	3.2	4.1
CaO	2.0	6.8	11.2	6.6	2.3	2.5	10.6	9.0
Na_2O	3.5	3.4	2.6	3.7	5.9	8.4	9.7	4.4
K_2O	4.2	2.2	0.9	3.8	5.2	5.5	2.3	2.3

Analyses given are those of the plutonic rocks. Names of the approximately equivalent volcanics are shown in brackets.

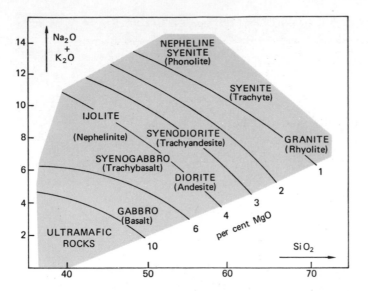

1. 13 **Alkali-silica diagram for some important igneous rocks. Names shown in capital letters refer to the coarse-grained (plutonic) varieties; names of equivalent fine-grained (volcanic) rocks are given in brackets. The shaded area indicates the field in which the great majority of igneous rocks plot. An indication of the approximate content of MgO is given by the contour lines.**

by fine-grained rocks, respectively *rhyolite*, *andesite* and *basalt*, which are characteristically found as surface lava flows (extrusions), though in the course of geological time these may become deeply buried by other rocks deposited over them. These six rock types, together with near relatives, between them account for well over 90 per cent by volume of all the igneous rocks known at the surface of the Earth. Granite is by far the most abundant intrusive rock, while basalt is the dominant extrusive composition. However, to add further significance to our study we must consider the additional types noted on Figure 1. 13. I must stress, however, that these have been chosen to illustrate some of the most important aspects of variation, and are not intended to constitute in any way a comprehensive list of the multitude of variants available.

Figure 1. 13 shows the positions of the rocks if we plot their total content of $Na_2O + K_2O$ against SiO_2. Igneous rocks occupy a somewhat restricted field on this diagram. There are two very important features to note. Firstly, the two parameters chosen are highly distinctive for each rock type. In other words we find, in general, that two rocks which plot in about the same place are comparable in their other chemical features. For example, a silica content of about 70 per cent with a total alkali content of about 7–8 per cent is characteristic of granites. The other 22–23 per cent of a granite analysis is usually made up of

about 12–16 per cent of Al_2O_3 and there are minor amounts of other constituents (see Table 3). We do not find rocks composed of, for example, 70 per cent SiO_2, 7–8 per cent total alkalis, and say 20 per cent of FeO or 20 per cent of CaO etc.

The second very important point to note is that the whole of the indicated field of Figure 1. 13 can be filled by analyses of individual rocks. In other words, igneous rocks form a compositional continuum. An important additional chemical feature is indicated by the contours for MgO shown on the figure. The behaviour of this constituent is closely matched by that of CaO and to a lesser extent by the iron oxides. If one considers a large number of analyses it is possible to contour the diagram for these constituents too, albeit in a somewhat approximate fashion. This means in practice that given the SiO_2 and $Na_2O + K_2O$ contents of an igneous rock it is possible to make an approximate prediction of the contents of MgO, CaO and iron oxides. All this is simply a way of demonstrating the extraordinary *coherence* of the chemical characteristics of igneous rocks. The variables such as SiO_2 content, CaO content, and so on, are by no means independent and behave according to a very rational pattern.

Amongst the major constituents only Al_2O_3 has so far not been considered. Nearly all igneous rocks contain somewhere between 13 and 20 per cent Al_2O_3 and the variation is thus very much less than in most other parameters. Only in the

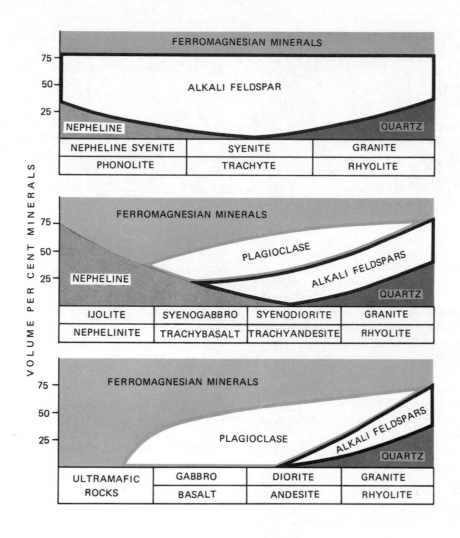

1. 14   Generalized mineralogical compositions of igneous rocks shown in Figure 1. 13.

ultramafic rocks, that is to say the rocks composed entirely or predominantly of ferromagnesian minerals, does Al_2O_3 fall to significantly lower values than those quoted.

Before considering the reasons underlying the chemical variation, it is important to see how this variation is expressed mineralogically. In Figure 1. 14 I have selected three 'traverses' across the continuum of compositions shown in the preceding figures. These have been selected for reasons of convenience, simply to illustrate some of the possibilities of mineralogical variation. I shall be considering later the question of *series* of igneous rocks which occur together in distinct

provinces, and in which the individual rock types are genetically related. Only one of the series shown in the figures, that joining the ultramafic rocks to the granites via the gabbros and diorites, is a likely genetic series—the others should not be taken as such.

The way in which the continuum of chemical compositions we have discussed generates the mineralogical variation illustrated in the figures can be demonstrated quite easily. In fact what follows applies equally well to cumulitic rocks as well as to the products of straightforward crystallization of liquids. Rhyolites and granites, for example, contain quartz, and this can be seen to be

1. 15 **Thin section of dolerite (medium grained equivalent of gabbro) showing some typical igneous features. Small phenocrysts of olivine (OL) show good form characteristic of early crystallization. Plagioclase laths (PL), pyroxene (PY) and magnetite (opaque) form the groundmass, the plagioclase having begun to crystallize before the pyroxene. (×33)**

1. 16 **Thin section of andesite showing large phenocrysts of plagioclase and small dark phenocrysts of pyroxene set in a fine-grained groundmass. (×12)**

inevitable because no silicate mineral contains the 70 per cent or so of SiO_2 which is characteristic of these rocks (see Table 2). Similarly, rocks such as ijolite, with their high content of alkalis, particularly soda, must almost inevitably contain a mineral of the feldspathoid group, in this case nepheline. The rocks approaching basalt and gabbro in composition are inevitably rich in the ferromagnesian minerals because of their high contents of magnesium, calcium and iron. The reader may find it an interesting exercise to continue this sort of reasoning to see why the rocks of the syenite-granite range are rich in alkali feldspars, whereas the gabbroic rocks are rich in calcic plagioclase. It helps to have the additional information that quartz is not compatible with any of the feldspathoidal group, nor is it compatible with the Mg-rich olivine so common in many basaltic rocks (quartz and nepheline if brought into contact at magmatic temperatures react to give albite; similarly quartz and olivine give an orthopyroxene).

The Crystallization of Igneous Rocks

Volcanic rocks are the only igneous rocks we can actually observe in their liquid state, and are commonly erupted as lavas which carry a certain proportion of crystals suspended in the liquid. After eruption the rapid cooling in the surface environment promotes quick crystallization of the liquid fraction of the lava and generates a characteristic texture known as *porphyritic* in the resulting solid rock. This texture consists of larger crystals (the crystals present before eruption) termed *phenocrysts* set in a fine grained *groundmass*, which in some instances may be finely crystalline and in others glassy (see Figs. 1. 15 and 1. 16). Experiments designed to study the crystallization of molten igneous rocks show that most examples crystallize over a *temperature interval*, and there may be a range of tens or hundreds of degrees centigrade in which the material consists of co-existing liquid and crystals. Studies of natural lava lakes in Hawaii by the U. S. Geological Survey have also provided direct evidence of this phenomenon. Porphyritic texture indicates that the majority of magmas are erupted with temperatures lying within this partly-solid, partly-liquid range. Indeed, it is likely that most magmas remain in this condition throughout most of their evolution, that is to say when they are first generated by partial fusion of existing rocks, and during their subsequent migration towards the surface.

With this knowledge we can discuss the most important mechanism known to produce variation in igneous rocks, namely *fractional crystallization*. This mechanism depends on the very strong tendency, already mentioned, of crystals to sink in magma. N. L. Bowen of the Geophysical Laboratory in Washington demonstrated this in

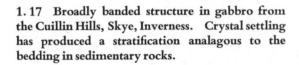

1. 17 Broadly banded structure in gabbro from the Cuillin Hills, Skye, Inverness. Crystal settling has produced a stratification analagous to the bedding in sedimentary rocks.

1. 18 Thin section of a cumulitic gabbro from Ardnamurchan, Argyllshire. Well-formed, in-equidimensional plagioclase crystals (PL) have taken on a preferred orientation as a result of crystal settling. Pyroxene started to crystallize at a later stage in the interstices between the plagioclase crystals. (×6)

experiments made in 1915 and was able to point to J. V. Lewis's description of the olivine-rich layer near the base of the Palisade sill, New Jersey, as a natural example. However it was not until 1939, when L. R. Wager and W. A. Deer described the rocks of the Skaergaard intrusion, East Greenland, while at about the same time H. H. Hess described the Stillwater intrusion of Montana, that detailed evidence of crystal settling in natural rocks was obtained. The Skaergaard is only one example of numerous *layered intrusions* which have subsequently been described. The rocks are entirely igneous but have been formed by a process of sedimentation of crystals on the floor of a magma chamber. Such rocks are generally termed cumulates, as has been mentioned, and they show a variety of structures more commonly associated with sedimentary rocks (see for example Figs. 1. 17 & 1. 18).

The study of layered intrusions enables the course of crystallization of a particular magma to be followed. Those minerals which are found in the lower part of the intrusion were clearly of early crystallization, and conversely the upper part of the intrusion is formed by minerals which have crystallized from the magma later, and at lower temperatures. The precise sequence of minerals observed does of course depend largely on the initial composition of the magma which fills the chamber. However, most layered intrusions are formed from essentially basaltic parental magmas, and as a generalisation we can say that the ferro-magnesian minerals, particularly olivine and the pyroxenes, start to crystallize early, as does calcium-rich plagioclase. More sodic plagioclase, alkali feldspar, feldspathoids or quartz, in general appear at lower temperatures and may therefore be prominent in the upper levels of such intrusions.

So far I have been talking mainly about crystals and the accumulations of them that are found in intrusive bodies. Let us now look at things from another point of view and consider what is happening to the composition of the liquid in a magma chamber while the process of crystal settling is in progress. The early removal of olivine, pyroxene and calcic plagioclase will tend to enrich most types of magma in alkalis and silica. Concomitantly the magma is impoverished in magnesium and calcium, and in many cases also iron. Suppose now that liquid is drawn off or expelled from the magma chamber at various stages, either to give rise to surface lavas or to higher level intrusions. Clearly a great variety of derived magmas can be produced in this way, and we may, for example if we study volcanoes, never see the cumulates which accompany their formation (though fragments of cumulitic rocks are occasionally found as inclusions in lava). The variation in the products of a single volcano are illustrated in Figure 1. 19, where some of the

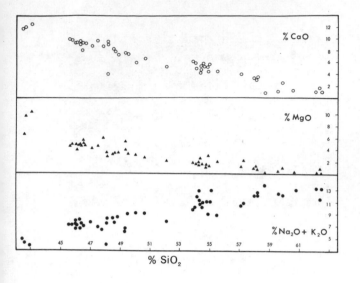

1. 19 Chemical variation in volcanic rocks from Tristan da Cunha, South Atlantic. Note the steady rise of silica and alkalis accompanied by a steady fall in calcium and magnesium.

chemical features of analysed lavas are displayed graphically. Although in cases such as this no cumulates are available for inspection, one can infer the operation of fractional crystallization in other ways. For example the minerals which form phenocrysts in the lavas indicate which phases are crystallizing at any given time and enable the chemical effects of fractionation to be predicted and to be compared with the natural variation in the series.

The general conclusion to be drawn from studies of lavas and cumulitic intrusions is that many igneous provinces owe their variety of rock types to the fractionation of basaltic parental magmas. It is inferred for a variety of reasons that the parental magmas are formed by the partial fusion of ultramafic rocks (rocks rich in olivine and pyroxene) within the upper part of the Earth's mantle (see Chapters 3 and 5). There are however several important exceptions to this pattern.

Other Igneous Processes

Liquids of granitic composition can be formed by the advanced fractionation of certain types of basaltic magma when the processes of silica and alkali enrichment mentioned previously are carried to extreme lengths. Some granitic intrusions and rhyolitic lavas are certainly formed in this way. However, granitic liquids exist at relatively low temperatures and are in equilibrium with quartz

and alkali feldspars, the major minerals which characteristically precipitate from them on cooling. Thus if crustal rocks containing these two minerals are heated, either on a regional scale during episodes of mountain building or on a local scale, for example, as hot basaltic magma is introduced into the crust from below, they may begin to melt and give rise to granitic magmas as partial melting products. This process is known as *anatexis* and it is certain that some granites are formed in this way rather than by the fractionation of basalt. The Tertiary intrusive complexes of Scotland (Skye, Mull, Rhum, Ardnamurchan) for example include gabbros and granites in abundance, but intermediate types are rare. It is this latter point which makes it seem unlikely that the granites are derived from basic magmas, for in that case we should expect to find at least moderately abundant rocks of composition between granite and gabbro.

A related problem, again concerned with the relative abundance of rock types, is presented by the lavas known as *andesites*, which have compositions intermediate between basalt and rhyolite. They are typical of the world's greatest volcanoes, lying in areas of active mountain building above Benioff zones (see Chapter 19). Although they are accompanied by basalts and rhyolitic rocks it is the andesites which are erupted most copiously. A hypothesis involving crystal fractionation of basalt must suppose that residual liquids, such as andesite, are produced in relatively small quantities compared with the amount of parental material available, and one can therefore see that an interpretation in terms of this hypothesis has some difficulties, at least in certain areas. Other hypotheses, for example involving the production of primary andesite magma by the melting of wet peridotites or the production of andesite magma by the contamination of basalt, have been put forward. This particular problem has not been satisfactorily solved, and is the subject of a considerable amount of current research (see Chapter 21 for further discussion).

SEDIMENTARY ROCKS

Almost everywhere on the surface of the Earth we see either the processes of erosion, which provide the raw materials of the sedimentary rocks, or we see the deposition of these materials in progress. On the land it is mostly the erosion of which we are conscious, and it is under water, either in the sea, in rivers, or in lakes where the eroded materials largely come to rest. Many of the agents which act to break down solid rocks into easily transported sedimentary particles are self-evident and I will mention them only briefly. These include the frost action which shatters the most solid rocks, particularly in mountain areas, the hydraulic action of the sea as cliffs are battered and worn back, and the slow attrition caused by the passage of sediment down streams and the swashing backwards and forwards on beaches. The organic world also plays an important part with the root-action of plants and the burrowing of organisms all contributing to the breakdown of the bedrock. Many of these processes contribute mainly to the physical disruption of solid rocks, and this is spoken of as *mechanical* weathering.

Chemical Weathering

Mechanical weathering is only part of the story, and *chemical* weathering, the breakdown of existing minerals and the formation of new ones, is equally important. Perhaps the best way to consider this process is to start with the concept of chemical equilibrium and ideas of mineral stability. Harking back for the moment to igneous rocks, the mineral assemblage we see in most fresh specimens we collect must have co-existed in equilibrium with a silicate melt at a very high temperature. Once an igneous rock has completely solidified, however, the absence of a fluid phase and the reduction of temperature make it very difficult for the minerals to re-equilibrate to new assemblages which would be stable at lower temperatures. Hence the assemblages we see in most specimens of igneous rocks are only *metastable* at normal temperatures and they owe their existence to the fact that the rates of the reactions which seek to change them are vanishingly low in a cold, dry, environment. I shall consider this question again with reference to the metamorphic rocks at a later stage, but for the moment it is sufficient to note that metamorphic assemblages formed at high temperatures are similarly only metastable at the low temperatures of the surface environment.

The main agent of chemical weathering is of course water, and this acts upon high-temperature minerals partly by providing a fluid phase and hence promoting the mobility of ions necessary for reactions, and partly because it acts as a new chemical component and makes possible the formation of hydrous minerals. Groundwater carries with it additional chemical components in solution, of which the most important is CO_2, and these also play their part in reactions by introducing new ions or simply by changing the oxidation or reduction potential and the acidity of the fluid.

If we consider sedimentary rocks as a whole we find that chemical weathering is very effective in producing a mineralogy which differs greatly from that of the igneous and metamorphic rocks. Whereas in the latter groups the feldspars are quantitatively very important constituents, in the sedimentary rocks their place is taken by the *clay minerals*. Amongst the common minerals of the igneous and metamorphic rocks, only quartz appears to be completely stable in the environment of chemical weathering, though even quartz is slowly dissolved, and passes into solution as the very weak silicic acid, H_4SiO_4. The *silicate minerals* are in general decomposed by the process of *hydrolysis*, of which the following reaction of forsterite (magnesian olivine) is a simple example:

$$Mg_2SiO_4 + 4H_2O \rightarrow 2Mg^{++} + 4OH^- + H_4SiO_4.$$

When the aluminosilicates are involved, one of the products of hydrolysation is usually a clay mineral. The following equation, showing the hydrolysis of orthoclase to give kaolinite is probably an oversimplification but gives a general idea of the sort of reaction involved:

$$4KAlSi_3O_8 + 22H_2O \rightarrow$$
$$4K^+ + 4OH^- + Al_4Si_4O_{10}(OH)_8 + 8H_4SiO_4.$$

Reactions of this type are aided by the presence of dissolved CO_2 which increases acidity by the formation of the HCO_3^- ion with the consequent release of H^+ from the water. Acidity also of course greatly increases the solubility of *carbonate minerals*, which are not common constituents of most igneous and metamorphic rocks but are relatively abundant in sediments.

Hydrolysis attacks practically all the minerals present in rocks subjected to weathering, but the different reactions proceed at a variety of rates and hence it is permissible to think of *relative* mineral stabilities in the weathering environment. If we consider, for example, the minerals of the igneous rocks we find a considerable similarity between the order in which they tend to appear during magmatic crystallization and their relative stability in weathering. Olivines and pyroxenes, which are mainly formed at rather high magmatic temperatures, decompose readily whereas amphiboles tend to be more stable, and biotite, which generally crystallizes at rather low temperatures, is the most stable of all the common ferromagnesian minerals. Similarly, amongst the feldspars the general sequence of appearance during magmatic crystallization is calcic plagioclase first, then more sodic plagioclase, lastly potash feldspars such as orthoclase and microcline. Of these the calcic plagioclase decomposes most rapidly under weathering conditions and potash feldspar is the most stable (see Fig. 13. 5).

With relative mineral stabilities in mind it is possible to introduce the useful concept of *maturity* into the discussion of sediments. By a mature sediment we mean one which has suffered the effects of weathering and other processes, such as transport, to an advanced degree. It will show a high degree of mineralogical change relative to the parental material. Conversely an immature sediment will approach the nature of the source material more closely. To take a hypothetical example consider a source area which consists of the igneous rock granodiorite, a type intermediate between diorite and granite and characteristically composed of hornblende or biotite, plagioclase, potash feldspar and quartz. Let us also imagine (and this is a rather artificial assumption) that sediment can be traced from this area to considerable distances and that the processes of sedimentation and transport do not have a very significant effect upon sorting the different minerals into different deposits. Very close to the source area we may expect to find hornblende and biotite amongst the detrital grains and we may also find a little plagioclase, though it is likely that much of this will have decomposed even before the parent rock is mechanically sufficiently disrupted to provide sediment at all. Also in this area we shall expect to see quite significant amounts of potash feldspar together with quartz. As we proceed away from the source the ferromagnesian minerals and plagioclase probably disappear completely, and the ratio of potash feldspar to quartz falls. Ultimately the only important detrital original constituent left in the sediment will be quartz. In principle we could use such features as the quartz/potash feldspar ratio as a quantitative measure of the maturity of the sediment at different localities, and this sort of thing is useful in the reconstruction of ancient sedimentary environments. I perhaps

need hardly add that only in rather favourable circumstances, for example when it is reasonably certain that only one source is involved, that the rocks are free of post-depositional mineralogical changes etc., would it be possible to make useful application of this principle (for further discussion see Chapter 13).

I would like to turn now to consider the chemical composition of sedimentary rocks in a more general way, and here it is useful to do a sort of 'book-keeping' exercise to find out what has happened to the various chemical constituents as source rocks are weathered and their materials are dispersed. Most of the discussion so far has been concerned with the original minerals of the source rock and their response to weathering. These minerals, while they still remain recognizable in the sediment, are referred to as the *resistate* fraction. However, in addition to providing a resistate fraction to potential sediment, the source rock will also give up materials to *aqueous solutions*, while both resistates and solutions will give rise to the production of *new minerals*, of which the clay minerals are a notable example. Thus at any given moment all the chemical constituents of a former source rock must be found somewhere in one of these three guises, resistate, solute, or new mineral. With this in mind let us consider the fate of some individual elements derived from igneous source rocks.

The *alkali metals*, sodium and potassium, are held largely in feldspars in the source material and, as we have seen, these minerals are not particularly stable in the weathering environment and the alkali metals therefore pass into solution. The relatively low stability of plagioclase relative to the potash feldspars means that sodium passes into solution more rapidly than potassium. One of the characteristic features of igneous rocks is that their sodium content is usually greater than, or approximately equal to, their potassium content (see Table 3), and thus the effect of chemical weathering is to increase the potassium/sodium ratio in the resistate fraction. At the same time, potassium enters rather readily into newly formed clay minerals while sodium tends to remain in solution almost indefinitely. Thus most sediments are rich in potassium relative to sodium, either because they contain clays or because they contain a resistate fraction enriched in potash feldspar relative to plagioclase, while conversely the waters of the seas and oceans are relatively strongly enriched in sodium. The analyses of sedimentary rocks given in Table 4 illustrates their generally

TABLE 4 : CHEMICAL COMPOSITIONS OF SOME TYPICAL SEDIMENTARY ROCKS (AFTER COX, PRICE AND HARTE)

	(1)	(2)	(3)	(4)	(5)	(6)
SiO_2	74.14	78.14	99.14	58.10	55.02	7.61
TiO_2	0.15	—	0.03	0.65	1.00	0.14
Al_2O_3	10.17	11.75	0.40	16.40	22.17	1.55
Fe_2O_3	0.56	1.23	0.12	4.02	8.00	0.70
FeO	4.15	—	—	2.45		1.20
MgO	1.43	0.19	nil	2.44	1.45	2.70
CaO	1.49	0.15	0.29	3.11	0.15	45.44
Na_2O	3.56	2.50	0.01	1.30	0.17	0.15
K_2O	1.36	5.27	0.15	3.24	2.32	0.25
H_2O^+	2.66	0.64	0.17	5.00	7.76	0.38
H_2O^-					2.10	0.30
CO_2	0.14	0.19	—	2.63	—	39.27
C	—	—	—	0.80	—	0.09
	99.81	100.06	100.31	100.14	100.14	99.78

(1) Typical greywacke

(2) Typical arkose.
    Torridonian, Scotland

(3) Devonian orthoquartzite

(4) Silty Clay

(5) Residual clay from
    weathering of gneiss

(6) Allochemical limestone

potassic nature, the only exception being the grey-wacke, which is an example of an exceedingly immature sediment.

Aluminium in contrast is one of the most insoluble elements in the sense that it is present in only very small amounts in solutions derived from weathered rocks. Much aluminium is taken up directly by clay minerals which in some cases form virtually *in situ* by the degradation of the crystal lattices of silicates. The aluminium contents of sedimentary rocks are often therefore rather high (though it follows that the *average* aluminium content of sedimentary rocks must be virtually the same as the average aluminium content of igneous rocks). There are essentially two ways in which this can come about, either by the physical processes of sedimentation which cause clay minerals to be separated from other constituents and deposited to form clay-rich rocks (analysis 4 of Table 4), or by the formation of residual deposits (analysis 5 of Table 4). The latter are formed in areas where chemical weathering is intense while the rate of removal of the resistate fraction is low. Such conditions are best reached in tropical areas of low topographic relief and high rainfall, the ultimate

product being *laterite*, a red, earthy rock containing in extreme cases almost nothing but oxides and hydrated oxides of aluminium and iron, even the silica having been removed by intense leaching.

Iron shows some similarity to aluminium because of the extreme stability and insolubility of its oxide Fe_2O_3 (haematite) and hydrated oxide, goethite. Much weathering takes place under oxidising conditions, so that iron released by the decomposition of silicates is almost immediately precipitated in these forms. It is this of course which accounts for the abundance of iron in the laterites, and may lead to the formation of valuable residual iron ores in situations where the source rocks are poor in aluminium.

Calcium and Magnesium are taken into solution during weathering and tend to be present only in small quantities in most sediments. Sea water, however, appears to be approximately saturated with dissolved $CaCO_3$ and thus slight physical and chemical variations in the marine environment may lead to precipitation of calcite or aragonite. In addition to this, however, large amounts of calcium are removed from solution by the activities of organisms such as shell fish and corals.

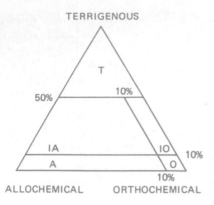

TERRIGENOUS

T

50%   10%

IA   IO   10%

A   O

10%

ALLOCHEMICAL   ORTHOCHEMICAL

1. 20   **Classification of sedimentary rocks:**
T — terrigenous rocks.
A — pure allochemical rocks.
O — pure orthochemical rocks.
IA — impure allochemical rocks.
IO — impure orthochemical rocks.

Thus both inorganically and organically precipitated carbonates can contribute to the formation of the common sedimentary rock, limestone. Magnesium appears in sediment mostly as dolomite, the double carbonate of magnesium and calcium.   Many dolomites appear to be formed by the alteration of limestones after their deposition.

Sedimentary Processes

   Most sedimentary rocks consist largely of transported fragments which are known by the general name of clasts. *Clasts* may be rock fragments, grains of resistate minerals, or grains and aggregates of new minerals such as clays or carbonates (e.g. shell fragments). Clasts are transported largely in water, but also to some extent by wind, and in the course of transportation two important things happen. Firstly the clasts tend to be reduced in size and angularity as a result of abrasion and secondly they tend to become sorted according to their grain size (see Chapter 13). Thus for example a sediment derived from a granitic source might be expected to consist of relatively large angular fragments of quartz and potash feldspar near the source area. Further away we might expect to find sandstones free of feldspar and composed of smaller, more rounded, quartz grains, while complementary deposits of finer-grained material such as clay might be found at an even

greater distance from the source, having become physically separated from the quartz grains as a result of their relative ease of transport. If we bear in mind the potential of the sorting and transporting processes, and the effects of chemical weathering, and if we add to this the possibility that new minerals may be formed actually in the area of deposition or in the sediment after it has been deposited, we have a good indication of all the processes which can interact to produce the large range of sedimentary rock types we observe.

Classification of Sedimentary Rocks

   The constituents of sedimentary rocks may be grouped under three headings to form the basis of a broad genetic scheme of classification:

   1. *Terrigenous constituents*, resistates and clay minerals derived from a source outside the depositional area.

   2. *Allochemical constituents*, that is chemically precipitated material formed within the depositional area but showing evidence of some degree of transport. Shell fragments are obvious examples.

   3. *Orthochemical constituents*, material chemically precipitated within the depositional area and showing no evidence of post-depositional transport.

Few rocks are composed exclusively of one of these types of constituent, and the classification is therefore based on relative proportions of constituents as shown in Figure 1. 20. The terrigenous rocks are the most abundant sediments (approximately 75 per cent of all sediments) and include mudstones (or clays, shales and siltstones) and sandstones in decreasing order of abundance. It may appear surprising that in fact the fine-grained terrigenous rocks such as clays, mudstones and shales are the most abundant of all sediments, but this is because they are generally rather soft rocks and do not form the imposing cliffs and other outcrops that are so often seen in sandstones and limestones. *Allochemical rocks* comprise about 15 per cent of all sediments and include shelly limestones and oolitic limestones among other types. The *orthochemical rocks* are relatively rare (about 10 per cent of all sediments), the most notable examples being some of the very fine-grained limestones and rock salt and related deposits.

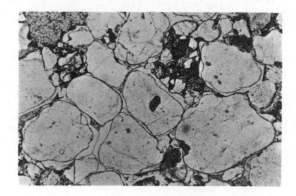

1. 21 Thin section of greywacke. Note particularly the angularity of the clasts, the most obvious of which are quartz, the variation in grain size (poorness of sorting), and the relatively large amount of fine-grained matrix. ($\times 25$)

1. 22 Thin section of sandstone composed predominantly of rounded quartz clasts. Original grain boundaries are shown by lines of dusty inclusions, and there is subsequent overgrowth by quartz to fill the inter-grain voids almost completely. ($\times 33$)

Terrigenous Rocks

The primary classification of the terrigenous rocks is by grain size into three groups, the *rudites* (average grain size >2 mm), *arenites* ($\frac{1}{16} - 2$ mm) and *argillites* (average grain size $< \frac{1}{16}$ mm). The rudites are represented by such rock types as *conglomerate* (rounded rock fragments, Fig. 13. 4b), *breccia* (angular rock fragments, Fig. 13. 4a) and such special types as *tillite* which are fossil boulder clays of glacial origin. The arenites (sandstones) are mainly rocks rich in quartz grains and amongst them the *greywackes* are probably the most abundant type. These are immature sediments with important amounts of clay matrix, feldspar grains and rock fragments (see Fig. 1. 21). The classification of the arenites is shown in Table 5. The argillites are represented by soft *clays* (e.g. the London, Gault and Oxford clays of southern England), compact *mudstones* and *siltstones* (which are transitional towards arenites) and by the common fissile rocks generally known as *shales*.

In terrigenous rocks especially it is important to consider the *cementing material* which binds the clasts together. This is mainly formed after deposition and converts unconsolidated sediment into more or less hard, coherent rock. This *lithification* is part of the wider process called *diagenesis* which covers all post-depositional changes of mineralogy and texture (excluding those generally thought of as taking place in response to increased temperature during metamorphism). Minerals formed during diagenesis are spoken of as *authigenic*, and an example is shown in Figure 1. 22 where we see quartz clasts overgrown by authigenic quartz in a sandstone.

Allochemical and Orthochemical Rocks

The most important rocks within this group are the limestones (rock consisting of more than 50 per cent $CaCO_3$) and the *dolomites*. These rocks are widely variable in appearance and mode of formation, and because of the complexity of the diagenetic changes they undergo—mainly recrystallization and dolomitisation—they are not an easy group to study. However, when diagenetic alteration is slight it is often possible to recognise both allochemical and orthochemical constituents. The *allochemical* constituents consist mainly of *fossil* and *shell fragments* which are sometimes abundant enough to justify the use of such terms as foraminiferal limestone, coral limestone etc. *Ooliths*, which are sub-spherical carbonate aggregates having a radial or concentric structure formed by crystallization round nuclei such as sand grains, are also common allochemical constituents, particularly for example in the Jurassic limestones of southern England. Ooliths are formed in environments where calcite is being precipitated and there is strong and continuous wave action.

TABLE 5: CLASSIFICATION OF ARENITES (BASED ON PETTIJOHN)

Matrix of fine, terrigenously derived silt or clay	Matrix prominent (>15%)	Matrix sparse or absent (<15%)	
SAND FRACTION Feldspar exceeds rock fragments	FELDSPATHIC GREYWACKE	ARKOSIC SANDSTONE	ORTHOQUARTZITE
Rock fragments exceed feldspar	LITHIC GREYWACKE	LITHIC SANDSTONE	
Quartz content	Variable; generally <75%	75–95%	>95%

The orthochemical constituents are represented in rocks by the fine-grained calcite aggregates called *micrite*, and coarser aggregates termed *sparry calcite cement*. The common distribution of limestone types is represented in Figure 1. 23. The absence of micrite from the sparry allochemical limestones suggests that microcrystalline calcitic ooze may have been removed by relatively strong current action. Some limestone types are illustrated in Figures 1. 24 and 1. 25.

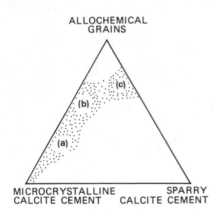

ALLOCHEMICAL GRAINS

MICROCRYSTALLINE CALCITE CEMENT   SPARRY CALCITE CEMENT

1. 23 The three main types of limestone: (a) microcrystalline limestones (micrites); (b) microcrystalline allochemical limestones; (c) sparry allochemical limestones. The shaded areas represent main fields of limestones when plotted in terms of the three end members.

Apart from the carbonate rocks the siliceous rock *chert*, iron rich sediments such as the *chamositic iron ores* (rocks containing ooliths of the iron mineral chamosite, e.g. as in the bedded iron ores of the Midlands and Inner Hebrides), and the *evaporites* are the most important chemically formed rocks. The evaporites are formed by intense concentration of sea water with the result-ant precipitation of such minerals as gypsum ($CaSO_4.2H_2O$), anhydrite ($CaSO_4$), rock salt or halite ($NaCl$) and a variety of potassium salts. These deposits form in present-day hot desert areas such as the margins of the Persian Gulf and the Red Sea. They are also important within the geological column and form many economic mineral deposits, as for example in the Triassic rocks of north-east England and Cheshire. Evaporites are prone to recrystallization and be-come very mobile under pressure, flowing in a way analogous to that of igneous magmas only much more slowly. Intrusive masses of salt rising through the overlying sediments are known as *salt domes* and have economic importance in form-ing structures suitable for trapping oil.

METAMORPHIC ROCKS

We have already discussed two sorts of change which can happen to rocks essentially in the solid state. One of these is chemical weathering, and the other, those post-depositional changes in sediments which were referred to as diagenetic. Both these processes involve mineralogical and textural adjustments in response to changes in the physico-chemical environment of the rock, but the changes concerned take place at low temperatures at or near the surface of the Earth. If we consider what happens at higher temperatures, for example when rocks are heated by the intrusion of magma nearby or by deep burial, we enter the field of

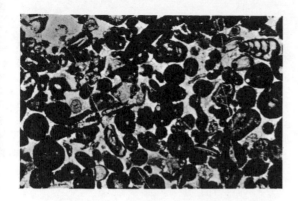

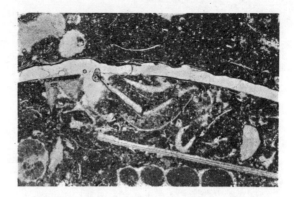

1. 24 Thin section of oolitic limestone. Ooliths and shell fragments (dark) are set in a sparry calcite matrix (light). This is a representative of the group of sparry allochemical limestones shown in Figure. 1. 23. (×10)

1. 25 Thin section of microcrystalline limestone (micrite) containing shell fragments. (×8)

metamorphism. It is not, however, possible to divide metamorphism sharply from diagenesis and weathering. In fact it would be perfectly permissible (though it is not very usual) to think of the latter processes as metamorphic changes of a particularly low-temperature type.

There are three principal agents of metamorphism, namely heat, stress (confining pressure and shearing stress), and the mobility of chemical constituents. It is not very profitable to discuss which of these factors is the most important, but the effects of heat are perhaps the most striking (certainly in a mineralogical sense), so I should like to discuss them first.

In talking about chemical weathering I pointed out that mineral assemblages which are stable at high temperatures are very commonly found in a metastable state at low temperatures. This applies to any igneous rock which we collect at the surface of the Earth, the failure of the assemblage to re-equilibrate at the lower temperature being due to the extreme sluggishness of the reactions involved. The process of re-equilibration, as we saw, is materially aided by the introduction of water in the environment of weathering. The general conclusion to be drawn from this, however, is that reactions between minerals in response to falling temperature take place only with relative difficulty. The converse is not true, for if an assemblage stable at a low temperature is heated it will react with comparative rapidity to produce a new assemblage more nearly in equilibrium at whatever higher temperature is reached. With subsequent cooling,

which must of course have affected all the rocks we actually collect, the sluggishness of reactions will again become important so that the rock may retain its high-temperature mineralogy just as most igneous rocks do. Reactions are also inhibited during this stage by the comparative dryness of the rocks, water having been driven off during the heating stage. Thus if we consider the case of a sedimentary rock which is heated up for a period and then cooled, the mineralogy we see is most likely to be characteristic of the highest temperature the rock has been subjected to during the metamorphic process. This type of metamorphic result is by far the most common and is termed *prograde*, that is metamorphic change in response to rising temperature. The term *grade* is often used to refer to the temperature attained during metamorphism as indicated by the mineral assemblage, a rock showing a high-temperature assemblage being termed 'high-grade' and a rock with a low-temperature assemblage being termed 'low-grade'.

Some metamorphic changes do however involve re-equilibration to lower temperatures, and these are called *retrograde*. A high-grade metamorphic rock may for example be metamorphosed a second time and held for a substantial period at a temperature lower than that of the first metamorphism. Any changes occurring would then be referred to as retrograde. Alternatively the rate of cooling in a particular metamorphic episode might be abnormally low, so that there would be a better opportunity than usual for down-temperature reactions to take place. Similarly, the metamorphism of many igneous rocks involves the replacement of a very high-temperature original mineral assemblage by a metamorphic assemblage stable at a lower temperature. This type of change, though not strictly referred to as retrograde since the starting material is not a metamorphic rock, nevertheless has all the essential characters of retrograde metamorphism.

So far in talking about mineral assemblages I have neglected the pressure factor. Pressure and temperature in fact go hand in hand in controlling the stability ranges of minerals and mineral assemblages, but let us ignore this factor for the moment and look at some examples of temperature-controlled changes.

Thermal Metamorphism

The simplest examples we can choose belong to the field of *thermal* or *contact* metamorphism. These take place in the country-rocks adjacent to igneous intrusions, the ring of metamorphic rocks surrounding the intrusion being known as a *metamorphic aureole*. In such aureoles metamorphic heating is the dominant effect and there are usually relatively few complications introduced by shearing stresses, intense pressures, and the consequent deformation which these may cause. We find that the highest grade rocks lie closest to the intrusion, and as we go away from the intrusion the grade decreases until the aureole passes insensibly into unaltered country-rocks. Consider a typical sequence of changes taking place in argillaceous rocks intruded by a granite :-

(i) bedding becomes obliterated, and the rocks become hard and relatively coarsely crystalline. The rocks may take on a spotted appearance due to segregation of organic material into clumps and the appearance of crystals of chiastolite (andalusite, Al_2SiO_5) and cordierite.

(ii) microscopic examination reveals a coarsening of texture accompanying recrystallization. Low temperature minerals such as clays and chlorites disappear. Minerals such as micas (especially biotite) grow and may be accompanied by large crystals of chiastolite and cordierite. In extremely high temperature metamorphism another Al_2SiO_5 polymorph, sillimanite, may be found.

1.26 **Granoblastic texture in pyroxene-plagioclase rock formed by thermal metamorphism of original basalt. The mutual interference of the two minerals as they crystallized simultaneously in the solid state results in a texture very different from that shown by the dolerite in Figure 1.15. ($\times$25)**

The high temperature rocks formed in this way are generally termed *hornfels*. These rocks have a characteristic texture known as *granoblastic*, in which mineral grains tend to be equidimensional and randomly oriented. Figures 1.15 and 1.26 show an interesting contrast between typical igneous texture and granoblastic texture, developed in this case in rocks of basaltic composition.

The mineralogy of thermally metamorphosed rocks is highly variable because it depends on both metamorphic grade and the bulk chemical composition of the starting material. The marbles are of particular interest because of the great range of minerals which may be formed by the reaction of calcium and magnesium with impurities such as clay minerals and quartz.

Effects of Pressure

As mentioned previously, temperature-controlled changes are often fairly obvious and striking in metamorphic mineral assemblages. Pressure-controlled effects are less so because most reactions are not as sensitive to pressure change as to temperature change. In certain environments we do find rocks which have been metamorphosed under abnormally high pressure, and here we may find mineral assemblages which are in considerable contrast to those in rocks of the same composition formed at lower pressures.

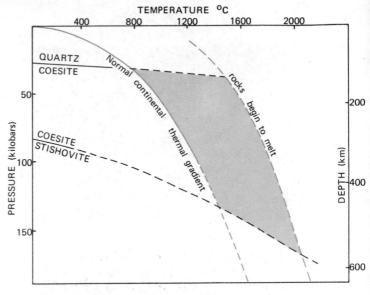

The most deep-seated rocks we have available for our inspection are the ultrabasic nodules brought up from great depths in the volcanic vents known as *kimberlite pipes* (named after the diamond mining area of South Africa). The nodules are mainly olivine-rich and contain lesser amounts of a number of minerals (including pyroxene and magnesium-rich garnet). A much rarer group of nodules, known as the eclogites, consist *essentially* of magnesium-rich garnet and sodium-bearing pyroxene with or without other minerals such as kyanite and, very rarely, diamond. Diamonds are found disseminated in the matrix of some kimberlite pipes, and a very few examples have been found in eclogites. The diamonds occasionally contain minute inclusions of coesite, a high pressure polymorph of quartz, but do not contain stishovite, another polymorph forming at even higher pressures.

Experimental work on the stability fields of diamond and the high-pressure polymorphs of quartz indicates that at least some of the material in kimberlite pipes originates from depths within the Earth of more than 150 km (see Fig. 1. 27) where the confining pressure is 50 kb or so.

Ordinary, crustal, metamorphic rocks, which are exposed at the surface as a result of uplift and denudation, are of course formed under conditions of much lower pressure (about 1 kb to 10 kb) than diamond and eclogite, but nevertheless considerable variations in mineralogy may result even within this range, as we shall see in the next section.

1. 27 PT diagram to illustrate the range of conditions under which natural diamonds probably form (coloured area). Two curves show the quartz-coesite and coesite-stishovite inversions. The temperature at depth is likely to lie between the other two curves, one of which represents probable minimum temperatures, and the other maximum temperatures. Temperature is unlikely to rise far above the curve showing the onset of melting because of buffering by latent heat of fusion.

Regional Metamorphism

Most metamorphism belongs to this type, characteristically developed over very large areas (often thousands of square kilometres) particularly in regions of mountain building. In contrast to the contact metamorphism already discussed, regional metamorphism is accompanied by more or less intense deformation caused by directed pressures and shearing stresses, and the resulting rocks show a number of distinctive structural and textural features.

Schistosity (see Figs. 1. 28 and 1. 29) is one of the most widespread metamorphic structures and refers to a parallel or sub-parallel orientation of inequidimensional mineral grains. Planar schistosity is most commonly seen in the metamorphosed argillaceous sediments known as *mica schists* where the structure is due to the parallel arrangements of tabular mica crystals. Linear schistosity may be seen in some rocks, and may be

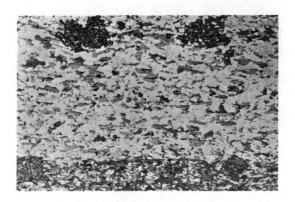

1. 28 Thin section of schist showing parallel orientation of micas (medium grey, elongated grains) giving schistosity. The foliation, shown by variations in the proportions of mica to feldspar and quartz (clear grains) and in the distribution of garnet (larger, dark grains), is due to original sedimentary bedding. ($\times$20)

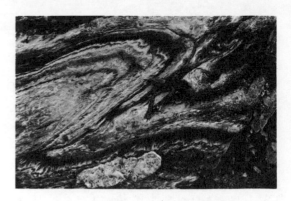

1. 29 Thin section of mica schist showing intense deformation with the development of minor folds leading to a cleavage oblique to the original schistosity. (×4)

1. 30 Field photograph of highly deformed gneiss. The hammer indicates the scale of the mineral banding.

produced by the parallel orientation of elongated minerals such as amphibole. Well-developed planar schistosity in fine-grained rocks is known as *slaty cleavage* and is characteristic of low grades of metamorphism where heating has not been sufficient to produce marked recrystallization. The rocks termed *phyllites* have characters intermediate between those of the low-grade slates and the higher-grade mica schists.

Foliation denotes the structure in which particular minerals are aggregated into layers, lenticles or streaks during the metamorphic process. Foliation and schistosity are commonly both present and are parallel to each other. Rocks showing a coarse foliation are referred to as *gneisses* (see Fig. 1. 30), though gneissose banding is also developed from original inhomogeneities in the rock such as bedding. Gneissose rocks, of course, usually belong to high metamorphic grades.

Metamorphic Zones and Isograds

In the field study of metamorphic rocks it is extremely useful to be able to compare the grade of metamorphism from one place to another and thus to map zones in which conditions of metamorphism have been approximately the same. Studies of this type were initiated by George Barrow working on the rocks of the south-eastern part of the Scottish Highlands in 1893. Barrow studied the mineralogy of *pelitic schists*, the term pelitic being used to refer to metamorphic rocks derived from original shales (as opposed to *psammitic*—derived from sandstones), so that the

question of variability of bulk composition was largely eliminated. In this area (see Fig. 1. 31) the pelitic rocks are represented by slates near the Highland Boundary Fault and gradually change to high-grade sillimanite-garnet-mica schists as they are traced to the north-west. Barrow divided this sequence into six zones, each based on the appearance of a characteristic metamorphic mineral. Zone boundaries are termed *isograds* and are surfaces along which metamorphic grade is approximately constant. The garnet isograd for example is the surface (line on the map) separating the garnet-bearing rocks of the garnet zone (high-grade) from the garnet-free rocks of the biotite zone (low grade). The surface ideally represents the locus of points all of which just reached the minimum necessary temperature for the formation of garnet in sediments of pelitic composition.

The characteristic mineral assemblages developed in the different zones are listed in Table 6. It must be stressed that here we are seeing mineralogical changes in response to rising temperature and pressure *in rocks of essentially uniform chemical composition*. The highest grade rocks are those which reached the highest temperatures, and were also metamorphosed under the greatest pressures. These factors were probably largely controlled by depth of burial during the metamorphism. However, Barrow's zones of the south-eastern Highlands represent a specific geological case, and in other metamorphic events the relationship between temperature and pressure is somewhat different. In the Buchan area of the Scottish

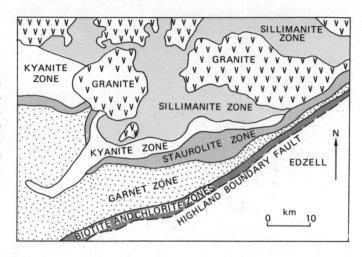

1. 31 Metamorphic zones near Edzell, Angus, along the border of the Scottish Highlands. The rocks to the south-east of the Highland Boundary Fault are younger, unmetamorphosed sediments and volcanics. The granites post-date the metamorphic rocks but belong to the same general episode of mountain building, the Caledonian Orogeny.

Highlands for example (Aberdeenshire, Banff), medium- to high-grade pelitic rocks characteristically develop andalusite and cordierite instead of garnet, staurolite and kyanite. Experimental studies (see Fig. 1. 32) on the polymorphs of Al_2SiO_5 (andalusite, kyanite, and sillimanite) lead to the conclusion that the rocks of the Buchan zones were formed at lower pressures than those of Barrow's zones, though over a similar range of temperatures. This has been interpreted as representing a lesser depth of burial in the Buchan area. In fact, though showing the structures characteristic of regionally metamorphosed rocks, the mineral assemblages produced in the Buchan type of metamorphism show considerable similarity to those produced during contact metamorphism at high levels in the crust. Conversely some metamorphic terrains show evidence of conditions of very high pressure without particularly high temperatures (see Fig. 1. 33), the characteristic rocks being schists carrying the soda-rich amphibole, glaucophane.

TABLE 6: SOME TYPICAL MINERAL ASSEMBLIES DEVELOPED IN PELITIC ROCKS OF BARROVIAN ZONES

ZONE	MINERAL ASSEMBLAGES
Chlorite	quartz-muscovite-chlorite-albite
Biotite	quartz-muscovite-chlorite-biotite-albite
Garnet (almandine)	quartz-muscovite-chlorite-biotite-garnet-albite or oligoclase★
Staurolite	quartz-muscovite-biotite-garnet-oligoclase-staurolite
Kyanite	quartz-muscovite-biotite-garnet-oligoclase-kyanite
Sillimanite	quartz-muscovite-biotite-garnet-oligoclase-sillimanite

★Oligoclase is a sodic plagioclase containing 10–30 per cent of the anorthite (calcium) end-member.

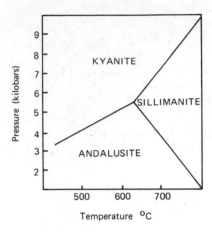

1.32 PT diagram showing stability fields of the polymorphs of Al_2SiO_5.

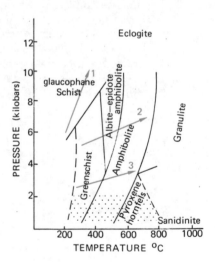

1.33 PT diagram showing metamorphic facies and three standard types of metamorphism. (1) Glaucophane schist facies type. (2) Barrovian type. (3) Abukuma type. Note that the 'triple point' of the Al_2SiO_5 polymorphs (Fig. 1.32) lies between curves (2) and (3). The stippled area indicates the range of pressures and temperatures realized in contact metamorphism.

Metamorphic Facies

Zonal classifications such as have been discussed above do not lend themselves particularly well to the comparison of different metamorphic terrains unless it is possible to consider not only differing pressure-temperature relations, as we have above, but also the complications introduced by variations in bulk composition. A metamorphic facies consists of a group of rocks of varying composition, all of which have been metamorphosed under similar conditions. A facies classification was first proposed by Eskola and depends on the study of large numbers of examples of metamorphic rocks of differing chemical compositions which are closely associated in the field and may therefore be assumed to have been metamorphosed under similar conditions. For example by studying metamorphosed basic igneous rocks associated with original pelitic sediments it is possible to see what mineral assemblages are developed in them under the same P-T conditions. The study of original limestone bands associated with pelites or with metamorphosed basic igneous rocks enables the comparison to be extended, and so on.

Because of the very large number of reactions possible when a whole range of starting compositions is considered it would be theoretically possible to define a large number of facies. For example the reactions which produce, successively, biotite, garnet and staurolite in pelitic rocks will not in general take place under the same pressure-temperature conditions as mineralogical changes in, say, associated calcareous rocks or metamorphosed basalts. Thus garnetiferous pelitic schists could be assigned to numerous different facies on the basis of the mineralogy of a whole range of associated rocks, and this would result in a rather fine-scale subdivision of the P-T conditions of metamorphism. However, a relatively sophisticated classification of this sort becomes difficult to apply, and for these practical reasons Eskola proposed only eight facies, the names of which are largely derived from the mineral assemblages developed in metamorphosed basic igneous rocks. Each facies encompasses a relatively large range of mineral assemblages and covers a wide range of P and T conditions. The wide range covered by each facies implies that certain bulk compositions which are sensitive to P-T changes show several different mineral assemblages within one facies. For example the rocks of the staurolite, kyanite and sillimanite zones in the Barrovian type of metamorphism are all of the amphibolite facies.

Facies boundaries are approximately shown on the P–T diagram of Figure 1. 33. Also shown are three curves representing different P–T relations in what Myashiro recognized as 'standard' type of metamorphism. At the high-pressure/low-temperature end of the scale we have the glaucophane-schist facies type, followed by the Barrovian and Abukuma (a locality in Japan) types at successively low pressures. The Abukuma type approximates to the Buchan metamorphism already discussed, and its close relationship to the conditions realised during contact metamorphism is illustrated.

FURTHER READING:

K. G. COX, N. B. PRICE AND B. HARTE 1967. An Introduction to the Practical Study of Crystals, Minerals and Rocks. *McGraw–Hill, London.*

A. HARKER 1960. Petrology for Students. *Cambridge University Press.*

Until the 17th century rocks, and the fossils contained in them, were interpreted as being the products of the biblical deluge, and so little attempt was made to determine their relative ages, let alone their absolute ages. The beginning of a more rational approach to dating rocks was made in 1669 by a Dane, Nicolaus Steno, who formulated the *Law of Superposition*, recognising that rocks higher in a sequence of sediments were younger than those below. Thus a relative order of events could be compiled in any one area. But it was not until the early 19th century that William Smith, the 'Father of English Geology', realised that strata could be equated from place to place if they contained similar fossil assemblages.

Studies using the law of superposition and fossil faunas and floras could only produce a *relative* time scale, and efforts were made during the 19th and early 20th centuries to quantify it. These methods tried to determine the age of the Earth from calculations concerning rates of sedimentation or the accumulation of dissolved salts in the sea. Many of these results gave an age of around 100 million years.

In 1897, the eminent physicist Lord Kelvin calculated the age of the Earth on the assumption that it had cooled from an initial molten state; he also made estimates based on assumptions concerning the origin of the Sun's heat, and proclaimed that both methods gave an age of 20 to 40 million years. So strong was the impact of these calculations and so great the magic of a precisely determined mathematical figure, arrived at by making seemingly valid assumptions, that attempts were made to compress the geological complexities of the Earth's history into this very short period of time.

What Kelvin did not know was that certain elements are unstable and, with time, decay to form other elements. This process of radioactive decay involves the liberation of energy in the form of heat—so Kelvin's answer for the Sun and the Earth had to be wrong. It is pertinent to note that the geologists were on the right track. If they had not been 'brain-washed' by Kelvin's mathematical arguments, they would undoubtedly have arrived at a figure in the right order of magnitude.

It was realized early in the 20th century by Rutherford and Holmes in England, and Boltwood in America, that the decay of unstable elements to produce radiogenic isotopes could be used to date the minerals and rocks of the Earth's surface. However, at that time, analytical techniques were nowhere near precise enough to determine the minute quantities of the radiogenic isotopes present in rocks. So it was not until the 1950's, when the mass spectrometer was developed, that rock dating became commonplace. Even now, it is a complex process fraught with difficulties. In the following article, Dr. Stephen Moorbath, of the geochronological laboratories in the Department of Geology and Mineralogy at the University of Oxford, describes the methods used to date rocks and minerals, the applicability and limitations of the methods, and the uses of radioisotope age data in geological and geophysical research.

2 Measuring Geological Time

by STEPHEN MOORBATH

The age of rocks and minerals can be determined by measuring the accumulation of the products of radioactive decay in them. The result is a time scale of fundamental significance to most branches of the Earth sciences.

IN RECENT YEARS enormous progress has been made in the study of the historical aspect of all branches of the Earth sciences by the development of precise and accurate methods for measuring the age of geological materials. These methods, which are based on the radioactive decay of certain nuclides with very long half-lives, are applicable over virtually the entire range of geological time (ca. 4600 million years) and to a wide variety of commonly occurring rocks and their constituent minerals.

The types of geological events which can be dated by these methods are (i) the crystallisation of igneous rocks from a magma, (ii) the recrystallisation of pre-existing rocks, of any type, to form metamorphic rocks, usually with a totally new mineral assemblage, (iii) uplift, cooling and erosion of fold mountain chains, and (iv) deposition of a sedimentary rock, provided that one or more new minerals are formed during, or very soon after, sedimentation.

Direct dating, expressed in millions of years measured backward from the present, can clearly yield fundamental information on the timing and duration of some of the major events and processes which have shaped the surface of the Earth in the geological past and given it the features which we observe in the geological present. This will be evident from many of the other contributions to this book.

The science of dating rocks in all its aspects is termed *geochronology*. In using the basic equations of radioactive decay* to calculate the age of a geological sample from the measured parent/daughter ratio and the appropriate decay constant or half-life, it is necessary to remember that the calculated age will yield the time of crystallisation

*The fundamental law of radioactive decay states that the number of atoms disintegrating per unit time ($-dN/dt$) is proportional to the total number of radioactive atoms, N, present. Hence we can write $-dN/dt = \lambda N$, where the proportionality factor, λ, is termed the *decay constant* and has a characteristic value for each radionuclide. It represents the probability that an atom will disintegrate in unit time (year for long-lived radionuclides). On integrating the above equation, we get $N = N_0 e^{-\lambda t}$, or $N_0 = N e^{\lambda t}$, where N_0 indicates the number of radioactive atoms at some original time t, and N the number present to-day. This equation forms the basis for calculating a geological age from the analytical data; the number of radiogenic daughter atoms, N_D, formed during a period of time t is given by

$$N_D = N_0 - N = N e^{\lambda t} - N = N(e^{\lambda t} - 1).$$

Therefore, $N_D/N = (e^{\lambda t} - 1)$, and solving for t, we have

$$t = \frac{1}{\lambda} \log_e \left[1 + \frac{N_D}{N} \right].$$

Several varieties of this basic equation are widely used in geological age calculations. Note that the commonly used term *half-life* ($T_{1/2}$) i.e. the time required for the radioactivity of a given amount of radionuclide to decay to one-half of its initial value is related to the decay constant (λ) by the following equation: $T_{1/2} = 0.693/\lambda$.

or recrystallisation of the rock only if (i) there have been no losses or gains of either daughter or parent nuclide by processes other than radioactive decay of the parent, (ii) the half-life of the parent nuclide is accurately known, (iii) precise correction can be made for any daughter nuclide incorporated into the mineral or rock at time of crystallisation, (iv) the mineral or rock was formed in an interval of time that is short when compared with its actual age.

When more than one dating method is applied to a mineral, or to a cogenetic mineral assemblage, and the resulting dates agree within analytical error, the above conditions are clearly obeyed. Such dates are termed *concordant* and are found in many rocks which have had a relatively simple geological history. However, *discordant* dates are also frequently found in certain types of geological environment. They can result from the failure of one or more of the above conditions in regions with a more complex geological history. As will become evident later, such discordant age-patterns can be extremely useful when several geological events have been superimposed on one another in a single area.

Direct measurements of long half-lives are obviously basic to geochronology and have been carried out in many physics and radiochemistry laboratories. Nevertheless, precise measurements of geological specimens are made difficult by the low-energy radiation and slow disintegration rates of long-lived radionuclides. In practice, radionuclides and their daughter nuclides can be respectively distinguished from stable and non-radiogenic isotopes of the same element by isotopic analysis with a mass spectrometer, which is by far the most important analytical tool of the geochronologist and which can also be used to measure precisely and accurately the frequently minute absolute quantities of parent and daughter nuclides in rocks and minerals. The actual laboratory techniques for determining so-called *radioisotope*, *isotopic* or *radiometric* ages of rocks (which used to be erroneously called *absolute* ages) are extremely complex. Further details of some of the principal techniques used by the fifty or more geochronology laboratories throughout the world may be found in the references at the end of this chapter.

Published age determinations are usually quoted with an error which is most frequently somewhere between two to five per cent of the actual age, e.g. 1 200 ± 40 million years, or 50 ± 2.5 million years. These errors refer only to uncertainties in the analytical measurements and represent a probability—usually expressed as a 66 or 95 per cent confidence level — that a replicate measurement on the same sample would fall within the quoted limits. The errors do not take into account uncertainties in the half-life of the relevant radionuclide, or in the geological interpretation.

POTASSIUM-ARGON (^{40}K – ^{40}Ar or K – Ar) METHOD

Potassium is a common and widespread element in many rocks and minerals (see Chapter 1). Of its three isotopes, namely ^{39}K, ^{40}K and ^{41}K, only ^{40}K is radioactive, forming a constant proportion of the total potassium of only 0.0119 atom per cent. ^{40}K decays by two entirely different processes, each with its own characteristic half-life, as follows:

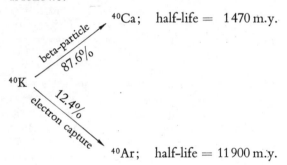

The ^{40}K – ^{40}Ca method is only rarely used for dating, because most common rock-forming minerals already contain so much primary calcium, which is mainly ^{40}Ca, that the comparatively minute amounts of radiogenic ^{40}Ca cannot be determined with anything like sufficient precision for geological age work.

The K-Ar method can be used to date a wide variety of rock types over almost the entire geological age range, including meteorites (ca. 4 600 million years old) as well as terrestrial rocks older than 3 000 million years to younger than 1 million years. However, rocks younger than about 0.1 million years can be very difficult to date because of the minute amounts of radiogenic ^{40}Ar present.

Commonly occurring igneous or metamorphic minerals suitable for K-Ar dating include biotite, muscovite, hornblende and nepheline. High-temperature forms of the potassium-feldspars (e.g. sanidine) and of the sodium-calcium (plagioclase) feldspars are generally suitable, particularly in Tertiary and Quaternary igneous rocks. In contrast, the most common potassium-feldspars, such as orthoclase and microcline, are unsuitable for K-Ar dating because they can lose argon readily even at atmospheric temperatures over long time intervals, due to the existence of gross structural defects in the crystal lattice and at boundaries between individual grains.

In many cases, whole-rock samples can be dated as they are found without mineral separation. (The term *whole rock* means just what it says: a piece of rock as it is found cropping out on the Earth's surface.) This is particularly useful for fine-grained volcanic rocks (lavas) such as basalts, andesites and rhyolites, and also for fine-grained, essentially micaceous, metamorphosed sediments, such as slates and phyllites. The marine sediments, limestone and sandstone, sometimes contain the potassium-bearing mineral glauconite, which is formed at time of sedimentation and can be dated by the K-Ar method.

Measured K-Ar dates do not necessarily relate to the time of crystallisation, because diffusive loss of radiogenic ^{40}Ar can occur at temperatures well below those of igneous and medium-to-high grade metamorphic crystallisation (ca. 1 200–1 000°C and 700–400°C respectively). This is because the radiogenic argon atom does not easily fit into the potassium lattice site in which it was formed and can escape out of the crystal lattice by diffusion, eventually finding its way upwards to join the reservoir of atmospheric ^{40}Ar which makes up nearly one per cent of the Earth's atmosphere. The so-called 'blocking' or 'diffusion-threshold' temperatures below which different minerals become effectively closed systems to argon diffusion are not known precisely, but are probably about 150–250°C for biotite, 250–350°C for muscovite and sanidine, and 400–500°C for hornblende. Furthermore, the basic laws of diffusion show that the larger the grain size of a particular mineral, the higher will be the effective blocking temperature. It is now known that very large, deep-seated igneous intrusions and regionally metamorphosed parts of the Earth's crust which have gradually been uplifted from great depths and have cooled extremely slowly from the initial crystallisation temperatures may yield K-Ar dates younger than the true age of crystallisation by several tens, or even hundreds, of millions of years. Furthermore, different types of minerals may become closed systems to argon diffusion at different times, because of the respective differences in their block-ing temperatures. For a given grain size the measured dates will clearly be in the order hornblende date > muscovite date > biotite date. These dates relate to various stages in the uplift and cooling history of the rocks. On the other hand, K-Ar dating of a lava-flow, volcanic ash or a small body of igneous rock emplaced high up in the Earth's crust may give the true age of crystallisation within analytical error, because such rock types cooled within a very short time interval compared to their actual age. Furthermore, K-Ar dates approaching the true age of metamorphism may be obtained from fine-grained, metamorphosed micaceous sediments, such as slates and phyllites, which recrystallised completely at only about 200–300°C, leading to complete expulsion of radiogenic argon inherited from the original sediment, but at temperatures only slightly above that at which subsequently formed radiogenic argon could begin to accumulate.

Cases have been recognised where mineral samples yield anomalously old K-Ar dates — sometimes greater than the accepted age of the Earth—because they have apparently incorporated radiogenic ^{40}Ar into their crystal lattices during crystallisation. This comparatively uncommon phenomenon tends to show up in potassium-poor and/or geologically young samples which contain very little intrinsic radiogenic argon, although it has also been observed in old micas with high potassium contents. This so-called 'excess' argon is probably produced by heating and outgassing of ancient, potassium-bearing rocks at some considerable depth and pressure in the Earth's crust, where some of it became incorporated into crystallising igneous and metamorphic minerals.

RUBIDIUM-STRONTIUM (^{87}Rb -^{87}Sr or Rb -Sr) METHOD

The two isotopes of natural rubidium are ^{85}Rb and ^{87}Rb, of which the atomic abundances are 72.8 and 27.2 per cent respectively. The decay scheme is as follows:

$$^{87}Rb \xrightarrow{\text{beta particle}} {}^{87}Sr;$$

half-life = 47 000 or 50 000 million years.

Unfortunately, there is still no agreement on the exact value for the half-life of ^{87}Rb because this is so difficult to determine by direct radioactivity measurements. Most workers nowadays seem to prefer the higher value because better agreement is obtained with K-Ar and U-Pb ages on identical, or other, minerals from the same rock. However, both half-life values are still widely used and it should be noted that ages calculated with the 47 000 million year half-life are 6 per cent lower than those calculated with the 50 000 million year value. Although this half-life uncertainty is much greater than for the other radionuclides used in rock dating, the advantages and potentialities of the Rb-Sr method far outweigh it.

The rare element rubidium forms no minerals of its own, but is always found in potassium minerals where it can easily substitute for the chemically closely related potassium in the crystal lattice because of similarity in ionic radius and charge. Most of the suitable minerals for Rb-Sr dating contain between about 100 to 1 000 parts per million of rubidium.

The Rb-Sr method can be used to date such common rock-forming minerals as muscovite, biotite, and (in contrast to the K-Ar method) all types of potassium-feldspars, including orthoclase and microcline. Hornblende is not suitable because it contains too little rubidium. The sedimentary mineral glauconite has also been successfully dated. Rb-Sr dating may also be carried out on whole-rock specimens without prior mineral separation, particularly those types of igneous and metamorphic rocks that are rich in potassium-feldspar and mica minerals, such as granites and granitic gneisses. Basic igneous and metamorphic rocks are usually not suitable. In whole-rock work, a sample of rock weighing several pounds is completely crushed to a fine powder, after which small aliquot samples are taken for rubidium, strontium and strontium isotope analyses. The Rb-Sr method may be applied to rocks of almost any geological age, although the comparatively slow decay rate of ^{87}Rb can create analytical difficulties for material younger than about 20–30 million years, unless it has an exceptionally high ^{87}Rb content.

In Rb-Sr dating it is usually necessary to make a correction for any so-called 'common' or 'normal' strontium which was incorporated into the mineral or rock when it crystallised. Although most common strontium follows the geochemically closely related calcium, a small amount is usually also incorporated into calcium-poor, rubidium-bearing potassium minerals. For example, micas and potassium-feldspars commonly contain between about 5 and 300 parts per million of common strontium, which may actually exceed considerably the amount of radiogenic strontium formed since crystallisation. Common strontium consists of isotopes of mass 84, 86, 87 and 88, in which the relative abundance of ^{87}Sr to the others can be quite variable in different crystallisation environments. It is therefore essential to distinguish between common ^{87}Sr incorporated on crystallisation and subsequently formed radiogenic ^{87}Sr. This can sometimes be done by isotopic analysis of a coexisting, cogenetic calcium mineral such as plagioclase which characteristically has an extremely low Rb/Sr ratio and in which the strontium isotope composition has remained unchanged since crystallisation, regardless of the age of the rock. The non-radiogenic isotope ^{86}Sr is customarily used as a measure of the common ^{87}Sr component. The basic equation of radioactive decay can then be used in the form:

$$ t = \frac{1}{\lambda} \log_e \left[\frac{(^{87}Sr/^{86}Sr)_P - (^{87}Sr/^{86}Sr)_0}{^{87}Rb/^{86}Sr} + 1 \right] $$

where t is the age of the rock, λ is the decay constant of ^{87}Rb, $(^{87}Sr/^{86}Sr)_P$ is the measured isotope ratio in the dated mineral, $(^{87}Sr/^{86}Sr)_0$ is the isotope ratio at time of crystallisation (i.e. in the plagioclase with extremely low Rb/Sr ratio). ^{87}Rb/^{86}Sr is computed from the analyses for total rubidium and strontium in the sample.

As stated above, the ratio $(^{87}Sr/^{86}Sr)_0$ varies significantly in different types of rocks, depending upon the pre-crystallisation history of the rock, particularly as regards the age and Rb/Sr ratio of its source region. For most igneous rocks derived from deep-seated sources in the lower crust or upper mantle the $(^{87}Sr/^{86}Sr)_0$ ratio varies from a minimum value of about 0.700 to about 0.715; for metamorphic rocks with a lengthy crustal pre-crystallisation history and with relatively high Rb/Sr ratios it can be much higher. The $(^{87}Sr/^{86}Sr)_0$ ratio, usually referred to as the *initial* strontium isotope ratio, is an important parameter for studying the ultimate origin of igneous rocks.

In Rb-Sr dating it is common practice to analyse several minerals from the same rock, or several whole-rock specimens from a given body of granite or granite-gneiss, and to interpret them by a graphical method based on the above equation,

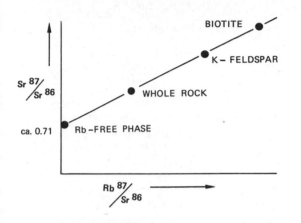

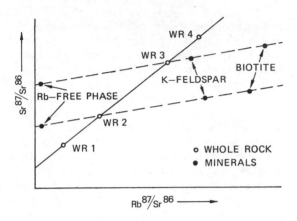

2.1 Rb–Sr isochron plot for a whole-rock specimen of granite and the minerals separated from it. The slope of the isochron is proportional to the age of the rock, whereas the intercept gives the isotopic ratio $^{87}Sr/^{86}Sr$ of common strontium when the rock crystallized.

2.2 Whole-rock and mineral Rb–Sr isochrons for a metamorphosed granite. The slope of the whole-rock isochron is proportional to the age of igneous intrusion. The slope of the mineral isochrons is proportional to the age of metamorphism.

which is termed an *isochron* plot. The equation can be rewritten as follows:

$$(^{87}Sr/^{86}Sr)_p = (e^{\lambda t} - 1)\,^{87}Rb/^{86}Sr + (^{87}Sr/^{86}Sr)_0$$

This is clearly the equation of a straight line of the type $y = mx + b$, where y and x are the axes, m is the slope and b is the intercept. The age of a cogenetic mineral assemblage can be calculated from the slope of the isochron, whereas the intercept represents the initial strontium isotope at time of crystallisation. An isochron plot for a hypothetical granite sample is shown in Figure 2.1. Clearly, the principal requirement for successful application of the isochron method is the availability of mineral and/or whole-rock phases with different Rb/Sr ratios from within a given rock sample or rock unit respectively. It is frequently found that whole-rock samples from a given rock unit exhibit considerable variation in their Rb/Sr ratios, even where the rock unit appears to be petrologically quite uniform over large areas.

Because of the geochemical dissimilarity between rubidium and strontium, radiogenic ^{87}Sr can diffuse out of rubidium lattice sites at temperatures well below those of igneous and metamorphic crystallisation. Rather surprisingly, perhaps, the diffusion behaviour of radiogenic ^{87}Sr is very similar to that of radiogenic argon. Blocking temperatures are probably in the general range

200–500°C, depending upon type of mineral and grain size. The blocking temperature order for a given grain size is potassium-feldspar > muscovite > biotite, and this is frequently the observed order of measured dates in large bodies of granite or gneiss in orogenic belts which underwent slow uplift and cooling after crystallisation. In such a situation whole-rock specimens of granite or granite-gneiss and of some types of feldspathic and micaceous schists are more retentive than individual minerals. This is because radiogenic ^{87}Sr that diffuses out of rubidium-bearing minerals down to relatively low temperatures can easily enter the lattice of the almost invariably present surrounding calcium minerals such as plagioclase or apatite and homogenise and equilibrate with the common strontium there, so that a hand-specimen of whole-rock can remain a closed system with regard to radiogenic ^{87}Sr.

To carry this one stage further, it is often found that a granite intrusion may remain a closed system with regard to radiogenic ^{87}Sr (as well as common strontium and rubidium) even during a subsequent metamorphism of sufficient intensity to convert it into a gneiss. Exactly as in the simple cooling situation described above, radiogenic ^{87}Sr diffuses out of micas and potassium-feldspars during the metamorphism but then mixes and homogenises with common strontium in neighbouring grains of calcium minerals, so that individual hand-specimens of the whole rock remain closed systems. Whole-rock Rb-Sr measurements on a body of granite-gneiss, even one occupying an area of tens or hundreds of square kilometres, can then yield a close approximation to the time of original crystallisation and intrusion of granite magma, whilst Rb-Sr and K-Ar mineral dates will record either the time of subsequent metamorphism, or the time of cooling and radiogenic closure after metamorphism, in the order of the mineral blocking temperatures for radiogenic ^{87}Sr and ^{40}Ar. A hypothetical example is presented in Figure 2. 2. The whole-rock isochron (points WR) yields the age of crystallisation and emplacement of the original granite, whilst the dashed lines show the isochrons for different minerals (all assumed to have become closed systems at the same time for the sake of simplicity) separated from whole-rock specimens WR-2 and WR-3, in which the slope is proportional to the time of metamorphism or radiogenic closure and the intercept yields the initial strontium isotope ratio at that time. The absolute differences between the initial $(^{87}Sr/^{86}Sr)_0$ ratio obtained from a whole-rock isochron and the corresponding mineral isochrons depend, of course, on the length of the time interval between igneous intrusion and metamorphism (or radiogenic closure), as well as the Rb/Sr ratio in the individual systems.

The practical application of Rb-Sr whole-rock isochrons has been of the greatest importance to geochronology and has contributed an immense amount of fundamental information on the age of igneous and metamorphic provinces throughout the world.

URANIUM-LEAD $(^{238}U - {}^{206}Pb, {}^{235}U - {}^{207}Pb)$, LEAD-LEAD $(^{207}Pb/^{206}Pb)$ AND THORIUM-LEAD $(^{232}Th/^{208}Pb)$ METHODS

The decay schemes, omitting details of the complex chains of intermediate nuclides, are:

$$^{238}U \xrightarrow[\text{6 betas}]{\text{8 alphas}} {}^{206}Pb; \text{ half-life} = 4\,500 \text{ m.y.}$$

$$^{235}U \xrightarrow[\text{4 betas}]{\text{7 alphas}} {}^{207}Pb; \text{ half-life} = 713 \text{ m.y.}$$

$$^{232}Th \xrightarrow[\text{4 betas}]{\text{6 alphas}} {}^{208}Pb; \text{ half-life} = 13\,900 \text{ m.y.}$$

Uranium and thorium frequently occur in the same mineral and it is therefore possible, in principle, to make three independent age determina-

tions on one mineral sample. The basic equation for calculating an age, t, is:

$$t = \frac{1}{\lambda} \log_e \left[1 + \frac{\text{radiogenic } {}^{206}Pb}{{}^{238}U} \right] \text{ etc.}$$

Correction for common lead incorporated into the mineral during crystallisation is usually necessary, although the correction is mostly quite small. The isotopic composition of lead in nature is very variable. Common lead consists of the isotopes ^{204}Pb, ^{206}Pb, ^{207}Pb and ^{208}Pb. Of these, only ^{204}Pb is entirely non-radiogenic and is used to correct for the common lead component as already described for the analogous case of strontium, i.e. by means of a coexisting uranium- and thorium-free lead mineral such as galena (PbS), or by isochron methods.

A semi-independent age can be obtained from the isotopic analysis of lead alone—without accompanying analysis for total amounts of uranium, thorium and lead in the mineral—namely from the $^{207}Pb/^{206}Pb$ ratio. This ratio is itself time-dependent, because of the different decay rates of ^{238}U and ^{235}U. After correction for any common lead the age, t, may be calculated from the equation:

$$^{207}Pb/^{206}Pb = \frac{1}{137.8} \frac{(e^{\lambda_2 t} - 1)}{(e^{\lambda_1 t} - 1)}$$

where λ_1 and λ_2 are the decay constants of ^{238}U and ^{235}U, and 1/137.8 is the present-day atomic ratio $^{235}U/^{238}U$. This equation is solved for t by a trial and error method, or from specially prepared tables in which $^{207}Pb/^{206}Pb$ is given as a function of t.

It is of passing interest to note from the half-lives that since the formation of the Earth approximately 4600 million years ago, about 50 per cent of the original ^{238}U has decayed to ^{206}Pb, 99 per cent of the original ^{235}U to ^{207}Pb, and 20 per cent of the original ^{232}Th to ^{208}Pb.

These methods can be applied over the greater part of the geological age range, although their usefulness decreases sharply for rocks younger than about 100–200 million years, partly because the rate of production of ^{207}Pb has become so low. The methods have a much more restricted range of application than the K-Ar and Rb-Sr methods but they may be used to date rare minerals in

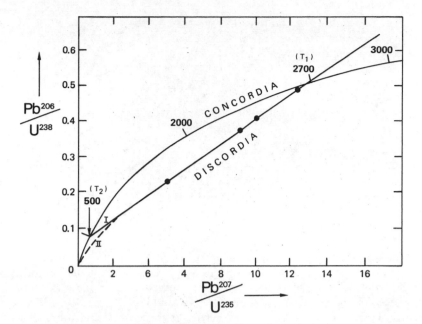

2.3 Concordia-Discordia Diagram showing (I) the effect of a single period of lead loss, so that samples of age T_1 (2700 million years) that lost variable amounts of lead at time T_2 (500 million years) fall on a chord connecting T_1 and T_2, (II) the hypothetical effect of continuous lead loss by diffusion from samples of age T_1.

which uranium and thorium are major constituents. Most frequently, however, the methods are applied to minerals such as zircon (zirconium silicate) and sphene (calcium titanium silicate) which occur in accessory amounts of up to about one per cent in many igneous and metamorphic rocks, and which have uranium and thorium contents ranging from a few hundred to a few thousand parts per million.

It should be noted that the U–Pb methods are of much greater practical value than the Th–Pb method. The latter is nowadays only rarely used because it often yields anomalous ages which can be difficult to interpret.

In some cases $^{238}U/^{206}Pb$, $^{235}U/^{207}Pb$ and $^{207}Pb/^{206}Pb$ dates for a given mineral are found to be concordant within analytical error and thus yield a valuable geochronological reference point. More frequently, however, the three measured dates are grossly discordant outside any plausible limits of analytical error. A commonly observed pattern is:

$$^{207}Pb/^{206}Pb > {}^{235}U/^{207}Pb > {}^{238}U/^{206}Pb.$$

Such discordant patterns result from lack of closed-system behaviour for the radioactive and radiogenic nuclides, including the intermediate members of different decay chains. The degree of discordance itself usually varies in different samples of the same mineral from a given rock unit.

Close study of discordant U–Pb age patterns shows that, in many cases, they result either from lead loss from a mineral during a metamorphic event long after crystallisation, or from continuous diffusion of lead out of the mineral over its entire geological history. Extremely useful geological information may be obtained from such discordant age patterns, particularly from the mineral zircon. This is illustrated by means of the so-called *Concordia-Discordia* diagram (Fig. 2.3), in which the measured $^{206}Pb/^{238}U$ and $^{207}Pb/^{235}U$ values (after correction for common lead) are plotted against each other. The Concordia curve represents the locus of all points for which the $^{238}U/^{206}Pb$ age equals the $^{235}U/^{207}Pb$ age. When, as usually happens, different zircon samples from a rock unit define a straight Discordia line, the upper intercept with Concordia gives the time of crystallisation, t_1, of the zircon. The lower intercept with Concordia indicates the time, t_2, at which a metamorphic heating event removed variable amounts of lead from zircons from different parts of the rock unit, whilst the actual position of a sample on Discordia is determined by the amount of lead lost by that sample at time t_2. This situation is referred to as *episodic* lead loss. In other cases, however, there is simply no evidence from independent K–Ar and Rb–Sr measurements on associated rocks for a metamorphic event at the time given by the lower, extrapolated intercept of Discordia on Concordia. An alternative mechanism of continuous diffusion of lead out of the

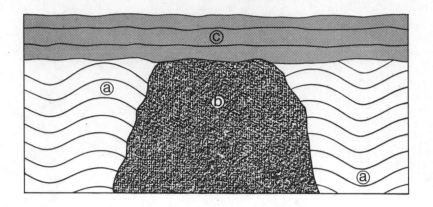

2.4 Cross-section of a 'bracketed' igneous intrusion. Provided that sediment (a) is not much older than sediment (c), the radio-isotopic date of the intrusion (b) can be used as a calibrated point in the time scale.

mineral has been postulated in such cases in which the extrapolated Discordia bends down towards the origin, as shown in Figure 2. 3. (No samples actually falling on the curved region have yet been reported.) Whichever mechanism is assumed for lead loss, extrapolation back to time of crystallisation, t_1, remains valid, but the assumption that lead loss was episodic must be verified by independent K-Ar and Rb-Sr measurements on the same rock unit.

GENERAL DISCUSSION OF DIFFERENT TYPES OF AGE PATTERNS

Concordant Age Patterns

When all measured mineral dates from a given igneous or metamorphic rock unit agree within analytical error, they either represent the time of crystallisation, or the time at which the entire rock unit cooled rapidly enough for all minerals to become closed systems to radiogenic nuclide diffusion at the same time. In those cases where whole-rock Rb-Sr ages are in agreement with the mineral ages it can be concluded that the time of crystallisation is being dated. Mineral and whole-rock dates from small igneous intrusions, volcanic lavas and ash-beds (*tuffs*), from the sedimentary mineral glauconite, as well as from low-grade metamorphosed sediments, are mostly concordant and usually give the age of crystallisation.

Such dates on rocks of known stratigraphic-palaeontological assignment are essential for fixing the post-Precambrian *Phanerozoic* time-scale. The time of deposition of glauconite-bearing sediments can be dated directly. This is best for Mesozoic and Tertiary sediments, since older glauconites are rather susceptible to loss of radiogenic nuclides due to the temperature effect of deep burial. A somewhat more widely applicable, indirect, method of dating sediments is by means of volcanic rocks such as lava flows and tuffs which were deposited rapidly and which are often interstratified with fossiliferous sediments of known stratigraphic age. Suitable minerals from lavas and tuffs include feldspar, biotite and zircon, whilst whole-rock specimens of acid type can be dated by the Rb-Sr method. Another related, indirect method is to date a small igneous intrusion such as a dyke, sill or stock ((b) in Fig. 2. 4) which intruded a fossiliferous sediment (a) and which has been unconformably overlaid by a fossiliferous sediment (c). For example, if (a) is Upper Silurian and (c) is Lower Devonian, then dating the intrusion (b) will obviously give an age value for the boundary between the two systems. This is called the method of *bracketed intrusives*.

Using the above principles, first pioneered by Arthur Holmes over fifty years ago, a Phanerozoic time-scale calibrated by radioisotope dates has been produced. More and more relevant dates from all over the world are being published each year, so that the time-scale is constantly increasing in refinement and sophistication. Several hundred points covering the entire Phanerozoic are now available, but the Mesozoic and Tertiary parts of the time-scale are more satisfactorily calibrated than the Palaeozoic.

Discordant Ages Produced by Cooling

According to the principles outlined earlier, in a deep-seated igneous or metamorphic terrain from the heart of a major orogenic belt one can usually expect a significant interval between the time of crystallisation and the time when different minerals reached their respective blocking temperatures for the radiogenic nuclides. In geological terms this corresponds to gradual uplift and cooling of the orogenic belt through a vertical distance of at least 15–30 km and also to gradual reduction of the thermal gradient at depth (see Chapter 5). The following sequence of measured dates would be quite typical from such an environment, although inversions and irregularities may well occur because of the effect of grain size, so that coarse-grained, pegmatitic minerals become effectively closed before fine-grained minerals of the same type:

Rb–Sr (whole-rock granite, or granite-gneiss)
> $^{207}Pb/^{206}Pb$ (zircon)
>   Rb–Sr (potassium-feldspar)
>   K–Ar (hornblende)
>   Rb–Sr (muscovite)
>   K–Ar (muscovite)
>   Rb–Sr (biotite)
>   K–Ar (biotite).

Of course, not all these minerals are necessarily, or even usually, present in a single area, and one must also remember that in the more ancient rocks an intrinsically discordant age pattern may be masked by the inherently greater absolute error of individual age measurements. However, the oldest dates from such a sequence are clearly the ones approximating most closely to the true age of crystallisation, whilst the youngest dates relate to various stages in the uplift and cooling of the orogenic belt and also to the time of erosion and deposition of the eroded detritus in bordering sedimentary basins and geosynclines.

Detailed study of age patterns in the Precambrian Grenville orogenic belt of Canada, the Lower Palaeozoic Caledonian belt of the British Isles, and of the Tertiary Alpine belt of Switzerland (see Chapter 20) indicates time intervals between crystallisation and attainment of the blocking temperature in fine-grained micas of approximately 150, 70 and 20 million years respectively. The actual differences are probably a consequence of the different depths of erosion exhibited by these three orogenic belts. With continuing uplift and erosion of the Western Alps, a geochronologist 30 million years from now, sampling surface rocks at that time, might well find a difference of 50 million years between time of crystallisation and cessation of diffusion of radiogenic nuclides.

At any given structural level, consistent differences of the type muscovite date > biotite date of about 12 million years for K-Ar, and about 8 million years for Rb-Sr, have been reported from the Caledonian belt of Scotland and from the Swiss Alps respectively. The spatial distribution of 'cooling' dates within orogenic belts is becoming an important tool for tectonic and structural investigations, because both the overall pattern and the rate of uplift can be assessed.

Discordant Age Patterns produced by Overprinting

These are commonly observed where igneous or metamorphic rocks which crystallised at time t_1 were reheated at a much later time t_2 but did not undergo complete recrystallisation at this time. The surviving minerals lose variable amounts of the accumulated radiogenic nuclides depending on mineral type and grain size, as well as on the intensity of the metamorphic heating event at time t_2. A whole range of so-called *overprinted* or *mixed* dates can result, falling between times t_1 and t_2, or more strictly between the times at which the minerals became closed systems after the earlier and later events. By using the various dating methods on different types of minerals and whole rocks close limits can frequently be set for times t_1 and t_2. The most retentive and coarsest-grained minerals formed at time t_1 yield the oldest dates, often very close to time t_1, whilst the least retentive and finest-grained minerals give the youngest dates, often very close to time t_2. Some of the mineral dates will fall between times t_1 and t_2 and give individually meaningless or so-called *apparent* dates. The actual order of radiogenic nuclide retentivities is, of course, the same as that mentioned previously with reference to discordant age patterns produced by cooling. It must not be forgotten that certain new minerals may be formed at time t_2, but it is often possible for the field geologist to decide which minerals have survived from the first event and which are the new minerals formed during the second event. A hypothetical overprinted age pattern is shown in Table 1, such as would be expected in the above situation. It corresponds closely with reality, although it is unlikely that so many suitable minerals and rocks would be available from a single geological terrain. In each case the geochronologist has to do the best with the available material. Note in Table 1 how whole-rock Rb-Sr measurements typically give the oldest dates, simply because individual hand-specimens may

TABLE 1

SAMPLE	AGE METHOD	MEASURED AGE (*million years*)	INTERPRETATION
Fine-grained biotite from country rock	K–Ar	300	Igneous and meta-morphic episode at 350 million years ago, followed by gradual uplift and cooling until at least 300 million years ago
	Rb–Sr	310	
Fine-grained muscovite from country rock	K–Ar	310	
	Rb–Sr	320	
Biotite } Muscovite } from a coarse-grained late pegmatite Potassium-feldspar }	K–Ar	340	
	Rb–Sr	350	
	K–Ar	350	
	Rb–Sr	350	
	Rb–Sr	350	
Hornblende from country rock	K–Ar	800	Partially overprinted, 'mixed' dates
Potassium-feldspar from country rock	Rb–Sr	900	
Biotite } Muscovite } from a coarse-grained early pegmatite Potassium-feldspar }	K–Ar	700	
	Rb–Sr	750	
	K–Ar	850	
	Rb–Sr	950	
	Rb–Sr	1 050	
Zircon from country rock	$^{207}Pb/^{206}Pb$	1 050	
	$^{235}U/^{207}Pb$	950	
	$^{238}U/^{206}Pb$	800	
	Discordia extrapolation on to Concordia	$t_1 = 1100$ and $t_2 = 350$	Igneous and metamorphic episode at 1 100 million years ago
Whole-rock granite-gneiss country rock	Rb–Sr	1 100	

Idealised, but realistic, discordant age pattern produced in rocks formed 1 100 million years ago, which underwent metamorphism and partial recrystallisation 350 million years ago. Note that from the age data alone it is not possible to say whether fine-grained biotite and fine-grained muscovite actually crystallised 1 100 or 350 million years ago.

remain closed systems with regard to radiogenic ^{87}Sr even during a subsequent metamorphism of sufficient intensity to convert granite into granite-gneiss.

Progress in recent years in recognising and interpreting discordant age patterns has greatly advanced the frontiers of geochronology. This is in great contrast to the situation of only a dozen years ago, when discordant age patterns were not understood and regarded as a severe limitation. In many orogenic belts and continental shield areas throughout the world geological histories extending over tens, hundreds and even thousands of millions of years have been elucidated by study of discordant age patterns. Particularly important for hypotheses concerning the origin, growth and structure of the continents is the clear recognition that many orogenic belts (at least those that are younger than about 3000 million years) involve the partial or complete reworking of crustal rocks formed at a much earlier time in the Earth's history. This clearly bears on the fundamental problem of whether the continents have grown with geological time, and if so, by how much.

It is not always easy to distinguish between a discordant age pattern formed by cooling or by overprinting, although close consideration of all associated geological, structural and petrological evidence can sometimes help to decide. Furthermore, whilst cooling patterns are probably limited by the nature of the orogenic uplift-cooling process to maximum ranges of about 100–200 million years, overprinted age patterns may span the entire interval between two (or more) thermal events.

SUMMARY OF PRINCIPAL APPLICATIONS

The aim of geochronologists is to date all igneous, metamorphic and, wherever possible, sedimentary episodes that together have shaped the Earth (and other accessible bodies within the solar system) throughout geological time. Below is given a list of some of the more specialised types of application in various branches of the Earth sciences:

(1) The timing and duration of events within a single orogenic belt, from the time of rock formation to the time of final uplift and cooling (Chapter 20).

(2) Study of the distribution of radioisotope dates in space and time within continental areas, related to problems of the origin, structure and evolution of the continents (Chapt. 15 & 22).

(3) Setting up a post-Precambrian (Phanerozoic) time-scale by dating rocks of known stratigraphical-palaeontological age (Chapter 22).

(4) Setting up a Precambrian time-scale and correlating Precambrian geological events all over the world. (The oldest reliably dated terrestrial rocks in several continental shield areas are about 3 300 million years old, and granitic gneisses recently discovered in West Greenland are about 3 800 million years old and are the oldest terrestrial rocks known in 1972.)

(5) Testing the hypothesis of continental drift, with particular reference to (a) matching geological age provinces on pre-drift reconstructions of the continents, (b) measurement of the rate of ocean-floor spreading, by dating oceanic island and ocean floor rocks in relation to their respective distances from mid-ocean ridges (Chapters 15, 16 & 21).

(6) Combined geochronological and palaeomagnetic measurements relating to (a) determination of the rate of polar wandering during the

2.5 **Close-up of the high-vacuum argon extraction system. Two rock samples are loaded in metal crucibles inside the two large glass furnaces. The crucibles can be heated to about 1300°C by means of the radiofrequency induction coil, causing the rock to melt although the glass remains quite cool. The remaining parts of the apparatus comprise vacuum gauge, metal valves, as well as various gadgets for purifying the argon, since several chemically reactive gases are also released.**

geological past (Chapter 15), (b) determination of the frequency and rate with which the Earth's magnetic field has reversed its polarity in the geological past (Chapters 16 & 17). This involves mainly the K-Ar dating of Tertiary and Quaternary basaltic lavas.

(7) Dating geologically relatively young events such as the age and rate of evolution of primitive man, and the age of the Pleistocene glacial and interglacial periods (Chapter 18).

(8) Dating extra-terrestrial materials such as (a) meteorites, which is widely accepted as an indirect approach to dating the formation of the Earth and the solar system, and (b) the lunar samples (Chapters 7 & 8).

FURTHER READING:

ERIC I. HAMILTON 1965. Applied Geochronology. *Academic Press.*

G. B. DALRYMPLE AND M. A. LANPHERE 1969. Potassium-Argon Dating. *W. H. Freeman & Co.*

E. I. HAMILTON AND R. M. FARQUHAR (ed.) 1968. Radiometric Dating for Geologists. *Interscience.*

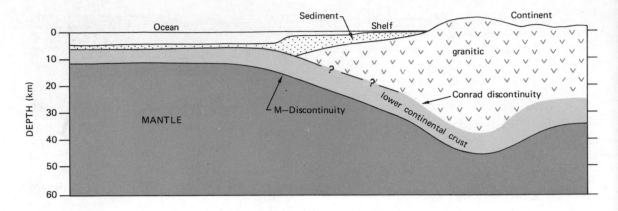

3. 0   Schematic section showing variations in thickness and composition of the continental and oceanic crust.

The factual evidence we have concerning the composition of the Earth is restricted to its surface and to samples taken from mines and bore-holes, none of which penetrate more than 8 km into its interior.   Geological processes can expose rocks created at some 20–25 km depth, and volcanoes erupt pieces of rock that may once have been part of the Earth's upper mantle. Apart from these scanty data, we have no direct evidence concerning the composition of the Earth's interior.

However, geophysical studies, particularly in seismology, provide physical parameters such as density, rigidity and compressibility for the various layers within the major structural units of the Earth.   Using these parameters in conjunction with other evidence, such as the composition of meteorites and the abundance and distribution of elements in the solar system, it is possible to decide on the most likely composition for the various parts of the Earth. Furthermore, very pertinent to its present-day composition is the way in which the Earth was created and how it evolved through time.   How was it created—by the cold accretion of particles within a gaseous nebula or dust cloud, or by some catastrophic event that 'tore' it away from a parent sun? Has it been cooling down or heating up since its creation?   Was it ever totally molten?   Has it always had oceans and an atmosphere?

In this article, Professor P. G. Harris, Chairman of the Department of Earth Sciences at the University of Leeds, summarizes the evidence available and discusses the most likely compositions for the Earth's crust, mantle and core before tackling the fundamental problems concerning its origin and evolution.

3 The Composition of the Earth

by PETER HARRIS

Models for the interior of the Earth are necessarily based on inferences from a variety of disciplines. We draw on physics, chemistry, solar system astronomy and on several branches of the Earth sciences to decide how the Earth may, and may not, be constituted.

LIKE OTHER TRIBAL GROUPS, geochemists tend to abide by well-established traditions and conventions. Because this chapter adheres to some of these conventions, it is as well to set them out in advance.

For example, chemical analyses of the major components of rocks are recorded and reported in terms of percentage of the constituent oxides, usually in the sequence SiO_2, TiO_2, Al_2O_3, Fe_2O_3, FeO, MnO, MgO, CaO, Na_2O, K_2O, H_2O ($-105°C$), H_2O ($+105°C$), P_2O_5, CO_2. The expressions H_2O ($-105°C$) and H_2O ($+105°C$) refer to water given off by the rock sample on heating up to $105°C$, and to much higher temperatures respectively. The halogens F, Cl, Br, I and the trace components present at less than 1 000 ppm (parts per million) are reported normally as elements, not as oxides or some other compounds.

Geologists and geochemists use some chemical terms in a misleading way. In igneous rocks, the silica contents have been used as a basis of classification. Those with more than 65% SiO_2 were termed *acidic*, those with 52-66% *intermediate*, those with 45-52% *basic*, and those with less than 45% *ultrabasic*. None of these rocks are acidic or basic in the true chemical sense. The terms acidic, basic and ultrabasic etc. are being modified in their usage, and are gradually being replaced by terms such as *silicic*, *mafic* and *ultramafic* although these are not exactly synonymous. The word 'mafic' is manufactured from magnesian and ferric, because mafic rocks and their constituent minerals are rich in magnesium and iron.

Rocks high in sodium or potassium are called alkaline or alkalic, again without any implications of alkalinity or pH.

COMPOSITION OF EARTH LAYERS

The surface of the Earth is readily sampled and analysed, so a good deal is known about its composition. However the deepest hole in the Earth, a drill hole, is only about 8 km deep, and there is no method of sampling from greater depth. Because there is no way of finding the chemical composition of unseen or inaccessible material, ideas on the composition of the deeper regions of the Earth must be based on indirect evidence, and consequently are somewhat speculative. The physical properties of the inaccessible regions of the Earth can be measured or estimated more readily. From a knowledge of the Earth's mass and volume its average density can be calculated. The change in density with depth can be estimated from the velocity of earthquake waves through the Earth, and from the Earth's moment of inertia. Most recently, the densities at different depths have been calculated from the free oscillations — the natural vibrations of the whole Earth — induced by very large earthquakes. If they are to be acceptable, models for the chemical composition of Earth layers must be consistent with the densities and other physical properties calculated for the layers.

The most important of these methods in locating Earth layers and defining their physical properties is *seismology*, the study of earthquake or seismic waves. On the basis of seismic investigations, the Earth can be divided into three major zones. The outer superficial layer is called the *crust*. This extends down to 30 or 40 km beneath continents, and even further beneath some mountain regions, and to about 10 km beneath sea level in oceanic regions. At the bottom of the

TABLE 1: THE EARTH'S INTERNAL LAYERING

Name of layer	Depth (km)	Density (10^3 kg m^{-3})	Pressure (kilobars)*
	0		
Crust		2.8	
	33	3.0 / 3.3	9
Upper Mantle		↓	
	700	4.3	260
Lower Mantle		↓	
	2890	5.5 / 10.0	1350
Outer Core		↓	
	5150	12.3 / 13.3	3340
Inner Core		↓	
	6371	13.6	3700

*The bar is an international unit of pressure equivalent to 0.987 atmospheres. 1 kilobar is 1 000 bars.

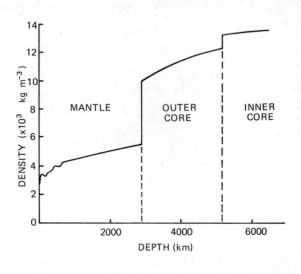

3.1 The variation of density with depth in the Earth.

crust, earthquake waves undergo an abrupt increase in velocity as they enter a denser layer, the *mantle*, the crust-mantle boundary being called the *Mohorovicic Discontinuity*, often shortened to the 'M' discontinuity or 'Moho'. The mantle itself has been divided into two layers, an upper and lower mantle, the boundary between these normally being taken at 1 000 km depth, although the most recent studies suggest a depth of about 700 km to be more appropriate.

The mantle is a solid layer. At about 2890 or 2900 km is the boundary with the third major layer, the *core*. The core is liquid, at least in the outer part, as can be shown from its inability to transmit shear waves, the *S* waves of earthquakes. The inner core, from about 5 150 km to the Earth's centre at 6 371 km, is solid. The core-mantle boundary is marked by an abrupt reduction in the velocity of compressional or *P* earthquake waves, as well as by the disappearance of the *S* waves. (See Chapter 24.)

These layers and the probable densities and pressures within them are shown in Table 1, while Figure 3.1 shows the change of density with depth in the Earth. It should be emphasised that the densities and the depth of the boundaries or discontinuities are not precisely determined, and the values are subject to frequent revision.

THE CRUST

Seismic evidence indicates that in continental regions, beneath a zone of superficial sediments, the crust can be divided into two layers, an upper crust often considered to be granitic or granodioritic in composition, and a lower crust probably of more basic composition. The boundary between the upper and lower crust, sometimes called the *Conrad Discontinuity*, is known only from the seismic evidence and is often difficult to define and variable in nature and in depth. In the floors of ocean basins, beneath the superficial sedimentary material, the 'granitic' layer is missing.

The Upper Continental Crust

When examined by geological and chemical techniques, the rocks exposed at the land surface show great heterogeneity and regional variation. For example, the younger margins of continents consist largely of sediment derived from the continued erosion of the continental surface and transported to the coast, where most of it is deposited in shallow water on the continental shelf. Such sediments may form accumulations many kilometres thick. On the other hand, the oldest continental regions, the *Precambrian Shields*, are often dominated by igneous rocks such as granite or by highly metamorphosed rocks such as gneiss.

Even within a restricted region of a few square kilometres there are likely to be great changes in chemical composition, e.g. from granite to basalt or gabbro or even to ultramafic rocks such as peridotite. Among sedimentary rocks the chemical differences may be even greater. Sedimentary processes are even more effective than igneous ones in causing the sorting out or separating from one another of chemical components. For example, in the chemical breakdown or weathering of surface rocks and the removal of the debris and decay products such as soil, high proportions of the cations Ca, Mg, Na, and K, which are soluble under these conditions, are removed in solution to the sea. Here Ca will be deposited as limestone, perhaps almost completely pure $CaCO_3$. In special circumstances of strong evaporation from enclosed basins and lagoons of salt water, $CaSO_4$ or NaCl or other salts may be deposited. The silica and alumina of the decomposed parent rock will be carried away in suspension in running water. Some of the silica will be hard, resistant sand grains of quartz, and the alumina with some silica will be very fine plate-like particles of clay. These two phases, sand and clay, are very effectively separated from each other by deposition or sedimentation from water under different settling conditions. Processes of this sort may cause a physical segregation or sorting-out of minerals which results in a very effective chemical separation (see Chapter 13). Some idea of the possible ranges in composition of common rock types is given in Table 2.

To calculate the average composition of the Earth's crust from such varied sources would require that one knew the volume of each rock type and its chemical composition. Because of the wide variation in composition of sedimentary rocks, and to simplify the calculations, earliest estimates of the composition of the Earth's surface assumed that all sedimentary and metamorphic rocks were derived ultimately from the breakdown of igneous rocks. So if a mean composition of all the igneous rocks exposed at the Earth's surface could be found, this should be close to the average composition of the whole surface. The best known early estimates of the composition were made in this way by Clarke and Washington, geochemists with the United States Geological Survey, who, from 1903 onwards, compiled all the published chemical analyses of igneous rocks

TABLE 2: TYPICAL CHEMICAL COMPOSITIONS OF COMMON ROCK TYPES (*Percentages*)

	1	2	3	4	5	6
SiO_2	95.4	55.1	5.2	72.3	50.8	43.5
TiO_2		0.9	0.1	0.3	2.0	0.8
Al_2O_3	1.1	16.3	0.8	14.0	14.1	4.0
Fe_2O_3	0.4	4.2	} 0.5	0.9	2.9	2.5
FeO	0.2	1.9		1.5	9.1	9.8
MnO		0.05	0.05	0.05	0.2	0.2
MgO	0.1	2.5	7.9	0.5	6.3	34.0
CaO	1.6	4.7	42.6	1.4	10.4	3.5
Na_2O	0.1	0.7	0.05	3.1	2.2	0.6
K_2O	0.2	3.0	0.3	5.1	0.8	0.2
H_2O+	0.3	5.2	0.5	0.6	0.9	0.8
H_2O-			0.2			
P_2O_5		0.11	0.04	0.15	0.2	0.05
CO_2	1.1	4.0	41.6			

1. Quartzitic sandstone, a sedimentary rock consisting chiefly of quartz grains.
2. Shale, a sedimentary rock consisting chiefly of clay.
3. Limestone, largely $CaCO_3$ formed from the shells of animals, or by precipitation from sea water.
4. Granite, an acid plutonic rock.
5. Basalt, a basic volcanic rock.
6. Peridotite, an ultrabasic plutonic rock.

The values quoted are averages of large numbers of analyses.

and then calculated from these an average composition.

These early estimates have many unsatisfactory features. For example, because of their novelty, unusual rock types tend to be analysed frequently while common rocks like granite or basalt that have enormous volumes may be represented by relatively few analyses. So any average of analyses is biased towards the rare and unusual. Also most analyses of rocks relate to Europe and North America, with few from Africa or Asia. If regional differences in composition exist, these might be ignored or overemphasized depending on where they occur.

For these and other reasons, the more recent calculations of the composition of the Earth's crust are based on estimates of the relative volumes of the different types of sedimentary, igneous and metamorphic rocks and the composition of each. The results of one such calculation are shown in Table 3. It can be seen that this composition of

TABLE 3:
COMPOSITION OF THE CONTINENTAL CRUST

	(1) (per cent)			(2) ppm
SiO_2	61.9	Barium (Ba)		425
TiO_2	0.8	Strontium (Sr)		375
Al_2O_3	15.6	Zirconium (Zr)		165
Fe_2O_3	2.6	Copper (Cu)		55
FeO	3.9	Scandium (Sc)		22
MnO	0.1	Lead (Pb)		12.5
MgO	3.1	Uranium (U)		2.7
CaO	5.7	Mercury (Hg)		0.08
Na_2O	3.1	Silver (Ag)		0.07
K_2O	2.9	Gold (Au)		0.004
P_2O_5	0.3			

(1) Average content of major constituents.

(2) Average content of some of the 'trace' elements.

the whole continental crust is intermediate between granite and basalt. Estimates of the composition of the granitic layer alone indicate that in fact this is not granitic, but is also intermediate in character.

In addition to the major components listed in Table 3 every rock can be assumed to contain each of the naturally occurring elements, though perhaps only at extremely low concentrations. Some of these 'trace' elements are shown in the table. It can be seen that some of the more abundant of these 'trace' elements are relatively uncommon in everyday use, while many of the metals of importance to society are extremely rare in nature. Usage is related in part to the industrially desirable properties of the metal or element, but also to the difficulties of extracting the metal. For example lead, at about £100 a ton, is a low-cost (and therefore much-used) metal because geological processes have concentrated it into easily mined veins and deposits of galena (PbS).

Scandium is more abundant in the Earth and has many desirable properties as a light metal of high melting point, perhaps ideal for super-sonic aircraft. But scandium does not occur in segregated workable deposits. The cost of extracting 50 ppm of scandium from a basalt would be prohibitive, and so the current price remains at £10 per gram; at this price it is only a scientific curiosity.

The composition of some types of sedimentary and igneous rocks in Table 2, and of the continental crust in Table 3, might suggest at first sight that silicon is the dominant element in the Earth's crust. This is not true. If the analyses are recalculated to show the concentration of each *element* instead of its oxide, it is seen that oxygen is the major constituent (46.5%) of the Earth's crust, with silicon second. This superiority of oxygen is even more enhanced if the relative *numbers* of atoms of each element are calculated. Nearly two-thirds of the atoms of the Earth's crust are of oxygen. Geochemists find it convenient to think of silicate minerals as consisting of Si^{4+} and O^{2-} ions, though it must be emphasised that this is just a useful convention. But if we accept this convention, then the oxygen ions in minerals are relatively large (radius of $O^{2-} = 1.40$ Å) compared with other major elements like silicon and magnesium (radius of $Si^{4+} = 0.42$ Å, $Mg^{2+} = 0.66$ Å). On a volume basis, oxygen is then 94% of the Earth's crust (see Table 4). The relatively tiny silicon ions, each of which is only about a fortieth of the volume of an oxygen ion, make up less than 1% of the volume of the crust. In effect, silicate minerals can be thought of as a stack or framework of oxygen ions, with the other much smaller ions fitting into the interstices and spaces between. As Chapter 1 makes clear, the structure of a silicate mineral is determined largely by the geometry of the best regular arrangement of a predetermined ratio of atoms of differing sizes.

TABLE 4: THE RELATIVE IMPORTANCE OF MAJOR ELEMENTS IN THE CRUST

	Weight per cent	Atoms per cent	Ionic Radius (Å)	Volume per cent
O^{2-}	46.5	62.1	1.40	94.07
Si^{4+}	28.9	22.0	0.42	0.88
Ti^{4+}	0.5	0.2	0.68	0.04
Al^{3+}	8.3	6.5	0.51	0.47
Fe^{3+}	1.8	0.7	0.64	0.09
Fe^{2+}	3.0	1.1	0.74	0.25
Mg^{2+}	1.9	1.6	0.66	0.26
Ca^{2+}	4.1	2.2	0.99	1.15
Na^+	2.3	2.1	0.97	1.07
K^+	2.4	1.3	1.33	1.71

The margins of the continents beneath sea-level, that is the continental shelves, consist of sediments derived from the erosion of the land surface and transportation of the solid debris in suspension in running water. Presumably these sediments are of much the same composition as the continental land surface itself, apart from those cations leached out in solution.

However, the floors of the deep oceans, far removed from land, have little addition of land-derived sediment unless this is transported by sea-floor currents, or as wind-blown dust. The bulk of deep-sea sediments are derived from the sea-water itself by chemical precipitation. Precipitates such as iron and manganese hydroxides, or silicate minerals like feldspars and zeolites, may form directly on the sea-floor as crystal growths. More important in many regions are the skeletons of marine organisms. These can be calcium carbonate, the shells or envelopes of microscopic organisms like foraminifera or the shells of larger animals, or the siliceous (SiO_2) skeletons of micro-organisms like diatoms or radiolaria. These build up calcareous or siliceous oozes on the deep-sea floor. The final composition of such a sediment is determined not only by the quality of life in the upper ocean, but also by the effects of solution as the skeleton sinks to the bottom. In very deep water, calcium carbonate is sufficiently soluble for nearly all calcareous skeletons to dissolve before they reach the bottom, and the accumulating skeletons are dominantly siliceous. On the floors of the deep ocean basins, these sediments are, on average, about half a kilometre thick. They accumulate only very slowly, normally at the rate of only a few millimetres in a thousand years. The floors of the ocean basins are relatively young (see Chapter 16) so there has been only a short time, mostly of less than a hundred million years, for accumulation.

Beneath this cover of sediments, and sometimes interbedded with them, deep-sea drilling reveals basaltic horizons. On the slopes and crests of the mid-ocean ridges (see Chapter 16) where sedimentary cover is thin or absent, the exposed rocks sampled by deep-sea dredging are also basalts. Most of these basalts are similar to one another in chemical composition, and are poorer in potassium and richer in aluminium than the basalts of the land surface. They are called *oceanic tholeiites* (Analysis 1 in Table 5). Sometimes, especially on steep surfaces exposed by differential movement

TABLE 5: POSSIBLE COMPOSITIONS OF OCEANIC CRUST AND MANTLE

	(1) per cent	(2) per cent	(3) per cent
SiO_2	49.2	44.5	45.2
TiO_2	1.4	0.1	0.7
Al_2O_3	15.8	3.1	3.5
Fe_2O_3	2.2	1.2	0.5
FeO	7.2	6.7	8.0
MnO	0.16	0.1	0.1
MgO	8.5	39.1	37.5
CaO	11.1	3.2	3.1
Na_2O	2.7	0.25	0.57
K_2O	0.26	0.04	0.13
P_2O_5	0.15		
Cr_2O_3		0.4	0.4
NiO		0.24	0.20

1. Average composition of tholeiitic basalts dredged from the mid-Atlantic ridge and probably representative of oceanic crust.

2. Composition of ultramafic xenoliths, thought to be representative of upper mantle.

3. Pyrolite model for the mantle. (Ringwood, 1966.)

and sliding of blocks of sea-floor, the rocks obtained by dredging are of serpentinite, a hydrated ultrabasic rock.

The ocean floors are studded with steep mounds. Some may be so high that they emerge above sea level as islands. Those that do not emerge are called *seamounts*. Some seamounts known as *guyots* are truncated and have flat tops, evidence that once they were eroded at sea level, though the top may now be 1 or 2 kilometres lower. In almost every case, these islands and seamounts are built of volcanic rocks and debris. All of the islands in the deep ocean basins are volcanic in origin, and are generally of alkali basalt, which differs from oceanic tholeiite, particularly in having a higher content of Na, K and Ti (see Chapter 21).

As a first approximation we may assume that in oceanic regions, below the superficial sediments, the crust of the Earth has the composition of oceanic tholeiite (Table 5). An alternative has been suggested, that the oceanic crust is dominantly serpentinite, the hydrated equivalent of mantle material. However, most marine geologists consider the serpentinite, sometimes dredged from the ocean floor, to be due to local injection of upper mantle material along planes of movement in the ocean floor, and not to be a major component of ocean-floor crust.

The Continental Lower Crust

Seismic data indicate that in continental regions the lower 15–20 kilometres of crust differ from the upper half in having higher seismic velocities. Presumably the lower crust also has a greater density (about 3.0×10^3 kg m^{-3}) than the upper crust (about 2.8×10^3 kg m^{-3}).

A density of about three is consistent with gabbro, the plutonic equivalent of basalt, and it was thought that the lower crust of continents was chemically similar to the basaltic layer beneath oceans. However, this is now doubtful. The pressure-temperature environment in the continental lower crust is such that gabbro would not exist in its normal form, but would be transformed to a dense high-pressure rock called eclogite. The density of this would be about 3.5×10^3 kg m^{-3} —too high for the lower crust.

Recently it has been suggested that the lower continental crust is of intermediate composition much like the upper crust, but that its minerals, formed at high pressures, have a greater density than the minerals of the upper crust, which formed at low pressures. The density difference could account for the seismic distinction between upper and lower crust. There is one chemical difference: compared with upper crustal rocks, those thought to be representative of the lower crust are depleted in potassium, uranium and thorium. This is important as these elements, because of their radioactive decay, are the present source of heat in the Earth.

THE MANTLE

At the base of the crust, earthquake waves increase abruptly in velocity, marking the Mohorovicic Discontinuity; this is by definition the boundary with the mantle. The mantle extends down to the boundary of the core at about 2 890 or 2 900 km. Although this is less than half the Earth's radius (6 371 km), the mantle forms 83% of the Earth by volume and about 68% by mass. Despite the fact that the mantle is physically so inaccessible, an understanding of its nature is extremely important for a number of reasons. For instance, it is very likely that early in its history the Earth had only one silicate layer, the primitive mantle, from which the crust developed during subsequent geological epochs by outward segregation, especially by addition of volcanic material. Also, the mantle is the source region of most of the Earth's internal energy and of the forces responsible for ocean floor spreading (Chapter 16), continental drift (Chapter 15), orogeny (Chapter 20) and major earthquakes. Because of this importance of the mantle to a better understanding of geological processes, the International Union of Geodesy and Geophysics, the body which speaks for geophysicists of all nations, initiated, in 1962, an international research programme. This Upper Mantle Project was devoted to the study of the upper mantle and its effect on the development of the crust. Much of our present knowledge of geological processes as discussed in this book stems from the stimulus of the project.

On the basis of seismic velocities, the mantle is usually divided into three regions—from the Mohorovicic discontinuity down to 400 km, from 400 to 1 000 km, and from 1 000 km to the core boundary. Very recently there have been suggestions that the secondary boundaries at about 400 and 1 000 km are in fact rather less deep than previously believed and should be located at about 200 and 700 km. Table 1 and Figure 3. 2 assume the latter values. The outer two regions constitute the upper mantle.

The shallowest mantle layer is one of gradual increase or perhaps in part even of decrease in seismic velocity and rock density. Then, from about 200 or 400 km, there is a rapid increase in velocity and density down to 700 or 1 000 km. Some calculations would show this region to be one of fairly uniform rate of increase, while others show it to be discontinuous, i.e. the velocity and density increase by steps as if there were a series of minor zones. From 700 or 1 000 km down to the core, velocity and density increase slowly but regularly (see Fig. 3. 1). The causes of these density changes are discussed later.

The nature and composition of the mantle is still uncertain. Earlier geophysicists had assumed the upper mantle to be equivalent to dunite, a rock composed almost completely of olivine $(Mg, Fe)_2SiO_4$. This was because the density and elastic constants of dunite were considered appropriate to those of the upper mantle.

This assumption has now been modified from the results of geological and geochemical studies. None of the surface rocks of the Earth can be firmly proved to have been derived direct from the mantle without change. However, some appear likely to be samples of mantle material. These are ultrabasic or ultramafic rocks consisting largely of olivine, with some ortho- and clino-pyroxene, enstatite $MgSiO_3$ and diopside $MgCa(SiO_3)_2$. Big masses of ultramafic rock which are exposed in regions of previous mountain building are thought to be slices of mantle that have been moved through the crust. Even so, during emplacement they may have undergone some unknown amount of change, for example by partial melting and loss of the liquid phase containing the fusible constituents. Other probable samples of upper mantle are *xenoliths*, blocks of foreign rock, caught up in some volcanic liquids. The xenoliths are ultramafic and of plutonic appearance, and occur in some erupted basalts and in kimberlites, the rocks that contain diamonds. The indications are that in their rapid passage upward through the upper mantle these volcanic liquids have torn off some of the solid rocks from the walls of the conduits, and transported them to the surface. There is a general similarity in chemical composition and mineralogy between these ultramafic xenoliths and some types of ultramafic rocks that supports a common origin for both—presumably an upper mantle origin. However, even though these rocks represent samples of parts of the upper mantle, some petrologists would claim that they do not represent average upper mantle. There is increasing geophysical and geochemical evidence that the mantle is laterally heterogeneous. The chief cause of heterogeneity is probably partial melting and removal of the liquid phase as basalt. If this melting was extensive enough, all that would be left would be the least fusible mineral, olivine. And indeed among the samples of probable upper mantle material there is a range of composition through to dunite, the rock consisting almost entirely of olivine. It has been suggested that in the samples of ultramafic xenoliths and rock supposedly representing the mantle, all have in fact undergone some degree of melting. The average composition of primitive or unchanged mantle would then be ultramafic xenolith plus some uncertain proportion of basaltic liquid phase. One popular view of upper mantle composition, due to A. E. Ringwood, considers it to be an amalgam of 3 parts ultramafic rock and 1 part basalt. This would imply that in the derivation of the ultrabasic rock from the mantle, one quarter was melted and removed as basalt. This mix of three to one

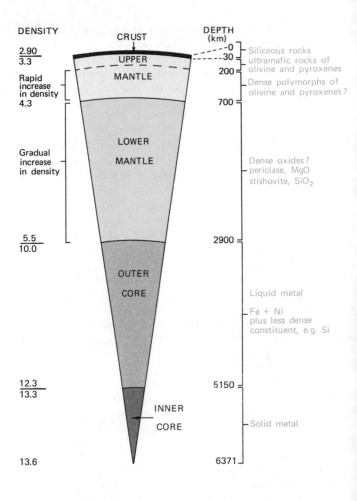

3.2   A schematic section through the Earth.

has been called *pyrolite*, a hypothetical primitive upper mantle material. The upper mantle composition given in Table 5, Analysis 2, is that of one group of ultramafic xenoliths of apparently primitive mantle composition, and does not make any allowance for possible prior depletion in liquid phase by melting. The third analysis in Table 5 is the 'pyrolite model' of Ringwood.

Another even more uncertain source of information on the mantle is provided by meteorites (see Chapter 8). The range of types of meteorite appears to provide a complete sampling of parts of the original parent bodies from which the meteorites were derived. If meteorites are specimens of asteroidal interiors, and if planets and asteroids were sufficiently similar to one another, then meteorites might tell us more of the Earth's interior than could crustal rocks from the Earth itself. However, it is now evident that meteorites differ markedly among themselves, indicating significant chemical differences between the parent bodies. It is likely that even greater chemical differences would have existed in the past between the parent bodies of meteorites on the one hand, and the Earth on the other.

Even though minor differences do exist between the different models for the composition of the upper mantle, estimates for the major components are all fairly similar. The average mineral composition of upper mantle material where this has not been subjected to partial melting is probably about 60 or 70% olivine and about 15 or 20% each of the orthopyroxene enstatite, and of the clinopyroxene diopside. Where melting and removal of liquid has been extensive, the proportion of olivine will be greater. This is true for depths down to about 60 to 100 km, the depths of origin of most basalts, and sampled by the ultramafic xenoliths carried by the basalts. The actual depth will vary with the local temperature environment. At greater depth, the three main minerals are joined by a fourth, a garnet close in composition to a variety called pyrope, $Mg_3Al_2(SiO_4)_3$. This region is represented by ultramafic xenoliths from kimberlite pipes. The conditions of pressure under which diamonds can form, and the existence of diamonds in kimberlite, require that these rocks come from depths of about 150 km in the Earth. Hence the xenoliths in kimberlite can include samples derived from greater depths in the Earth than the xenoliths in basalts.

Both groups of xenoliths have similar chemical compositions, although they differ in the minerals that they contain. We have no knowledge of the possible changes in chemical composition at deeper levels, and if changes do occur they probably affect the minor components of the mantle, like potassium or uranium, rather than major components like magnesium. However, some sort of major change must occur, for example between 200 and 700 km, to account for the changes in velocity of earthquake waves and the calculated increases in density at these depths. What are the changes in the deeper parts of the mantle? The present explanation of the physical differences between the three regions of the mantle are that these are due to changes in the minerals stable within each region.

An assemblage of minerals stable under one set of pressure-temperature conditions will become unstable and recrystallize to form new minerals in a different pressure-temperature environment. Le Chatelier's principle, in one of its many forms, considers that 'If some stress is brought to bear on a system in equilibrium, a change occurs such that the equilibrium is displaced in a direction which tends to undo the effect of the stress'. From this we would expect that if a rock containing an assemblage of minerals was compressed, any resultant change in the minerals would be to relieve the compression by reducing the volume. So with increasing depth and pressure within the mantle, any changes in the minerals should be such that their volumes are reduced and their densities increase.

These density increases in the mantle are of three types:

(a) Normal compression of solids results in a slight decrease in volume due to their compressibility. The equivalent property used by geophysicists is the *bulk modulus* (M_B) — the reciprocal of compressibility, and

$$M_B/V_1 = (P_2 - P_1)/(V_1 - V_2),$$

where P_1, P_2, V_1, V_2 are the initial and final pressures and volumes respectively. This increase in density requires no modification in the structure of the crystalline solid. Changes due to compressibility or bulk modulus are small and regular, such as those down to 200 km and those below 700 km.

(b) In some minerals, the atoms can be rearranged or repacked in a more compact way. That is, the mineral can recrystallize to different and denser crystalline structures. This property, *polymorphism*, occurs in the minerals of the mantle. For example, olivine Mg_2SiO_4 can recrystallize at pressures of over 150 kilobars to a new polymorphic form called a spinel structure. The increase in density due to this is about 8%. Equally the pyroxene $MgSiO_3$ can probably recrystallize at pressures in excess of 200 kilobars to a denser structure analogous to that of corundum (Al_2O_3). The resultant increase in density probably exceeds 10%.

(c) Under some conditions, minerals may react with one another to form new mineral species, or a single mineral may decompose or dissociate to form two new minerals. For example, at low pressures the reaction

$$MgO + SiO_2 \rightarrow MgSiO_3 \text{ (or } Mg_2SiO_4)$$

has its equilibrium point to the right, and $MgSiO_3$ forms at the expense of the oxides. However, at very high pressures the reaction may be reversed:

$$MgSiO_3 \rightarrow MgO + SiO_2.$$

MgO has a density of 3.6×10^3 kg m^{-3} whereas at pressures above about 150 kilobars SiO$_2$ exists as stishovite, a very dense (4.3×10^3 kg m^{-3}) polymorph of quartz. So MgO and stishovite would be much denser than MgSiO$_3$ (3.3×10^3 kg m^{-3}), and high pressures would favour their formation.

It has been suggested that the lower mantle consists of high density polymorphs of silicates, or of dense oxides, like stishovite and MgO, and that the intermediate layer from 200 to 700 km (or 400–1 000 km) is the transition zone between the high density and low density phases. The rapid and irregular increase of density in this zone is due then to these changes in mineral stability with increasing depth.

THE CORE

The Earth's core is the region extending downwards from the core-mantle boundary at 2 890 or 2 900 km. The boundary is sometimes called the *Gutenberg* or *Wiechert-Gutenberg Discontinuity*. The outer part of the core must be liquid because it does not transmit S waves, the transverse or shear type of earthquake waves. In 1936 it was shown that the core contains distinct inner and outer parts marked by rapid increase across the interface of the velocity of P waves, the longitudinal or compressional type of earthquake waves. Subsequently the inner core, extending from about 5 150 km down to the centre at 6 371 km, was shown to be solid with a probable density of about 13.

The boundary at 2 900 km between the mantle and core marks a sharp change in density from about 5.5×10^3 kg m^{-3} in the mantle to about 10^4 kg m^{-3} in the core, while at greater depth the core density increases to 12 or 13×10^3 kg m^{-3}. So the core on average is more than twice as dense as the mantle, and although it is only 16% of the Earth by volume it is about 32% by mass.

Normally it is assumed that the core consists of iron with some nickel. This is based largely on the analogy with meteorites. Iron meteorites and the metal phase in stony meteorites consist of iron with about 6% nickel. Also chemical considerations of the abundance of elements in the Sun and of the probable conditions during formation of the Earth make metallic iron the most likely major component of the core. Other possibilities have been suggested of course. One suggestion is that the core is of silicate but that at extremely high pressures, the electronic structure has broken down to give a metallic type of material of high density. A more extreme theory suggested that the core was hydrogen, but in a metal-like form at the core pressures of more than a million atmospheres. However, even if these transformations in silicates or hydrogen did occur at core pressures, there is no evidence that the resultant densities would be high enough to explain that of the core. One of the densest planets, Mercury, is the smallest and therefore the one with the least compression in its interior. So a transformation due to high pressures alone could not explain the high density of Mercury.

Although geophysicists and geochemists now accept that the Earth's core is metallic, and largely iron, there are complications. The latest evidence is that the density of the core, when corrected for the effect of pressure, is too low for pure iron, or iron and nickel, and that the core must contain an appreciable content of some less dense element. There has been much speculation on this. The presence of sulphides or carbides in the core, or the partial solubility of magnesium oxide in the metal phase at very high pressures, could all provide the necessary reduction in density. However, the favourite is silicon. It has been suggested that the Earth has been very strongly reduced during or after formation, so much so that not only were most iron compounds reduced to metal, but some silicates were also reduced to silicon which is now present in the core. The estimated quantity of silicon in the outer core is 20%, the other 80% being iron and nickel. However, this interpretation is still likely to be amended.

The solid inner core is denser than the liquid outer core. Usually it is assumed that the density change is merely that between liquid and solid metals of the same chemical compositions. This is unlikely. If the inner core grew gradually, it would incorporate the least fusible components, leaving the more fusible ones in the liquid phase. It is impossible to speculate on what chemical changes exist between inner and outer core.

Core-Mantle Relationships

If the material from which the Earth was formed had been subjected to some sort of smelting or reduction process, with the production of metal and silicate phases analogous to the iron and slag in a blast-furnace, then it should be

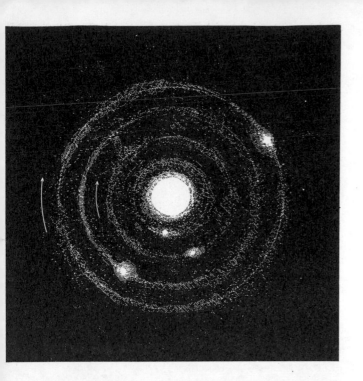

3.3 According to the Nebular Hypothesis, the solar system evolved from a rotating nebula which formed gaseous rings; these in turn condensed into planets and their satellites, while concentration of the central part gave rise to the Sun. The theory allows the planets to be at least as old as the Sun, but does not account for their high angular momentum.

possible to predict the distribution of elements between the two phases. The phases would in fact be the core and the mantle. The prediction is somewhat uncertain in that the effects of pressure are unknown, whether the initial reduction was in the cloud of gas and dust from which the planets formed or occurred in the interior of the planet after its formation.

Assuming that iron occurs both in the core as metal, and in the silicate mantle in an oxidized form, the behaviour of other elements can be assessed on the basis of whether they are more easily oxidized or reduced than iron. If the energy liberated by the oxidation of an element and its transfer from the metallic to the oxidized phase is greater than for iron, the element will be preferentially oxidized and occur in the silicate phase. If the energy evolved is less than for iron, the element will tend to be more easily reduced than iron and will be concentrated in the metal phase. On this basis, elements that are easily oxidized like Na, K, Mg, Ca, Ba, Al, Cr, Zr, U and Th should be almost completely concentrated

in the silicate layers of the Earth, the mantle and crust. The elements with a low affinity for oxidation, like Cu, Ag, Au and Pt, should be almost entirely in the metallic form and be preferentially concentrated in the core. This, of course, does assume no solubility of metals in the silicate phase and vice versa. Confirmation of the general distribution comes from meteorites, where for example the platinum content of the iron phase was found to be about 24 ppm but platinum could not be detected in the silicate phase, and probably was less than 0.01 ppm. The probable nickel content of the core can be explained similarly in that nickel is more easily reduced to metal than is iron and so should be enriched in the metallic core. In meteorites, the nickel content of the iron phase is about 6% whereas in the silicate phase it is only about 0.01%.

There are anomalies; for example in the mantle the nickel content is about 0.2%. Also if the core contains 20% of elemental silicon this suggests far more severe reduction conditions than would be indicated by the 8% of oxidized iron in the mantle. These suggest that in the formation of the Earth some material was very highly reduced, while other material was much less so.

In discussing the occurrence of elements within the core and mantle, it is convenient to ignore the possible existence of a third phase—sulphide. In meteorites there are three major phases— metal, silicate and about 5% of sulphide, so the distribution of elements is three-way. Some metals like copper, lead and zinc occur preferentially in the sulphide phase, rather than in the metal or silicate. In the Earth there is no evidence for a discrete sulphide layer, and it is convenient to assume that none exists, any sulphides being dissolved in the other two phases of metal and silicate.

ORIGIN OF THE EARTH AND PLANETS

The Earth is a planet, a member of the planetary system around the Sun, which, with the Sun, is usually called the solar system. For some purposes a better understanding of the Earth can be obtained by studying the other planets and considering the Earth in relation to them, rather than by studying the Earth in isolation. For example, more is known of the early heating and cooling histories of the parent bodies from which meteorites have formed (see Chapter 8) than of the initial temperature and heating of the Earth.

Most theories for the origin of the planets derive the planetary material from the Sun, or from a primitive cloud of dust and gas in our galaxy which condensed to form both the Sun and the planets. However, the composition of the Sun differs markedly from that of the inner planets; it consists very largely of the simplest elements, hydrogen and helium, only 1% of its atoms being of elements heavier (i.e. with atomic weights greater) than helium. In the inner planets hydrogen and helium are virtually absent, and 99.9% of the atoms present are of elements heavier than helium.

In general the theories for the origin of the solar system can be divided into two kinds:

(a) natural or evolutionary theories, which suggest that planetary systems form as part of the evolutionary history of some stars. If these theories are correct, planetary systems around stars should be relatively numerous;

(b) catastrophic theories, which imagine that the planetary systems form only by some special accident or catastrophe, such as the close approach or collision of two stars. However, because the stars are so far apart in the galaxy, the possibility of such a catastrophe is extremely remote, so that perhaps only about one in each ten thousand million stars could have acquired a planetary system by this mechanism during the whole life of the solar system (4.6×10^9 years). In fact, although planets around other stars cannot be directly seen by telescopes because their luminosity is so low, more than one in a thousand stars is thought to have a planetary system. A far higher proportion are *known* to be binary or multiple systems, in which two or more stars orbit around their common centre of gravity; indeed, binary systems are so common that they form part of the natural order of things, and a catastrophic explanation for their existence must be ruled out.

Any satisfactory theory for the origin of our planetary system must account for several of its observed features, in particular the uniform movement of the Sun and planets. The solar system can be thought of as disc-like in form, with the Sun at the central hub. Each planet moves around the Sun in the same direction, all of the orbits being almost circular and almost in the same plane, which is close to the equatorial plane of the Sun's rotation. In other words, the whole solar system 'disc' is rotating, though different parts within it move at different speeds. Most planets spin or rotate fairly rapidly on their own axes. For

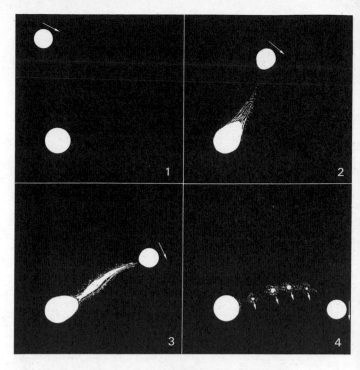

3. 4 Diagram, not to scale, to illustrate the Jeans-Jeffreys Theory. 1: a comparatively dense star approaches the Sun. 2: both bodies are tidally distorted. 3: a filament of gaseous material streams away from the Sun. 4: as the star recedes, the filament breaks up and the planets condense out of it, retaining some of the angular momentum imparted by the passage of the star.

the Earth the period of rotation, i.e. the time taken for a single complete rotation, is 23 hours 56 minutes, but for most of the large outer planets it is about 10 hours. (The Earth's mean solar day of 24 hours is slightly longer than the period of rotation because the direction of the Sun relative to the Earth changes as the latter moves along its orbit.) Where planets have satellites of their own, these secondary systems show the same general features of movement as the solar system, in that most of the moons move around their central planets in the same direction.

Planets are of two types. The four innermost planets are small and dense, their densities each being 4×10^3 kg m^{-3} or more. The outer planets are much larger, Jupiter having more than 10 times the radius of the Earth and 1000 times its volume. Also the outer planets are much lower in density, Saturn having a density less than that of water (10^3 kg m^{-3}). Pluto, the outermost planet, is an exception to this, being fairly small and dense. The actual sizes and densities of the planets relative to Earth are shown in Table 6.

There is some uncertainty in these values because the sizes of the smaller planets are difficult to measure accurately. This relationship—the small planets being of high and the large ones of low density—is contrary to normal expectation. The larger and more massive the planet, the greater its gravitational attraction and the greater the pressure within its interior. One would expect that the largest planet would be the most compressed and hence have the highest density. The fact that the large outer planets are of low density suggests that they consist mostly of substances like hydrogen and helium and of hydrogen compounds such as water, ammonia and methane. The small high-density inner planets consist almost entirely of silicates and metal.

Among the many theories for the origin of the solar system, perhaps the best known is that put forward by the Marquis de Laplace in 1796. The Laplace or 'Nebular' hypothesis suggested that the material that later formed the Sun and planets was originally a disc-shaped rotating nebula or cloud of hot gas. As the gas lost energy by radiation and became cooler, the cloud would have shrunk inwards and in doing so would have rotated more rapidly to conserve angular momentum, until at some period in the contraction the speed of the outermost rim of the disc would have become sufficient for the 'centrifugal force' to be as great as the inward gravitational attraction. At this position, material would be in stable orbit so that continued contraction of the cloud would have left behind a continuous ring or disc of material. Somehow this would then have been swept up into planets. The vast mass of the nebula would have continued to shrink and finally formed the Sun. In this way one could explain the regularity of movement within the bodies of the solar system. After an initial period of wide acceptance, this hypothesis fell into disfavour because it could not explain the energy distribution within the solar system. At present the planets have only 0.1% of the mass of the solar system, but 98% of the energy of movement, the energy of angular momentum. From the Laplacian hypothesis, the Sun would be expected to have much more angular momentum, i.e. a much higher speed of rotation. If the Sun rotates too slowly on its axis, it must have lost some of its rotational energy. Until recently, there seemed no obvious way for this energy to be lost. Because of this difficulty, the catastrophic theories became popular. These explained the very high energy of movement of the planets as being imparted not from the Sun but by the close approach of another star. If the star approached very close to the Sun, its tidal attraction on the Sun, coupled with the Sun's explosive or destructive tendencies, would have caused some of the mass of the Sun to

TABLE 6: PLANETARY DATA

	Mean distance from Sun (AU)	Orbital Period d = days y = years	Period of rotation on its axis (days)	Radius relative to Earth (= 6371 km)	Mean density (10^3 kg m^{-3})	Number of satellites
Sun			25.4	109	1.4	
Mercury	0.387	88d	58	0.38	5.50	0
Venus	0.723	225d	−243	0.95	5.27	0
Earth	1.000	365¼d	1.00	1.00	5.52	1
Mars	1.524	1.88y	1.03	0.53	3.95	2
Asteroids	ca. 2.7				3.5?	
Jupiter	5.20	11.86y	0.41	11.2	1.33	12
Saturn	9.52	29.5y	0.43	9.5	0.69	10
Uranus	19.16	84.0y	0.45?	3.7	1.7	5
Neptune	30.0	164y	0.66?	3.9	1.6	2
Pluto	40	247y	6.4	0.5?	4?	0
Moon	1.000	27.3d	27.3	0.273	3.33	

1 AU or Astronomic Unit is the mean distance of the Earth from the Sun of 1.496×10^8 km.

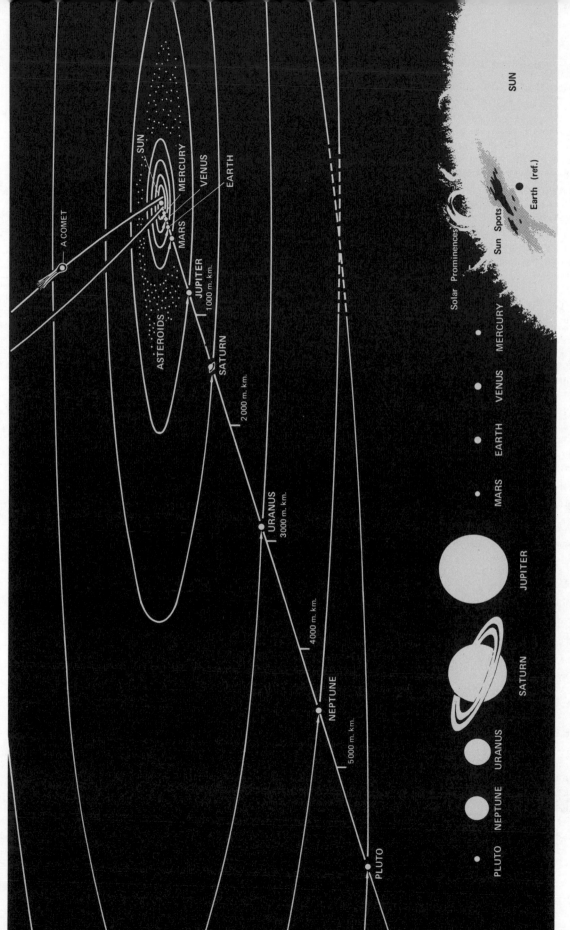

3. 5 A diagrammatic summary of the Solar System.   The Sun and the planets are drawn to scale in the lower half.   All the planetary orbits except those of Pluto and of Mercury lie in nearly the same plane as the Earth's orbit.   The asteroid belt and a highly eccentric comet orbit are shown.

be torn away from the surface. This ejected material, under the influence of the gravitational attraction of the other star, instead of falling back into the Sun assumed a path round it and on condensation formed the planets. Because the direction of movement was imparted by the passing star, all of the ejected material would have moved in the same direction around the Sun. There are a number of variants on this theory, those of Chamberlain and Moulton and of Jeans and Jeffreys being the best known.

Although variants of the two-body theories could account for the energy of angular momentum of the planets, there are many flaws. It is difficult to explain how individual clouds of hot gas of the temperature of the Sun and of the mass of planets could condense rather than disperse. Also the probability of a close approach of two stars is extremely unlikely.

Since about 1943, there has been a swing back to theories of a Laplacian type, in which the planets formed from a rotating gaseous nebula or dust cloud, either around the Sun and perhaps derived from it, or else parental to the Sun which formed during the condensation processes. Within the dust cloud, aggregation of the planets might have been initiated by local irregularities in the concentration of gas and dust, due to vortices within the cloud. The objection to Laplacian theories that the Sun's angular momentum is too small has now been overcome. At present the Sun rotates in a few days whereas the outer planets revolve around the Sun with periods of many years. In the original gas cloud, any viscous drag in the gases would have slowed down the rotation at the interior of the nebula and accelerated the movement at the edge. Magnetic forces could have achieved the same result. It is thought that originally the primitive Sun had a considerable magnetic field. The gas cloud around it would have consisted mostly of hydrogen, some of which would have been ionized. The hydrogen ions would have acted as magnets. A large central rotating magnet surrounded by small magnetic particles tends to drag them around until their period of rotation is the same as that of the central magnet itself.

In this way, a Sun rotating perhaps with a period of a few hours would tend to accelerate the hydrogen ions in the dust cloud in which it was enveloped. This acceleration of the gases would move them outwards away from the Sun, carrying the fine dust with them and leaving only larger solid masses in the region of the inner planets. The

transfer of momentum from the Sun to the gas ions would at the same time slow down the Sun's rotation. In this way Hoyle explains the necessary loss of the Sun's momentum and shows how the dust cloud would be segregated into two regions, an inner gas-free region of solid particles and an outer gas-rich one. The dense inner planets would have accreted in the inner region, the large low-density planets in the outer region.

The actual way in which the planets grew from the initial cloud of gas and dust is uncertain. Studies of the chemical nature of planetary accretion show that the inner planets formed from solid materials at relatively low temperatures. The relative paucity of the inert gases, helium, neon, argon, xenon and krypton in the Earth compared with the Sun, indicates that gases did not accrete to form the Earth. The present oceans and atmosphere of the Earth are secondary features due to the subsequent dewatering or 'sweating-out' of the Earth's interior. The original water would have accreted in the Earth not as gas, but in solid hydrated minerals such as amphibole or mica.

The temperature and gas pressure at the time of planetary formation are also uncertain, although extremely important. It appears that the parent bodies of meteorites accumulated over a range of temperatures down to as low as about 300°K, at which temperature magnetite and the hydrated silicates of carbonaceous chondrites could have formed. For the Earth the final temperature of accretion was probably substantially higher, though the retention of water in hydrated silicates does set an upper limit perhaps of 400–500°K. This temperature is important not only in relation to the Earth's early temperatures but also in deciding the chemistry of the accreting material. For example, in the gas-dust cloud, iron could have been present as metal or in an oxidized form.

In the simplest reaction

$$Fe + H_2O \;\rightleftharpoons\; FeO + H_2 \;,$$

in presence of hydrogen and water in the proportion estimated from the Sun's composition, the iron would be in the reduced form at temperatures above 270°K. When one allows for other possible compounds and equations, most of the iron is probably reduced at temperatures above 400–500°K.

From the probable conditions of condensation of the Earth, it seems likely that it accumulated as a mixture of silicates especially magnesium silicate, together with metallic iron and possibly some sulphides, mostly iron sulphide.

A good deal is known about the present temperatures within the Earth. Temperature and conductivity measurements in deep boreholes and mines and from sediments in the ocean floor provide satisfactory values for the increase of temperature with depth near the Earth's surface, and for the 'heat flow'—the rate of escape of interior heat from the Earth's surface (see Chapter 5). The temperature of the Earth's interior is defined to some extent by the requirement that the oxides of the lower mantle are still not molten, and the core changes from solid to liquid, i.e. melts, at about 5150 km or at over 3 million atmospheres pressure. By reference to laboratory experiments on the effect of pressure on the melting point of similar materials it seems that, the Earth's core is likely to be at temperatures of around 4000°K. Other sources of evidence support temperatures of this order in the core.

From a geological point of view it is important to know how long these temperatures have existed —did the Earth begin as a cool body and become heated, or has it always had temperatures of this order? This is particularly important because the convective movement of the mantle, which causes sea-floor spreading and continental drift, requires heat for its operation, and would increase in speed if the temperature increased and the viscosity of the Earth's interior diminished. Also, does the lack of igneous (or any other) rocks older than four thousand million years and the gap of nearly a thousand million years between that time and the origin of the Earth (4.6×10^9 years ago) result from the early temperatures being so low that igneous processes did not occur, or so high that rapid convective movement engulfed and destroyed the earliest surface rocks?

If, as suggested earlier, the Earth accumulated from solid particles at temperatures of about 500°K or less, how do we explain the presence of molten rocks in the volcanoes? This problem was one of the early reasons for supposing that the planets were derived from the Sun. The Sun was so hot that one could easily imagine the Earth's present heat to be a residue of this original solar temperature. We know now that the Earth has its own heat source, the radioactive decay of elements such as uranium and thorium and their daughter products. These are a considerable heat source. For example, in one year one kilogram of uranium and the equivalent amounts of its decay products would emit 3.1×10^3 joules of heat. Moreover, such elements as potassium and rubidium, not usually thought of as radioactive, are in fact very weakly radioactive because they contain small amounts of the isotopes ^{40}K and ^{87}Rb, and although the energy output of potassium is very low (1 kg of potassium emits about 0.1 joules yr^{-1}) the abundance of potassium in the Earth is so much greater than that of uranium that potassium is almost as important a source of radiogenic energy. Although there is some uncertainty about the total concentrations of radioactive elements within the Earth, there is no question that these can provide the heating necessary to explain the Earth's present temperatures.

There are other sources of heat. As the Earth grew and its gravitational attraction became greater so the energy released by an infalling body increased. At present the energy of infall of a meteorite or a space rocket is very large, enough to cause complete vaporization under some circumstances. For the whole Earth, the gravitational energy release during accumulation and formation is equivalent to 38×10^6 joules per kilogram, enough to raise the temperature of the Earth by more than 20000°C. However, if the Earth accumulated at a fairly slow rate the infalling particles could have lost most of their gravitational energy by radiation after impact.

Another source of energy is also gravitational. If the Earth accumulated as a homogeneous mix of metal and silicate, and subsequently the metal phase sank down to the centre to form the core, this redistribution of lighter and denser phases within the Earth would turn a considerable amount of potential energy into heat, again enough to cause melting of the Earth.

The actual models of the thermal history of the Earth (i.e. the changes in internal temperature since the Earth formed) produced by different authors vary greatly from one another, but nearly always assume a fairly cool beginning, a long period of heating to the Earth's present temperatures, and a subsequent period of relatively uniform temperature.

However, if one examines the only other available planetary material, meteorites, a completely different picture is given (see Chapter 8). Presumably the meteorites are formed from the break-up of asteroids. It seems quite definite that when these asteroid parent bodies were first formed, it was from low temperature material which is now embodied in carbonaceous chondrites. The bodies then became heated extremely rapidly, enough to cause local melting, then gradually cooled. The heating mechanism was by short-lived radioactivity due to the presence of radioactive isotopes with half-lives of about 10^6 or 10^7 years, which have all decayed long ago to non-radioactive isotopes. Although the radioactive parents no longer exist, some of their characteristic decay products can be isolated. The presence of such relatively short-lived radioactive species requires that the atoms were made by nuclear synthesis within a star, or in some stellar event like a supernova explosion, only a short time (10^6–10^7 years) before the formation of the planets.

If one applied the same mechanism to the Earth, it would require that the Earth became heated very early in its history by these short-lived isotopes and then, after their decay, settled down to a steady state in which heat generation from decay of uranium, thorium and potassium was more or less balanced by heat losses from the Earth's surface.

THE ORIGIN OF THE OCEANS AND ATMOSPHERE

The nature of the early oceans and atmosphere is extremely important to our concepts of the origin of life. For example, did the primitive atmosphere contain oxygen, thus permitting biological life to begin, or is the oxygen the result of biological activity?

There are two extreme possibilities for the origin of the oceans. If the Earth formed at very high temperatures, the water would be largely in the gaseous envelope around it, and as the Earth cooled and solidified, the oceans would be the last phase to condense. In this way, from the earliest times, the Earth would have had a primitive ocean of about the present volume. If the Earth formed as a cool body, the water would have been retained in hydrated minerals. Then, with gradual heating and partial fusion, the water could have escaped to surface in volcanic liquids and gases. In this case, the ocean would have been growing continuously by additions from the Earth's interior.

The cations Na^+, Mg^{++}, Ca^{++} etc. in ocean water are derived from leaching of the rocks of the land surface. The chloride anion, however, seems to have an origin similar to that of water,

being either a primitive condensate or an addition from volcanic gases. One must also account for the carbon dioxide in sea water, and all the carbon that has been removed from the seas and atmospheres by plants and animals in the past and is now contained in sediments (e.g. limestones, coal, petroleum, or carbonaceous shales).

For each square metre of the Earth's surface one has to find 3.3×10^6 kg of water, 1.8×10^5 kilograms of carbon dioxide, and 6×10^4 kilograms of chlorine. If all of these were condensed in a primitive ocean and atmosphere, then it would have been enormously rich in carbon dioxide, and one would expect limestones to be very abundant early sediments, falling off in abundance as the carbon dioxide became progressively removed from the oceans and atmosphere. In fact, limestones are formed at about the same rate in each period of geological time, with no large initial deposits. This suggests that the carbon dioxide renewal in the oceans and atmosphere is sufficient to replace the removal of carbon and carbon dioxide in sediments, i.e. that the carbon dioxide is being continuously evolved.

It is probable that in the Earth any early heating resulted in considerable dewatering and degassing of the interior, so that there was an early primitive ocean and atmosphere. However, this would have been subject to subsequent continuous addition and modification. Much of the ocean will be recycled in the ocean-floor spreading process discussed in Chapter 16. New ocean floor will become hydrated and the ocean floor sediments will contain sea water. Where sea-floor is consumed or ingested by the mantle, this water will be incorporated into the mantle, to be released in some subsequent volcanic episode. Some volcanic gases will contain recycled sea water, others will contain water that is escaping for the first time.

The composition of the primitive atmosphere is most speculative. Some consider that it was a reducing atmosphere of hydrogen, methane, ammonia and similar gases, whereas others would suggest that CO_2, nitrogen and water were dominant. The present oxygen content of the atmosphere is renewed by the photosynthetic activity of plants. Whether photosynthesis could have begun in an environment initially completely devoid of oxygen, or whether in the primitive atmosphere some very small amount of free oxygen had been already formed from the photo-dissociation of water vapour by ultraviolet radiation from the Sun, remains to be decided.

FURTHER READING:

B. MASON 1966.   Principles of Geochemistry.
John Wiley & Sons.

J. A. WOOD 1968.   Meteorites and the Origin of
Planets. *McGraw-Hill*

In one sense, geomagnetism is the oldest branch of geophysics. By 600 B.C., the Greeks had discovered lodestone, a highly magnetic natural ore of iron; and because lodestone forms part, albeit a small part, of the Earth's crust, the Greeks had thus discovered that the Earth is associated with magnetism. But they did not know that the Earth as a whole possesses a magnetic field. Nor did the Chinese, who by the first century had invented a simple compass in the form of a lodestone spoon rotating upon a smooth board. In fact, the compass had been in use for over a thousand years before anyone realized that the reason it points approximately north-south is the influence of a field produced by the Earth itself.

But once it had been discovered, the Earth's field became the source of a great deal of study and observation. Many of the earlier measurements were, of course, rather crude; but over the past 350 years, and especially over the past 150, scientists have managed to build up a remarkably detailed picture of the geomagnetic field and its changes with time; and they have done so, for the most part, in complete ignorance of how the field is produced.

What, then, is the source of the Earth's magnetic field? Surprisingly, in view of the antiquity of geomagnetic study, a viable theory for the origin of the field has only been formulated over the past twenty years. One of the earliest theories was that the Earth acts like one huge permanent magnet; but the interior of the Earth turned out to be too hot for permanent magnetism. A theory developed by Professor P.M.S. Blackett during the 1940's was that a magnetic field is a fundamental property of rotating bodies. Again, this idea had to be dropped when sensitive equipment failed to detect such a field in an experiment with rotating gold cylinders, and when a prediction of the theory — that the Earth's magnetic field should decrease with depth below the surface — was proved incorrect.

However, during the 1940's and 1950's, Professor W.M. Elsasser in America and Sir Edward Bullard in England developed a theory, similar in some ways to one which had first been applied to the Sun's magnetic field by Larmor in 1919. This was that perhaps the Earth's core of liquid iron, which is a conductor, is not static but contains convection currents. If there were a small magnetic field to start with, the moving liquid iron would cut the lines of force and thus give rise to electric currents. These currents would, under certain circumstances, then produce their own magnetic fields; and, as long as there was a source of energy in the core to keep the iron moving, there would henceforward always be a magnetic field.

This theory, the dynamo theory, has proved to be remarkably successful; and although it is difficult to work out in detail, it is now accepted by almost all Earth scientists. In this article, Sir Edward Bullard of Cambridge University, one of the two pioneers of the theory, shows how it works, and discusses possible sources for the energy which keeps the dynamo going.

4 The Earth's Magnetic Field and its Origin

by Sir Edward Bullard

If, as seems highly plausible, the Earth's liquid core acts as a dynamo, the geomagnetic field can be accounted for in a natural and satisfying manner.

THE NATURE OF THE FIELD AND ITS CHANGES

A PIVOTED MAGNET will swing round to point approximately to the north and, if it is free to tilt, will take up an inclined position, downwards to the north in the northern hemisphere and upwards to the north in the southern hemisphere. The force on the needle is conveniently described in terms of a *magnetic field*, which is a vector giving the direction and magnitude of the force on one of the poles of a magnet. We need three numbers to describe the force at a given point, and clearly the numbers can be chosen in more than one way. We might, for example, choose the forces in the downward, northward and eastward directions (Fig. 4. 1), or we might choose the magnitude of the force and let the other two numbers represent directions (conveniently the angle of dip of the force below the horizontal plane and the angle between the direction of the horizontal component of the force and the true north). Whatever system we choose, the three numbers must completely specify the force at a given place and time. A *line of force* is a curve whose direction gives the direction of the field at the points along it. The strength of the field and of its components are measured in units called weber metre^{-2}; the field varies over the Earth from 7×10^{-5} to 2.5×10^{-5} weber metre^{-2} (Wb m^{-2}).

Geomagnetism is a complicated subject because all three components of the magnetic field vary not only from place to place, but also with time. Part of the variation from place to place is due to the magnetisation of rocks near the surface of the Earth. For example, if molten rock forces its way to the surface and flows out as lava from a volcano, it will become magnetised in the direction of the Earth's field as it cools. Such magnetised rocks produce a local disturbance in the field which is important in prospecting for ores and in the interpretation of detailed magnetic surveys. Here such magnetised rocks concern us only because they record the history of the Earth's field and show that it has not only changed in the way that we have seen in the last 400 years, but has also reversed on numerous occasions at intervals of a few hundred thousand years. The reversals take place in a surprisingly short time, probably in one or two thousand years.

In a rough way the field outside the Earth resembles that outside a uniformly magnetised sphere or, what is the same thing, that outside a sphere with a dipole at its centre. Such a field is called a *dipole field* and is illustrated in Figure 4. 2. A dipole field has two magnetic poles where a magnetised needle stands vertical, and a magnetic

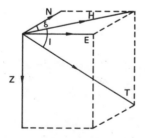

4. 1 The components of the magnetic field. The vector T shows the direction and magnitude of the field; it is called the total force. N is the north component, E the east component, Z the vertical component, H the horizontal force, I the angle of dip (also called the inclination) and δ the declination (called the variation by sailors). N, E, Z or H, δ, I or T, Z, δ are sufficient to specify the field but N, E, H or T, Z, I are not (because they are not three independent quantities; if we know N and E we know H).

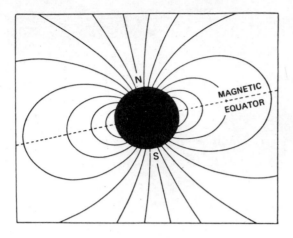

4. 2 The lines of force of a dipole field outside a sphere. The section shown contains the poles. The form of the field inside the sphere depends on how the field is produced.

equator where it lies horizontal. The dipole field that best fits the actual field of the Earth has its poles 11½° from the geographic poles, the northern one being in the north-west of Greenland. This resemblance of the Earth's field to the field of a uniformly magnetised sphere was first pointed out by William Gilbert of Colchester, who was physician to Queen Elizabeth I.

The fit of the best choice of dipole field to the Earth's field is only approximate. This can be seen in Figure 4. 3, which shows lines of equal horizontal force; if the field were that of a dipole, these would be lines of latitude relative to the poles of the dipole. In some places the difference between the actual field and that of the best-fitting dipole reaches 20% of the whole field. If the best-fitting dipole field is subtracted from each component of the actual field we get the *non-dipole field* which is shown in Figure 4. 4. Even when all the short distance changes have been smoothed out we are still left with a rather complex pattern. There are areas a few thousand kilometres across over which the field differs systematically from the dipole field, and these areas do not show any obvious relation to geography or geology. Some of the sets of closed contours in Figure 4. 4 are centred over the land, and some over the ocean; none of them follows

the coast line or the great mountain chains in any systematic way. The only exception to this lack of relation to geography is a tendency for the non-dipole field to be smaller over the Pacific than it is over the rest of the Earth. Whatever story this large-scale, smoothed picture of the field tells, it is not a story connected with surface geology. In looking for an origin for the field we should not look too near the surface in the region studied by geologists.

The variation of the field with time suggests the same conclusion. The pattern of Figure 4. 4 is temporary; 300 years ago it was quite different and, if present changes continue, it will be quite different again 300 years hence. The dipole itself changes more slowly and has hardly moved in the last 300 years. We know, however, from the magnetisation of rocks that it has moved in the more remote past and that if its position is averaged over a few thousand years it is very close to the geographic pole.

The changes in the field at a given place are quite rapid and quite large. The first measurements of the declination of the compass at London were made in 1580; the needle then pointed 11° to the east; in 1660 it pointed due north; by 1820 it was 24° to the west. Since then it has come back and is now only 7° to the west. These changes and that of the dip are shown in Figure 4. 5. Figure 4. 6 gives a world-wide picture of the rate of change of the horizontal force in 1942. The rates of change not only bear no relation to geology but also have a time scale very short compared to geological time. At London the compass swung through 35° in 240 years. At Cape Town the horizontal component of the field decreased by 30% in 100 years. These are large changes and, from a geological point of view, very rapid changes. It is inconceivable that large motions of material or large temperature changes could occur on a world-wide scale in the solid part of the Earth in such short times without cataclysmic consequences. Again we must get away from geology.

It is possible to show that the main features of the Earth's field, the dipole field, the non-dipole field and the changes on time scales greater than a few years all have their origin within the Earth. This result was stated by Gilbert, who deduced it from the analogy between the Earth's field and that of a sphere cut from a naturally magnetised lump of iron ore. The formal theory, which enables one to say with certainty whether the poles

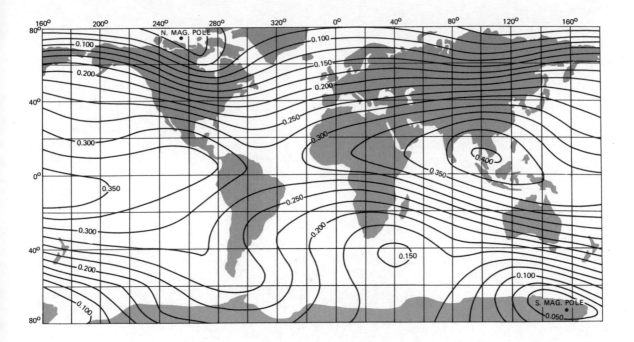

4. 3 Lines of equal horizontal intensity. If the field were exactly a dipole field these would be lines of latitude relative to the magnetic poles of the dipole, that is they would, except near the poles, be almost straight lines on this map.

or currents producing a field are within or without a closed surface, was developed by Gauss (1777–1855); all that is needed is measurements of the three components over the surface. There do exist small rapid variations of the field with periods from a second to a few years which are of external origin and are due to currents circulating in the upper atmosphere; in our broad picture we ignore these or imagine them smoothed out. We look at the field through an imaginary filter which smooths out variations of short period or wave length in both time and space.

Since the origin of the field is inside the Earth we must flee the geologists in a downward direction and not upwards to the upper atmosphere, into space or to the stars. Thus we look for an origin deep within the Earth in a place where rapid changes can take place. The obvious place to consider is the Earth's core, where the material is fluid and we can hope for motions that are much more rapid than those that can occur by creep in the outer, solid part of the Earth.

POSSIBLE THEORIES OF ORIGIN OF THE FIELD

How can a changing, reversing field be produced within a spherical body of fluid? There are two ways in which fields are commonly produced: by permanent magnets and by electric currents. A fluid cannot be a permanent magnet; at least no fluid magnets are known, and if they did exist the different parts would soon get mixed up and there would be no general magnetisation, only a chaotic lot of little magnetised regions jumbled together. The other way of producing a magnetic field, by currents, seems more promising. The core of the Earth has a density of 10 to 12×10^3 kg m^{-3}, which is typical of metals at the pressure there prevailing and is too high for any rock; it is usually supposed that the core is composed largely of molten iron and is therefore a good electrical conductor.

If electric currents flow in the core, is it necessary to suppose that there are e.m.f.'s to maintain them? Once a current is started in a conducting

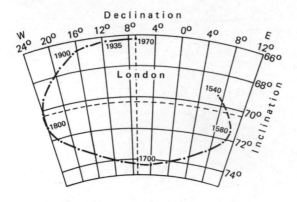

Declination

W 24° 20° 16° 12° 8° 4° 0° 4° 8° E 12°

London

Inclination

4. 5   Changes in field direction at London.

body it will run on for some time through the effects of self-induction. For laboratory-sized bodies this time is short; for example, for a copper sphere 0.1 m in diameter it is always less than 0.1 s, but it increases as the square of the size of the body and may therefore be large for the Earth's core. The actual time depends on how the currents are distributed. The smoother the distribution, the longer the time. For a sphere the maximum possible time for the current and field to fall to $1/e$ of their initial values is $\mu_0 k a^2 \pi^{-2}$ seconds, where k is the conductivity in ohm^{-1} metre^{-1}, a is the radius in metres and μ_0 is the permeability of the material (at the prevailing temperatures this will be the same as for a vacuum; that is $4\pi \times 10^{-7}$ henry metre^{-1}). The Earth's core has a radius of 3400 km and its conductivity is probably around 3×10^5 ohm^{-1} metre^{-1}; with these values the time to decay to $1/e$ is 14 000 years. This is a very short time compared to the age of the Earth $(4.5 \times 10^9$ years) and shows that the currents must be maintained against decay if the field is to last for any geologically significant time. The occurrence of reversals of the field also strongly suggests that there is some process controlling the currents and that they are not just coasting along, kept up by the inductance of the circuit.

One might suggest many ways of keeping electric currents flowing in the core of the Earth; one might imagine them driven by chemical e.m.f.'s (i.e. batteries) between the hot silicates and the surface of the core or by the thermo-electric effect. The core is so well protected from observation that it is difficult to say for certain that such

views are wrong. But it is difficult to imagine how the field could be suddenly reversed if it were produced in this way.

In recent years the hypothesis that the Earth's field is produced by a dynamo within the Earth's core has found many supporters. It cannot be said that it has been shown conclusively that this is how the field is produced, but it does seem to give a promising line of attack and to lead to a theory that should be capable of being worked out in some detail.

THE DISC DYNAMO

First we consider what we mean by a dynamo by considering the simplest possible example. Figure 4. 7a shows a disc rotating on an axle in a magnetic field perpendicular to the disc. As it turns, every radius of the disc cuts the field and generates an e.m.f. between the axle and the edge of the disc. This e.m.f. cannot drive a current since there is nowhere for the current to go; all that happens is that charge collects on the rim of the disc. If an external circuit is connected to the axle and the disc by brushes, as in Figure 4. 7b, then current will flow in the disc and around the circuit. The current will produce a field in the neighbourhood of the circuit. This is a dynamo. It produces a current and a magnetic field, but it requires a field supplied from outside to work it. However, if the circuit connected to the brushes consists of a coil going around the axle as in Figure 4. 7c, then the field needed to excite the e.m.f. in the rotating disc is itself supplied by the dynamo. We now have a *self-exciting dynamo* which produces a field whenever the disc is turned sufficiently rapidly. If the disc is turned too slowly, not enough current is produced to keep up the field and the dynamo will not work; if it is turned too fast the current and field will grow without limit. At a critical velocity between the two a constant current and field are produced. The paradox of the infinitely great current produced by a fast-turning dynamo is easily avoided by noting that the couple needed to turn the axle above the critical speed would grow indefinitely as the field builds up. If the axle is turned by a constant couple the disc can always turn at the critical speed and produce just enough current and field for their interaction to balance the driving couple. We then have a steady dynamo driven by a constant couple, turning at a constant speed

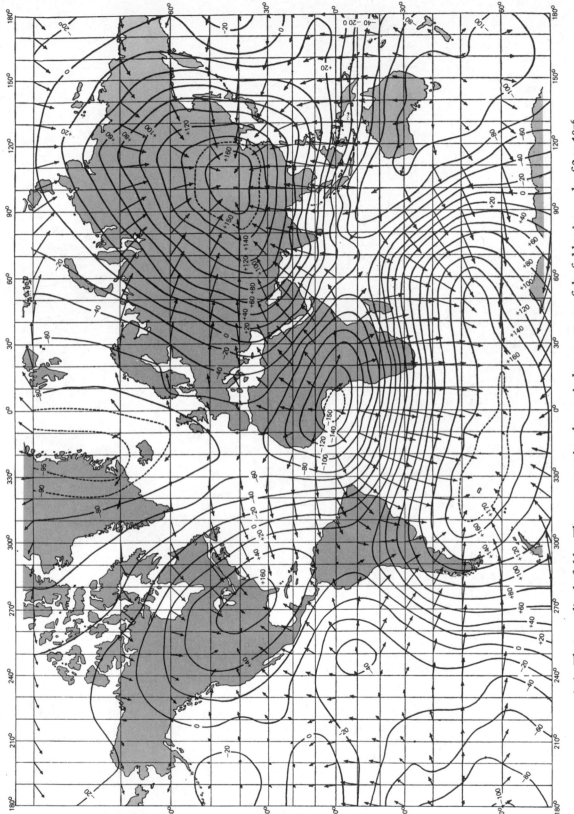

4.4 The non-dipole field.   The contours give the vertical component of the field at intervals of 2×10^{-6} Wb m^{-2}, the arrows give the horizontal component.

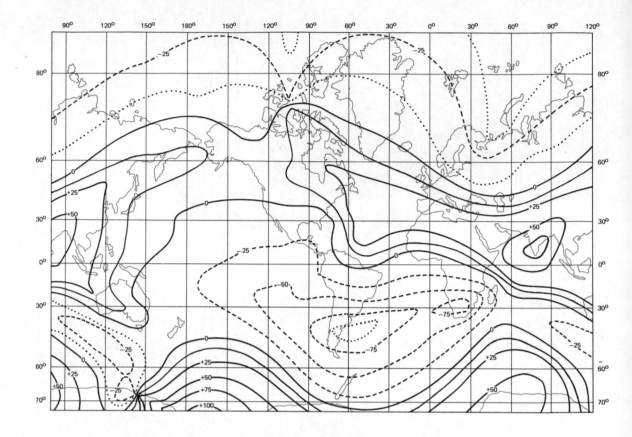

4. 6 The rate of change of the horizontal intensity of the Earth's magnetic field in 1942.   The units are $10^{-9}\ \text{Wb}\ \text{m}^{-2}\ \text{yr}^{-1}$.

and producing a steady current and field.   Power is supplied by the applied couple and dissipated as heat by the current in the coil.

It is clear from Figure 4. 7c that if *both* the current and the fields are reversed the dynamo will work as before.   That is, a dynamo that can produce a field in one direction can equally well produce a similar field in the reverse direction.   The same is not true of the velocity; if it is reversed the field from the coil will produce a current that gives a field opposing that already there, and the dynamo will not work.   Thus for the dynamo to work it must be turned in the right direction, but it may produce a field in either direction.   This is a most promising result—the field from a dynamo can be in either direction—but will it reverse if left to itself?   A man can stand on his head, but he does not necessarily do so in the ordinary course of business.   What we are asking here is whether, if

we disturb a steadily running dynamo, for example by changing the driving couple, we can make it flip over so that the current and field are reversed.   After not very difficult calculations it is found that the dynamo of Figure 4. 7c will not flip over but that there are systems that will; for example, two dynamos like that of Figure 4. 7c with the coil of each connected to the other's brushes do flip over repeatedly and irregularly.

In this account of a dynamo we have assumed that there was some field present and have shown that, if there is, it will build up and be maintained at a value appropriate to the driving couple.   But what if there is no initial field?   Why should not the couple just accelerate the disc without any field being produced?   Such a motion is possible but it is unstable.   The slightest current or field will be seized on and built up in a time that depends on the inductance and resistance of the coil.

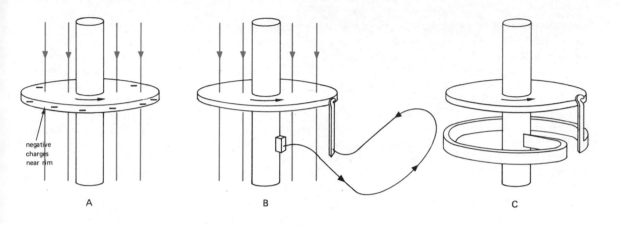

negative
charges
near rim

A

B

C

4.7 (a) Disc rotating in magnetic field, no current flows.

(b) External circuit enables current to flow.

(c) Current provides a field and gives a self-exciting dynamo.

DYNAMOS IN A SPHERE

The simple system described in the previous section illustrates the properties of dynamos, but it is clearly not a thing that we can expect to find in the Earth's core. For one thing it is not what the topologists call a simply connected body; it is like a Henry Moore statue: it has a hole in it. Also it is not, like the Earth's core, a spherically symmetric body. We have, therefore, a very serious and, as it turns out, a very difficult problem: are there any motions in a sphere of electrically conducting fluid that will cause it to act as a dynamo? Can we trade off complexity of motion for simplicity of structure? The disc dynamo has a very simple motion, just a rotation of the disc, but a rather complicated structure. Can we have a dynamo with a very simple structure, just a sphere of conducting fluid capable of quite a complicated motion? Such general questions are often difficult to answer, and for many years the answer to this one was unknown. Eventually, however, examples of motions in a sphere that could be shown to act as dynamos were produced, and the possibility of the process was demonstrated. In the other direction it has been shown that many kinds of symmetry in the motion prevent dynamo action. For example, the motion must have a radial component and must not be confined to planes at right angles to an axis. If we want a dynamo in a body as simple as a sphere we must have a fairly complicated motion.

To work out what will happen with any particular motion or with any particular set of driving forces is a problem of great difficulty which is only now becoming soluble; it is one of the outstanding problems of theoretical physics. It is, however, clear that there is a condition similar to that for the disc dynamo. If a given pattern of motion can work as a dynamo, it will only do so if the speed of the motion is fast enough. The condition is of the form

$$k \, l \, v \, \mu_0 = C$$

where k is the electrical conductivity (perhaps about 3×10^5 ohm^{-1}m^{-1} for the Earth's core), l is a typical distance in which the field changes by a large part of itself (this cannot be greater than the core's diameter and can perhaps be taken as 1000 km), v is the minimum velocity needed for dynamo action, μ_0 is the permeability of the material ($4\pi \times 10^{-7}$ henry m^{-1}), and C is a non-dimensional constant which would be given by a detailed theory for any specified pattern of motion. From the little that has been done, C seems to be of the order of 100. Thus

$$v = C/k \, l \, \mu_0 = 0.3 \times 10^{-3} \text{ m s}^{-1} \ .$$

At any rate such velocities are not absurdly large.

Of the forces that might drive the dynamo we know almost nothing. One possible view is to suppose that the motion is driven by thermal convection. Small amounts of radioactive materials, particularly uranium, thorium and potassium, are found throughout the universe; as these materials decay they produce heat and raise the temperature. If, as is to be expected, the core of the Earth contains small amounts of these materials, its temperature will be raised and its density reduced. If the material were incompressible the inner hotter parts would be less dense than the outer parts, the fluid would be unstable and motion would be produced, the hotter material rising and the cooler sinking. This process can be seen when water is heated in a saucepan. Here, the heat is applied at the bottom and it is pretty obvious what is happening. In the Earth's core the heat is produced throughout the material and it is not so obvious that the deeper material will get hotter; in fact it will, owing to the greater loss of heat by conduction from the outer parts. If the material is compressible a critical gradient, called the *adiabatic gradient*, must be exceeded before motion can take place. This sets a lower limit to the heat generation and the amount of radioactivity below which motion cannot take place and no dynamo can function. The actual radioactivity of the core is unknown. The radioactivity needed to produce motion is much below that of most rocks but above that of iron meteorites. There seems no reason to suppose that the process is impossible.

An alternative view ascribes the motion to the precession of the Earth. The attractions of the Moon and the Sun on the equatorial bulge of the Earth cause its rotational axis to trace out a cone similar to that described by an inclined top spinning on a table. The fluid material of the core may not be able to follow this motion exactly, and a complicated eddying motion may be produced. We have no way of choosing between these two theories and it is quite possible that both mechanisms play a part.

Motion in the Earth's core may not only account for the origin of the magnetic field but also for its fluctuations. When a conducting fluid moves through a field it produces an e.m.f., an electric current and a disturbance of the field. A complicated eddying motion will therefore produce a complicated and varying disturbance of the field which is just what is observed at the surface of the Earth. It is a characteristic of fluid motion that the motion is more complicated than its causes. We can see this in meteorology. The causes are simple; the Sun shines on the Earth and warms the air and the ground over one hemisphere; out of this grows all the infinite complexity of weather and climate. Looked at in this way, the Earth's magnetic field illustrated in Figures 4. 4 and 4. 6 has a hydrodynamical look about it. It looks about as complicated as a weather map; and a theory depending on motions in a fluid seems very appropriate. The time scale also seems reasonable. The field changes up to 30% in 100 years, and to produce this change the fluid in the core should move a substantial distance in this time. If the speeds are of the order of 0.3×10^{-3} m s^{-1}, the distance gone in 100 years would be about 1 000 km, which is of the right order of magnitude, neither very large nor very small compared to the size of the core.

In brief, the dynamo theory seems natural and unforced, and order-of-magnitude arguments are encouraging; but the difficulties of setting up a detailed theory which will predict details of the field from assumed forces are immense, and even if they were surmounted we should still have no direct proof that our mechanism was the true one. Perhaps conviction would be easier to attain for the Sun, where the motions at the surface of the fluid conductor can be observed as well as the magnetic field.

FURTHER READING:

W. M. ELSASSER 1958 (May). The Earth as a Dynamo. *Scientific American* (*Reprint No. 825*).

D. W. STRANGWAY 1970. History of the Earth's Magnetic Field. *McGraw-Hill*.

P. J. SMITH 1968 (April 4). The Earth's Fluctuating Dynamo. *New Scientist*.

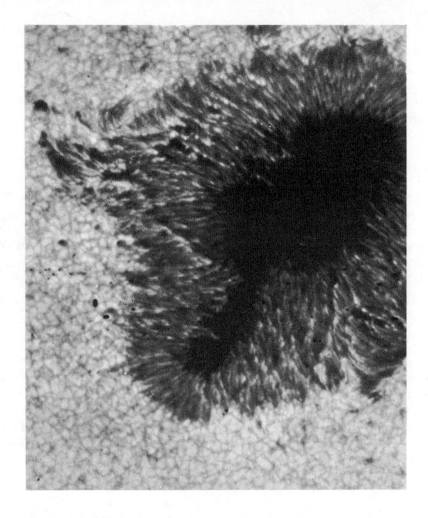

4. 8 A sunspot photographed with a stratospheric telescope. The picture shows the radiating pattern of the filaments in the penumbra, and the granulation of the photospheric surface which probably corresponds to a pattern of convection cells.

The Sun has a magnetic field which is specially strong in sunspots, where it reaches a strength of up to 5 000 times that at the surface of the Earth. The fact that a field can occur in a hot, molten or gaseous body is a powerful argument in favour of the dynamo theory. It is difficult to imagine any other mechanism that could work in such circumstances.

The heat we feel at the Earth's surface comes mainly from the Sun. However, the Earth itself is also a vast heat engine. Over 8×10^{20} joules of heat, produced mainly by the decay of radioactive elements in the interior, reach the Earth's surface each year. From the energy point of view heat flow is thus the most impressive of all terrestrial phenomena, being many orders of magnitude greater in terms of energy than all earthquakes, volcanic activity and tectonic processes put together. Furthermore, geothermal processes play an important role in almost all theories concerning the origin and development of the Earth.

The reason we are not usually aware of this internal heat is that, because the thermal conductivity of the Earth's crust is low, it flows to the surface only slowly. Very sensitive instruments are therefore required to measure it. But the low thermal conductivity is really a blessing because it means that the Sun's heat also penetrates the Earth very slowly. Thus below about 50 metres, seasonal variations in the solar heat have no effect at all; and there we can concentrate solely on the measurement of the Earth's internal heat. Below the 50 m level the temperature gradient averages about 30°C per kilometre.

The Earth's internal thermal state is the most important factor affecting physical and chemical processes operating at depth. So, we have to attempt to interpret the surface data in terms of conditions in the interior. The problems involved in making these interpretations, the general picture of the Earth's internal temperature that emerges, and its significance in the Earth sciences are discussed in the following article by Dr. J. H. Sass of the United States Geological Survey, a leading authority on the Earth's heat.

Parameter	Symbol	SI Unit	Working Unit	Conversion Factor (SI to Working Units)
Heat flow	q	W m^{-2}	μ cal cm^{-2} s^{-1}	23.9
Vertical temperature gradient	r	°K m^{-1}	°C km^{-1}	1000
Thermal conductivity	K	W m^{-1}°K^{-1}	mcal cm^{-1} s^{-1} °C^{-1}	2.39
Thermal capacity	C	J kg^{-1}°K^{-1}	cal gm^{-1} °C^{-1}	239
Thermal diffusivity	α	m^2 s^{-1}	cm^2 s^{-1}	10^4
Heat production	A	W m^{-3}	$\times 10^{-13}$ cal cm^{-3} s^{-1}	2.39×10^6

TABLE 1. SYMBOLS AND UNITS FOR THERMAL PARAMETERS

5 The Earth's Heat and Internal Temperatures

by JOHN H. SASS

A discussion of observational evidence and theoretical considerations on which models for the Earth's heat regime are based.

ON MOST OF THE EARTH'S SURFACE, the flow of heat from the interior is almost imperceptible. It is several thousand times smaller than the heat exchanged by radiation between the Sun and the Earth's surface. There are spectacular exceptions such as volcanoes and hot springs, but these occupy an extremely small fraction of the globe. Although it is small, about 0.06 W m^{-2} on the average, heat from the interior supplies most of the energy for volcanoes, earthquakes and mountain building. A knowledge of surface heat flow and its variation from place to place is therefore required for an understanding of deep-seated Earth processes. In particular, recent technological advances have made possible the laboratory study of physical and chemical processes in probable Earth materials at temperatures and pressures corresponding to a large range of depths within the Earth. The findings of these experiments can be applied to the real world only to the extent that the temperature and pressure at a given point can be determined. Because of the large uncertainties in the variation with depth of important variables, especially the coefficient of heat transfer and the radioactive heat productivity, the temperature at a given depth cannot be specified with nearly the same accuracy as can the pressure.

The only directly measurable quantity related to the Earth's internal temperatures is the surface heat flow. In this discussion we shall examine briefly the variation of heat flow over the terrestrial surface and its implications for thermal models of the Earth. Table 1 lists the symbols and units for the most common geothermal parameters. In addition to SI units, the table shows the most common working units used in the literature together with the appropriate conversion factors.

OBSERVED HEAT FLOW

A determination of heat flow requires two separate measurements: the rate of increase of temperature with depth (temperature gradient, r) and the thermal conductivity (K) of the rocks in which the temperatures are measured. Heat flow (q) is then calculated according to the formula:

$$q = Kr.$$

The temperature of the ocean bottom where the water is deeper than a few hundred metres is usually very stable, so q can be determined by measuring temperatures in the upper one to five metres of ocean-bottom sediments and by determining conductivity over the same interval. On continents, the effects of seasonal variations of surface temperature and movement of ground water extend to depths of 50 to 100 metres or more, so that the measurements must be made in oil wells, mines and tunnels that penetrate below this near-surface zone of variability.

Table 2 shows a recent summary of over 3 000 heat-flow results as tabulated by W. H. K. Lee. The standard deviation is included in the table as a means of comparing the scatter among the various data sets.

IMPLICATIONS OF THE OBSERVED HEAT FLOW

One of the most important and most unexpected results of the measurements is that the mean heat flow from continents (61 mW m^{-2}) is essentially equal to that from oceans. Until the first successful measurements of oceanic heat flow by E. C. Bullard and his associates in the 1950's, it was tacitly assumed that the mantle beneath continents was physically and chemically the same as that

TABLE 2.   MEAN* HEAT-FLOW VALUES FOR VARIOUS TECTONIC REGIONS

TECTONIC REGION	NUMBER OF DATA	MEAN HEAT FLOW (q) mW m^{-2}	STANDARD DEVIATION
All continents†	597	61.0	19
Precambrian shields	214	41	10
Post-Precambrian non-orogenic areas	96	62	17
Palaeozoic orogenic areas	88	60	17
Mesozoic–Cenozoic orogenic areas	159	74	24
All oceans†	2530	61.4	33
Ocean basins	683	53	22
Mid-oceanic ridges	1065	80	62
Ocean trenches	78	49	29
Continental margins	642	75	39
World†	3127	61.4	31

*The individual data were weighted according to quality, so that a poorly determined result makes a smaller contribution to the mean than a precisely determined value.

†To compensate for the uneven geographical distribution of data within these larger units, means are determined by giving equal weight to the average heat flow in quadrilaterals of equal area (3×10^5 km^2 or $5° \times 5°$ at the equator).

TABLE 3.   TYPICAL ABUNDANCES OF RADIOACTIVE ISOTOPES FOR VARIOUS TERRESTRIAL ROCKS

	U ppm	Th ppm	K %	HEAT PRODUCTION (A) μW m^{-3}
(a) *Crustal Rocks*				
Silicic igneous rocks	4.0	16.0	3.3	2.5
Mafic igneous rocks	0.5	1.5	0.5	0.3
Shales	4.0	12.0	2.7	2.1
Carbonates	2.2	1.7	0.3	0.7
Beach sands	3.0	6.0	0.3	1.2
(b) *Possible Mantle Rocks*				
Dunite	0.005	0.02	0.001	0.004
Eclogite	0.04	0.15	0.1	0.04
Oceanic Lherzolite	0.02	0.06	0.005	0.01

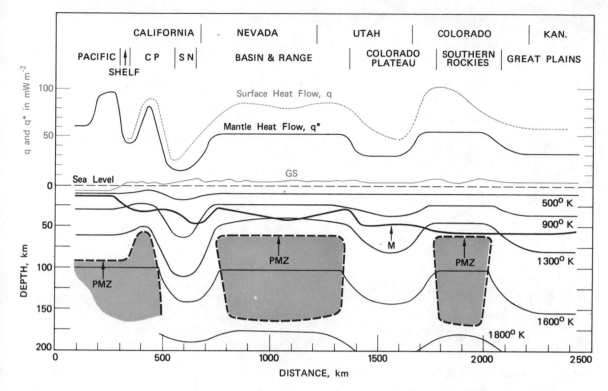

5.1 Cross section across the Western United States near latitude 38° N (modified from Roy and others, 1971). Abbreviations: CP, Coastal Provinces; SN, Sierra Nevada; GS, Ground Surface; M, Mohorovicic Discontinuity; PMZ, Partially Molten Zone.

beneath oceans. Because of this, it was thought that the oceans, with their thin (about 5 km) mafic crust and a lower concentration of radioactive elements would have a much lower heat flow than the thick (30 to 40 km) silicic continental crust.

Virtually all radioactive heat is produced by isotopes of uranium, thorium, and potassium. From the variations in radioactivity for main rock types given in Table 3 we can see that a 20 to 25 km thickness of 'silicic' rocks could generate all of the heat observed at the continental surface; on the other hand 5 km of 'mafic' rocks would generate only about 1.5 mW m^{-2}, or less than 3% of the observed oceanic heat flow.

The near-equality of continental and oceanic heat flow thus implies a fundamental difference between the mantles underlying these major topographic units. Heat flow from oceanic mantle is higher, on the average, than that from the mantle beneath the continents because of some combination of higher radioactivity and transient temperature rises associated with sea-floor spreading.

HEAT FLOW PROVINCES

Within both continents and oceans, a number of heat-flow provinces can be defined, and their boundaries can often be correlated with physiographic or tectonic boundaries. Thus, in the oceans, ridges are provinces of high but variable heat flow, basins are characterized by moderate, relatively uniform flow and trenches are low heat-flow provinces. On continents, Precambrian shields have low heat flow, and regions of Mesozoic-Cenozoic tectonic activity contain both low (Sierra Nevada) and high (Basin and Range) heat-flow provinces (Figs. 5.1 and 5.2).

The characteristic heat flow for these various provinces can be correlated with observations of such parameters as crustal thickness, seismic velocity, and magnetic intensity to yield information about variations in composition and temperature beneath the surface. Figure 5.1 illustrates such an interpretation by R. F. Roy, D. D. Blackwell and E. R. Decker for the observed heat flow near 38° N latitude in the western United States (see Fig. 5.2). The zones desig-

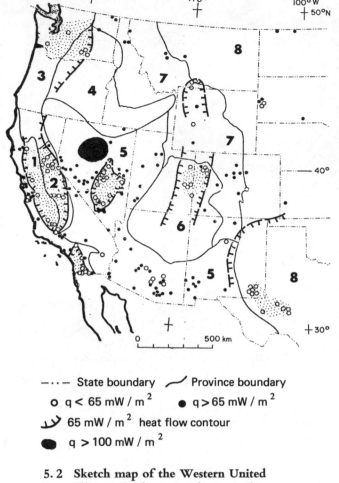

$$q = q^* + DA$$

where q^* and D are constants (within a given province) having the dimensions of heat flow and depth respectively. This relation was first discovered for the Appalachians (Fig. 5.2) by Francis Birch and his colleagues at Harvard, but the Appalachian parameters have since been found to apply in the stable continental region of North America between the Appalachians and the Rocky Mountains, and the linear relation has been observed in both the Basin and Range and Sierra Nevada provinces by both the Harvard group and A. H. Lachenbruch of the U.S. Geological Survey. For the provinces defined to date, D ranges from 7 to 10 km and q^* from 17 in the Sierra to 60 in the Basin and Range. The simplest interpretation of the observed relation is that the surface heat production A at any point within a province extends to a depth D and that the intercept q^* is the heat flowing from below that depth. q^* for any province is a combination of the heat flow from the mantle and the lower crust (below depth D) for which the mean heat production is laterally homogeneous. Lachenbruch pointed out that this interpretation requires that all plutons be the same thickness regardless of different rates of erosion for different plutons. He suggested an alternative interpretation in which the abundance of radioactive material decreases exponentially with depth within plutons. This is the only distribution that fits the observed relation and allows for differences in thickness. For either interpretation it is convenient to regard q^* as an upper limit to the mantle heat flow. Establishing q^* for a province, in effect, strips off the zone of variability in the upper crust and gives us another geophysical window on the Earth's deep interior (Fig. 5.1).

State boundary — · —   **Province boundary**

o $q < 65 \ mW / m^2$ ● $q > 65 \ mW / m^2$

$65 \ mW / m^2$ heat flow contour

● $q > 100 \ mW / m^2$

5.2 Sketch map of the Western United States illustrating the major physiographic provinces; 2 (Sierra Nevada), 5 (Basin and Range), and 8 (Stable Interior) are discussed in the text as heat-flow provinces. 1 and 3 are the coastal provinces; 4, the Columbia Plateau; 6, the Colorado Plateau, and 7, the Rocky Mountains.

nated as partially molten (PMZ) coincide with high surface heat flow and low seismic velocities in the lower crust and upper mantle.

For the granitic intrusive rocks within a given continental heat-flow province, heat flow and near-surface heat production appear to be closely related. Specifically, if heat flow, q, is plotted as a function of the surface heat production, A, calculated from measurements of the abundances of uranium, thorium, and potassium, a straight line with positive intercept on the q axis results. This line may be described algebraically as

THERMAL MODELS FOR THE CRUST AND UPPER MANTLE

The observed linear relation between heat flow and heat production in granitic plutons may allow precise estimates of temperature as far down as the base of the crust. Lachenbruch recently presented a formal mathematical treatment based on his own interpretation of the linear heat-flow relation in terms of an exponential decrease of radioactive heat production with depth. Subject to reasonable assumptions, the mathematical treatment leads to a simple geochemical view of the crust in which the exponential decay of heat production with depth reflects the upward concentration of radioactive elements within a zone of partial melting which (over a period of time) involved most of the crustal thickness. Thus the intrusive part of the crust within a given thermal province may be treated as a single system, and, after the actual heat of intrusion has been dissipated, the temperature-depth curve below any point can be calculated from a knowledge of the surface radioactivity, A, at that point. Assuming the validity of the model, the only uncertainty in temperature lies in the uncertainty of the thermal conductivity, K, which (for temperatures up to about 1 000°K) is reasonably well represented by the value of 2.5 W m⁻¹ °K⁻¹.

Figure 5.2 illustrates the location of the three well-documented heat-flow provinces in the United States. Some of Lachenbruch's crustal temperature profiles for these provinces are reproduced in Figure 5.3. It is seen that for the Basin and Range physiographic province, temperatures predicted by this model (or any other plausible model) are higher than for the other provinces. At the base of the crust the predicted temperatures for this province are in the range commonly thought to be high enough to allow partial melting. This is, of course, perfectly reasonable in view of the observed and recently apparent volcanic and hot spring activity in the Basin and Range. It is also consistent with the observed low seismic velocities in the upper mantle beneath this province. (See also Fig. 5.1.)

Observed heat flows are generally compatible with the hypotheses of sea-floor spreading and plate tectonics. Heat flow is high along ridges where hot material is supposed to be rising, and it is low near trenches where a cold plate evidently is descending.

Figure 5.4 is a diagram from the work of M. N. Toksöz, J. W. Minear and B. R. Julian, who made a theoretical study of the thermal effects of a descending slab. The figure shows the thermal perturbations due to an 80 km thick slab of oceanic crust and upper mantle descending at the rate of 8 cm each year for about 13 million years. The thermal effects of radioactive heating, phase changes (in the shaded regions), adiabatic compression, and frictional heating along sliding surfaces are represented in the diagram.

The near-equality of continental and oceanic heat flow places an important constraint on the concept of plate tectonics. There is an apparent paradox, because we have already concluded that

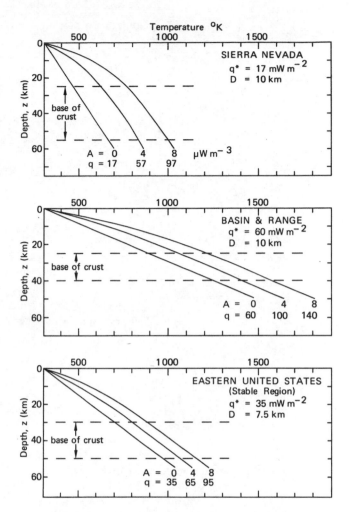

5.3 **Crustal temperature-depth profiles as a function of heat flow and surface radioactivity for various heat-flow provinces (after A. H. Lachenbruch). The symbols are defined in the text.**

the mantle beneath continents differs significantly from that beneath oceans; yet there is evidence that continents and sea floors are moving around, and that some parts of continents now overlie mantle previously covered by oceanic crust and *vice versa*. The paradox can be resolved by requiring that the individual *plate* (which includes the crust and several tens of kilometres of mantle) contains that portion of the mantle which is distinctively continental or oceanic, and that below the plates, the mantle (at least in terms of heat flow and abundances of radioactive elements) is laterally homogeneous.

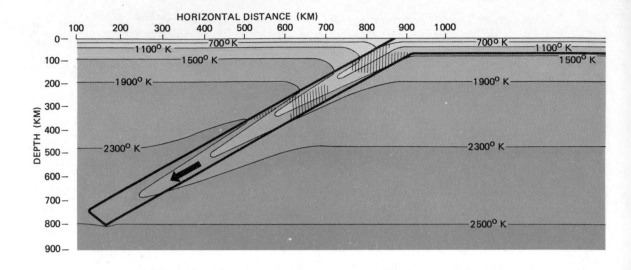

5. 4 Temperatures in the neighbourhood of a descending slab.

A factor in any discussion of the thermal implications of moving plates is the very slow rate at which a thermal wave is propagated through rocks. This velocity depends on the thermal diffusivity $(a = K/\rho\, C)$ which is in the range 0.5 to 2×10^{-6} m^2 s^{-1}, or, on a more geological scale, 15 to 60 km^2 per million years, for most rocks. Thus, thermal perturbations associated with tectonic activity which started upwards of 100 million years ago may still provide a sizeable component of the observed heat flow. Both the high temperatures beneath the Basin and Range and the low temperatures under the Sierra Nevada (Figs. 5.1 & 5.3) may be associated with the thermal effects of a slab like that illustrated in Figure 5.4, which ceased its downward movement ten or more million years ago.

THERMAL MODELS FOR THE WHOLE EARTH

The geotherm (temperature versus depth within the Earth) is a function of surface temperature and surface heat flow, and of the subsurface variation of both radioactive abundances and the coefficient of heat transfer. It is also dependent on the mode of origin and early history of the development of the Earth. We have reasonably detailed knowledge of all physical parameters near the surface, but at present there seems no satisfactory method of determining uniquely the variation of radioactive abundances with depth. The largest uncertainties, however, are in the coefficient of heat transfer. At moderate temperatures, up to about 1 000°K, heat transfer in rocks is almost entirely by 'lattice conduction' (the vibration of particles within the crystal lattices of individual minerals). The lattice conductivity of the various rocks tends to converge with increasing temperatures to something like 2 to 3 W m^{-1} °K^{-1}. At higher temperatures, heat can be transferred much more quickly by radiation and/or penetrative convection by molten rock. The coefficient of radiative heat transfer can be influenced by subtle changes in composition. The amount of heat transferred by the motion of molten rock is determined by the melting point which, in turn, is influenced by composition as well as by temperature and pressure.

Despite the difficulties involved in estimating relevant parameters, thermal models can provide valuable insights regarding the Earth's history and present internal constitution. Figure 5.5 shows a generalized temperature–depth curve and melting point curves for the present-day Earth. The absence of a numerical temperature scale should serve to emphasize that there are many uncertainties, depending on the assumptions that are made with respect to the physical and chemical properties of the Earth's interior.

5. 5 Generalized melting point curves for the mantle and core, and the temperature-depth relations within the Earth.

In the upper 1 000 km, we show two separate geotherms, one for stable continental plates (e.g., Precambrian shields) and another for ocean basins. The point of convergence is arbitrary, and it is possible that beneath some portions of continents the continental curve is above that for ocean basins. Irrespective of the model chosen, we can make a number of generalizations regarding the geotherm:

(1) Temperatures rise very rapidly in the upper 100 km or so. This reflects the strong upward concentration of heat-producing radioactive elements within the crust and upper mantle.

(2) In the upper few hundred kilometres the geotherm closely approaches the melting point curve; this is necessary to account for the layer of low seismic velocity in the upper mantle and the observed volcanic activity on parts of both continents and oceans. As stated previously, in continental areas of volcanic activity the geotherm will be above the 'stable continental' curve. In fact, it may even lie above the oceanic curve locally.

(3) The geotherms beneath various tectonic units converge with depth. For example, the stable continental curve is below the oceanic curve in the upper few hundred kilometres. To catch up, the temperature gradient must exceed that below oceans for a few hundred kilometres. This can be caused by differences in the distribution of heat sources, or by an increase in effective thermal conductivity beneath oceans due to partial melting

or an increase in radiative heat transfer. The conductivity increase seems a more attractive cause than the difference in heat sources in that it is compatible with plate tectonics.

(4) Below depths of 1 000 km or so variations of several hundreds of degrees in temperature can be obtained by making slight changes in the values of individual parameters. The only things that we can say with any certainty are determined from seismological evidence, especially the absence of shear wave transmission in regions that are predominantly molten:

(a) There is apparently a sudden drop in melting point at the core-mantle boundary. This is consistent with the theory that the core consists primarily of iron, and with the various estimates of the melting points of silicate rocks and iron at these pressures.

(b) The geotherm must rise at a slower rate with depth within the core than the melting point, because the inner portion of the core is solid.

FURTHER READING:

E. A. LUBIMOVA 1969. Thermal History of the Earth. *American Geophysical Union, Monograph 13* on 'The Earth's Crust and Upper Mantle'.

M. G. LANGSETH JR AND R. P. VON HERZEN 1970. Heat Flow through the Floor of the World Oceans. From 'The Sea', v. 4, *Interscience*.

R. F. ROY, D. D. BLACKWELL AND E. R. DECKER 1972. Continental Heat Flow. From 'The Nature of the Solid Earth'. *McGraw-Hill*.

Although the Earth sciences have made great advances over the past twenty years or so, there are still many geophysical phenomena—some of them discovered a long time ago—which have not yet been explained satisfactorily. One of the most fascinating of these is the Chandler wobble, the wobble of the Earth with respect to its axis of rotation, which has a period of about 14 months and an amplitude of about 0.5″. It was discovered in 1891; but even now we are unsure of its cause.

An idea currently receiving some support is that the Chandler wobble is excited by earthquakes. Some years ago it was shown that under certain circumstances one large earthquake could produce 10 per cent or more of the observed Chandler wobble amplitude; and more recent calculations show that, again under favourable conditions, a single earthquake of magnitude 8.5 could maintain the wobble for about a year. Is it possible, then, that all earthquakes put together could account for the whole of the Chandler wobble? It seems intuitively likely; but, as so often happens, the theory has its detractors—and at the moment we cannot say that it is firmly established.

In this article, Professor M. A. Chinnery of Brown University, Rhode Island, shows just what the Chandler wobble is, describes how it was discovered and discusses its possible causes with special reference to the earthquake origin theory.

6 The Chandler Wobble

by MICHAEL CHINNERY

The Earth has a measurable wobble with respect to its axis of rotation. Until recently, there were aspects of this phenomenon that appeared to defy any rational explanation.

WE USUALLY THINK OF THE EARTH as rotating uniformly, once a day, about an axis through the North and South poles; and so of course it does, to a first approximation. However, as we study the motion of the Earth in more detail, we begin to find complications. This is not surprising, because the Earth is not a perfect sphere, nor even a perfect oblate spheroid; this is obvious from the distribution of continents and oceans. Contained within the observations of this complex part of the Earth's motion is some very useful information about the mechanical properties of the Earth's interior, though our success at extracting this information from the data has been very limited as yet. One of the most interesting aspects of the Earth's motion, and one of the most mysterious, is the Chandler Wobble.

In order to describe this phenomenon, we define several axes. These have different physical meanings, but all pass through the Earth in a roughly North-South direction. Since these axes move relative to one another, it is sometimes simpler to consider the corresponding poles, which are the points where the axes intersect the surface of the Earth. In what follows, we shall only consider the North pole of each axis.

The first axis that we shall need is the geographic axis. This is no more than a straight line joining the cairns that have been erected at the North and South poles (Fig. 6. 1). It is an artificial axis, but it is important because it is tied to all of our maps of the Earth's surface. Also, it is fixed, and forms a convenient reference from which to observe other motions.

The next axis that we must consider is the total angular momentum vector of the Earth. Any motion of the Earth as a whole, or of parts of the Earth relative to other parts, can be associated with an angular momentum vector. We can

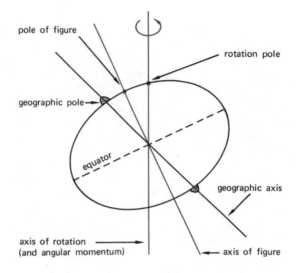

6. 1 **Highly exaggerated view of the geographic North-South axis, and the axes of rotation and figure. The actual distance between the geographic pole and the rotation pole averages about 6 metres (see Fig. 6. 6).**

add all of these vectors together (vectorially) to obtain a total angular momentum vector. The important point here is that angular momentum is conserved. Hence, *if* there were no external forces acting on the Earth, the total angular momentum vector would remain fixed in space, with a constant magnitude and direction. This is a useful property, for it means that we can eliminate the effects of external forces on the motion of the Earth by restricting our observations to those which are made relative to the angular momentum axis.

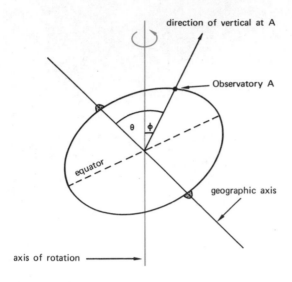

direction of vertical at A

Observatory A

θ φ

equator

geographic axis

axis of rotation

6. 2 We can determine the angular departure of the rotation axis from the geographic axis by observing the difference between the angles θ and Φ at an observatory. θ is simply the complement of the known latitude of the observatory. Φ, the angle between the vertical at the observatory and the rotation axis, can be determined very accurately from observations on stars. Φ is called the astronomical co-latitude, and is found to change with time. When several observatories measure Φ at the same time, it is possible to locate the position of the rotation pole on the Earth's surface.

In practice, it is very difficult to measure the direction of the axis of angular momentum. However, by studying the apparent motion of stars, it is very easy to determine the direction of the axis of rotation of the Earth. Fortunately, the angle between the two is very small, about 0.001 seconds of arc (this corresponds to a distance of about 2.5 cm between the pole of angular momentum and the pole of rotation), and we may treat these axes as being coincident (Fig. 6. 1). The reason for these axes being so close together is, of course, that by far the largest contribution to the total angular momentum arises from the daily rotation of the Earth.

We shall need one further axis, the axis of figure. This is rigorously defined as the axis of maximum moment of inertia, and its position depends on the distribution of mass throughout the Earth. Any change in the mass distribution will cause a change in the orientation of the axis of figure. In principle, every time I take a step there is a (minute) corresponding change in the axis of figure. Any such mass movement is said to excite the axis of figure, and the size of the change in the direction of the axis is a measure of the magnitude of the excitation.

We are concerned here with the effect of these internal mass movements on the motion of the Earth. The basic observational data for this study are obtained from astronomical measurements of latitude (Fig. 6. 2). By very carefully observing the passage of stars across the zenith at an observatory, it is possible to measure the astronomical co-latitude with an accuracy of about 0.01 seconds of arc. By combining the measurements from several observatories on many stars, the position of the rotation pole can be located on the Earth's surface to better than 0.3 metre, and by repeated observations any movement of the rotation pole can be followed. Any such movement will result in a variation with time of the measured latitude of an observatory, and this field of research is frequently referred to as the study of the 'variation of latitude'.

This variation in latitude indeed occurs, and it is not hard to understand why. First, let us consider the case of a rigid planet (by rigid we mean that the axis of figure has a fixed orientation with respect to the planet). If the planet rotates about its axis of figure, then we have the simplest case of completely stable rotation. There would be no latitude variation under these circumstances. However, suppose that the planet is set to rotate about some other direction. Now the motion is more complex. The planet has a tendency to align its axis of rotation with its axis of figure, but it cannot. The axis of figure starts to precess about the axis of rotation. If you live on the planet, you may prefer to say that the axis of rotation precesses about the axis of figure, as in Figure 6. 3, and the rotation pole will trace out a circular path with its centre at the pole of figure. In this case, we should observe a latitude variation at any point on the planet's surface.

The mathematical analysis of this phenomenon is quite simple, and was first given by Euler in 1765. It has since become known as *Eulerian nutation*, or free nutation. Euler showed that the rotation pole should describe one complete circle about the pole of figure in $A/(C-A)$ sidereal days, where the moments of inertia of the Earth are

C (about a polar axis) and A (about an equatorial axis). This ratio is known very precisely from a different astronomical phenomenon (the precession of the equinoxes), and the theoretical period turns out to be 305 days, or about 10 months. Notice that the theory only predicts the period of the motion; it can say nothing about the amplitude of the latitude variation, for this depends on why the Earth began to rotate unsymmetrically in the first place.

If the Earth were truly a rigid body, the latitude variation would be simple, and would follow Euler's theory. Unfortunately, however, the Earth is far from rigid, and mass movements of many kinds occur. These include movements of the atmosphere and oceans; changes in the surface mass distribution due to erosion, volcanic eruptions and earthquakes; motions in the fluid parts of the Earth's core; and probably other motions in other parts of the Earth that are only poorly understood. In such a mobile Earth, the axis of rotation will trace out a more complicated path, as it attempts to precess about each new position of the axis of figure.

There is one more effect that we must mention. If internal motions within a planet are possible, then a redistribution of mass is possible so that the axis of figure will align itself with the axis of rotation. In a completely fluid planet, this will happen very rapidly. If the planet offers some resistance to flow (i.e. has viscosity), there will be a certain time necessary for the adjustment. In this case we might expect the rotation pole to spiral in towards the pole of figure. Clearly, then, in addition to the concept of excitation of the rotation pole, we must include the idea of the damping of the excitation. The rate of this damping will give a measure of the viscosity of the Earth.

As a result of Euler's conclusions, many famous scientists, including Peters, Bessel and Maxwell, searched the astronomical records for a variation in latitude, with no success. The small variations they found were of doubtful significance, and had an amplitude of less than 0.1 second of arc. Though the true period is rather different from Euler's value of 10 months, it is still hard to see how these eminent men missed the effect. Lord Kelvin was undeterred. He calculated that the annual motion of air masses should give a measurable variation in latitude, and persuaded Newcomb to analyse the data for Washington, D.C. from 1862 to 1865, to find a 10 month periodicity. The best that Newcomb could do was to find a motion

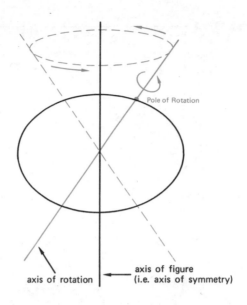

Pole of Rotation

axis of rotation

axis of figure
(i.e. axis of symmetry)

6.3 Eulerian nutation. If a rigid symmetrical planet is made to rotate about an axis that is inclined at an angle to the axis of symmetry, the planet will develop a wobble. To an observer on the planet, the axis of rotation will appear to precess (i.e. describe a cone) about the axis of symmetry. In this case, the rotation pole will trace out a circular path about the pole of figure.

with amplitude 0.05″ ± 0.03″, about which he was very sceptical. Kelvin, however, was overjoyed, and announced the discovery of the free nutation of the Earth during his Presidential Address before the British Association in 1876.

Kelvin's joy was somewhat premature. The true state of affairs was discovered by accident a few years later. In 1884 Küstner, in Berlin, was studying the constant of aberration. He was careful to remove all sources of error, and was surprised to find a residual, roughly annual, term which appeared to be a latitude variation of amplitude about 0.2″. When this result was announced in 1888, great interest was generated, and in 1891 a crucial experiment was performed. The latitudes of two stations, Berlin and Waikiki (which are 180° apart in longitude), were studied for a little

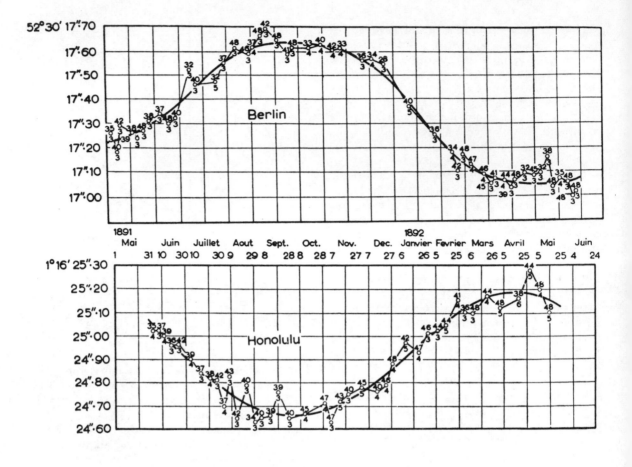

6. 4 The famous graph that resulted from the latitude observations in Berlin and Hawaii in 1891–92. A clear latitude variation, with amplitude about one half of a second of arc, was found. After Melchior.

over a year in great detail. The resulting plot (Fig. 6. 4) has now become famous. A very clear latitude variation, with amplitude about 0.5″ and a period of 14 months, was observed.

Meanwhile, Mr. Chandler had appeared on the scene. He was a prosperous merchant, in Cambridge, Massachusetts, with enough time on his hands to pursue a scientific hobby. He decided to repeat the analysis of latitude observations, in the search for a periodic variation. In 1891 he announced that he had detected both an annual component and another, with period 428 days. Initial reaction was hostile, but the results of the Berlin-Waikiki experiment confirmed his finding completely. His name went down into history,

associated with the 14-month wobble. The reason for the difference between this period and that predicted by Euler was soon discovered. Only a year later, Newcomb showed that elastic yielding of the Earth is enough to lengthen the calculated period to 14 months.

These discoveries led to the formation of the International Latitude Service (ILS), and since then a set of stations has made continuous observations of latitude. The number of active stations has varied somewhat, and was augmented to over 30 during the International Geophysical Year. The result has been a wealth of data, and a multitude of papers (one compilation refers to a bibliography of 1761 papers before 1957). Funnily enough, however, the net result of all this work has been minimal, and we really understand little more than did Newcomb in 1893.

When the variation in latitude is analysed in detail, it is found that energy is distributed over a range of period, and not just at the Chandler period of 14 months. This is best expressed by constructing a power spectrum of the data, and one example is shown in Figure 6. 5. Clearly a part of the variation has a period of exactly one year, and this can be nicely explained by seasonal movements of the atmosphere. The remainder, which is what we call the Chandler wobble, has periods that range from 13 months to 15 months. We are now faced with two difficult problems. First, we must account for the amplitude of the Chandler wobble, which means that we must find out what has caused, or is causing, the angular departure of the axis of figure from the axis of rotation. Second, we must explain the range of periods involved in the wobble, i.e., explain the width of the spectral peak.

A broad spectral peak, like that found for the Chandler wobble, can be ascribed to two very different physical mechanisms. It can result from the continuous excitation of a mechanical system whose natural period changes with time. This possibility has received a great deal of support from authors who include Chandler himself, and more recently Melchior and Nicolini. The variation in the Chandler period (about ±4%) is rather large, however, and it seems unlikely that changes in the physical state of the Earth large enough to produce this can occur in times of the order of a year.

The second possibility is that the Earth has a fixed Chandler period, and that random excitations of the system are subject to damping, in the manner described earlier. This is a much more reasonable idea, and it has been championed by authors who include Jeffreys, Munk and Mac-Donald. Recent studies of the motion of the rotation pole seem to support this idea. Figure 6. 6 shows the motion of the pole during 1957, and (with a little imagination) the rotation pole can be seen to follow circular arcs about different centres. The implication of this is that the pole of figure is jumping from one point to another in a jerky fashion. The required amount of excitation of the pole of figure is of the order of 0.1 seconds of arc (about 3 metres on the Earth's surface). If this is the mechanism, it is possible, from the data, to estimate the Chandler period and the damping

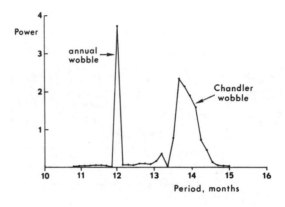

6. 5 **This graph shows how the energy in the latitude variation is spread over a range of periods. It was obtained from analysis of the International Latitude Service observations between 1900 and 1954.**

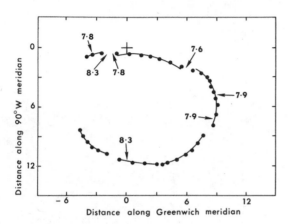

6. 6 Each point shows the position of the rotation pole, averaged over a ten-day period, determined from latitude observations by Bureau International de l'Heure, during 1957, after the removal of the annual wobble. The origin of coordinates is the geographic pole, and the rotation pole moves anti-clockwise. The times of occurrence and magnitudes of major earthquakes are shown. Distances in metres. (After Mansinha and Smylie.)

time (the time for the amplitude to fall to $1/e$ of its initial value). The period is found to be very close to 1.20 years, and estimates of the damping time range from 10 to 30 years. We now have to explain the cause of the random excitations of the wobble, and account for the rather large amount of damping.

First, let us consider the excitation, since it is here that progress seems to have been made. We have to find some mass motions in the Earth that occur almost instantaneously, and are distributed roughly randomly in time, amplitude and orientation. They must be large enough to displace the axis of figure by a tenth of a second of arc. An obvious candidate is the occurrence of earthquakes, and many authors have looked into this possibility. Munk and MacDonald, in their excellent monograph *Rotation of the Earth*, concluded quite definitely that earthquakes are much too small to be the source of excitation. They argued that even a very large earthquake, in which a block of the Earth's crust with dimensions $1\,000 \times 100 \times 30$ km is moved horizontally by 10 m, will not excite a wobble amplitude of more than 0.001 seconds of arc.

This observation, and others like it, have thrown the field into confusion. If earthquakes are not the cause of this phenomenon, what is? A great many suggestions have been made, though many of them suffer from the philosophical difficulty that they propose processes within the Earth which are not observable in any other way, and therefore are difficult to verify. Some recent suggestions include the following: Büsse has indicated that precession of the inner core of the Earth might be coupled to the mantle by pressure forces arising from its ellipticity, and so excite the Chandler wobble; Runcorn has re-examined the possibility of exciting the wobble by the effects of magneto-hydrodynamic turbulence in the fluid outer core; Hide has discussed the role of irregularities on the core-mantle boundary, and the possibility of a mechanical coupling between fluid motions of the core and the mantle; Colombo has suggested that the differential gravity torque exerted on the Earth by the Moon has a destabilizing action that might excite the wobble.

As a result of the work of two Canadian geophysicists, Smylie and Mansinha, the possibility that earthquakes may yet provide the excitation has been reopened. They showed that, according to their interpretation of the data, 'breaks' in the path of the rotation pole seem to correlate with large earthquakes (see Fig. 6. 6). In some instances the break in the path appears to occur shortly before the earthquake, which is a disturbing result.

The basic problem here is, how much mass movement accompanies an earthquake? In order to answer this question, since observation (except at the surface of the Earth) is impossible, we have to resort to theory. Unfortunately, the theory is very difficult, and our knowledge of the mechanical properties of the Earth is very incomplete. However, there are indications that the mass redistributions caused by earthquakes might be very much larger than previously thought. Intuitively it is very hard to believe that the amount of mass movement caused by an earthquake is 100 times that given by the example of Munk and MacDonald quoted earlier. If it is, then each time an earthquake occurs, appreciable permanent deformations occur throughout the Earth. To give an example, the calculations suggest that during the 1906 earthquake in San Francisco, the east coast of North America (3000 miles away) was uplifted by about 1 mm, and the European continent was raised by about half of this.

We are left, then, with a paradoxical situation. If we can demonstrate that the Chandler wobble is excited by earthquakes, we may have solved one problem, but we shall have introduced many new questions about our misunderstanding of earthquakes. It is this aspect of the problem that makes the study of the Chandler wobble so fascinating.

The damping of the Chandler wobble remains a mystery. The damping time, though not known precisely, does appear to be too short. In other words, the Earth somehow acts as if it were much more fluid than we had expected from the study of other phenomena. We may have to revise some of our present ideas about the mechanical properties of the Earth's interior before we can account for this observation.

FURTHER READING:

P. M. MELCHIOR 1957. Latitude Variation. Chapter 7 in 'Physics and Chemistry of the Earth', vol. 2. *Pergamon Press.*

W. H. MUNK AND G. J. F. MACDONALD 1960. The Rotation of the Earth. *Cambridge University Press.*

L. MANSINHA AND D. E. SMYLIE 1968. Earthquakes and the Earth's Wobble. *Science,* v. 161.

D. E. SMYLIE AND L. MANSINHA. The Rotation of the Earth. *Scientific American,* v. 225, December 1971.

So far, we have concerned ourselves with the Earth, our planet, its origin, evolution, composition and structure. The Earth, however, is not alone in space. In its immediate vicinity, measured in astronomical distances, there revolves around it a satellite—the Moon. Why do we include in a book entitled *Understanding the Earth* a section on the Moon? There are two reasons.

First, the Moon interacts with the Earth gravitationally, influencing its motion through space, its movement about its centre of mass and its axial rotation. Indeed, the length of our day is governed by the presence of the Moon and, as you will see, the changes in the length of the day and indeed the length of the lunar or sidereal month through geological time have been due to changes in the position of the Moon with respect to the Earth.

Second, and perhaps more directly pertinent here, the surface of the Moon, unprotected by any atmosphere and unaffected by erosive processes, presents an uninterrupted and unmodified record of all the events that have happened in this part of the solar system since its formation 4 600 million years ago. Although the Earth was formed at about the same time as the Moon, the oldest rocks exposed on its surface are no more than 3 500 million years old. There is no remaining record on the Earth of the first 1 000 million years of its history. Therefore the Moon rocks and structures provide invaluable evidence which spans these 'dark aeons' in the history of the Earth-Moon system and is of unique interest to astronomers and Earth scientists alike.

In this article, Professor Zdeněk Kopal of the Department of Astronomy at the University of Manchester, after presenting the fundamental properties of our satellite, discusses the structure and composition of the Moon's surface and how it was formed. He then describes the effect of the Moon on the Earth, before concluding with a brief statement concerning the Moon's origin.

7   The Earth-Moon System

by ZDENĚK KOPAL

The Earth and the Moon are companions in space. They affect each other strongly, and each bears traces which help to elucidate the history of the other within the solar system.

7. 1   A synoptic view of the visible face of the Moon, on which photographs of the first and last quarters are merged together.

The fundamental facts and figures which we shall have to keep in mind as we discuss the Moon are well known. To the astronomer in particular —and not only to him—the Moon has indeed been a friend of old standing; and an at least rudimentary knowledge of its motion goes very far back in the history of mankind on this planet. In prehistoric times the waxing and waning of lunar phases, and the light changes accompanying them, provided the first astronomical basis for the reckoning of the time. Whenever we go sufficiently far back in the history of almost any primitive civilization, we find it invariably dependent on the lunar, rather than the solar, calendar. The month became a unit of time long before the concept of the year emerged from accumulating observations, and the Moon, as the graceful carrier of this knowledge, thus gained entrance to the pantheons of most ancient nations as a female deity.

The *orbit* of the Moon around the Earth is approximately an ellipse of mean eccentricity equal to 0.055 which is slightly distorted by the attraction of the Sun and by the uneven attraction of the non-spherical Earth. The plane of the Moon's orbit is inclined to the ecliptic, the plane in which the Earth revolves around the Sun, by a little more than 5° 9'. The mean period of revolution of the Moon around the Earth is equal to 27 days, 7 hours, 43 minutes and 11.5 seconds. This is the so-called *sidereal month*, the time it takes for the Moon to return to the same place with respect to the Earth. However, because of the apparent motion of the Sun it takes 29 days, 5 hours, 5 minutes and 35.8 seconds for the Moon to attain the same phase. This period is termed the *synodic month*.

How far away is the Moon? This distance had been calculated by the Greeks, from the relative durations of different phases of lunar eclipses, to be equal to about sixty times the Earth's radius. More recently, astronomers have determined this distance much more accurately by triangulation and, still more recently, from the measured time-lag of radar echoes reflected from the Moon. The results show that the distance from the Earth to the Moon varies in the course of each month from 356 000 to 407 000 km. The mean distance in the course of a month is 384 400 km, or 60.267 times the Earth's equatorial radius. In other words, it is 0.00257 times the mean distance separating the Earth from the Sun. This represents less than one per cent of the distance separating us from our nearest other celestial neighbours, the planets Venus and Mars, at the time of their closest approach.

Knowing the size and form of the Moon's orbit, we can easily compute the mean velocity of its motion. It averages 3 681 kilometres per hour or about 1 023 metres per second, which corresponds to a mean angular velocity, as viewed from Earth, of about 33 minutes of arc per hour. This is just a little greater than the apparent diameter of the Moon itself.

The mean apparent diameter of the lunar disc in the sky is just over half a degree, which at a mean lunar distance of 384 400 km corresponds to a radius for the Moon of 1 738 km. The Moon is, therefore, about one quarter of the Earth in size. It rotates around an axis which is almost perpendicular to its orbital plane with a uniform angular velocity in exactly the same period as it revolves. It therefore always shows us the same face. However, since the angular velocity of revolution in an elliptical orbit varies inversely with the square of the ellipse radius, the rotation is sometimes ahead and sometimes behind the orbital motion by as much as 7° 7'. Hence, actually more than one half of the Moon's surface can be seen from the Earth. This phenomenon is known as the *libration in longitude*. In actual fact the axis of the Moon is not exactly perpendicular to the plane of its orbit. The angle between this plane and the lunar equator is about 6° 5', which means that in the course of one month we can again sometimes see more of one polar region and sometimes more of the other. This gives rise to a *libration in latitude*. Again, when the Moon is rising for the observer on the Earth, we look over its upper edge, seeing a little more of that part of the Moon than if we were observing it from the centre of the Earth; and when the Moon is setting, the converse is true. This *diurnal libration*—not of the Moon, strictly speaking, but of the observer —amounts to almost 1°. Taken together the librations enable us to see considerably more than one half of the Moon from the Earth. On the whole, not less than 59% of the whole lunar globe can be seen from the surface of the Earth at one time or another. Only 41% is permanently invisible and 41% always visible; the remaining 18% is alternately visible and invisible.

The next quantity of interest which should contain a clue to further fundamental characteristics of our satellite is its *mass*. The mass of the Moon, like that of any other celestial body, can be determined only by the effects of its attraction on another nearby body of known mass. In the case of the Moon, this was first our Earth, but more recently fly-by or impinging spacecraft have been used. As a spacecraft approaches the Moon, lunar attraction accelerates it. Accurate tracking of this acceleration recently led to a very precise determination of the Earth : Moon mass-ratio, which was found to be 81.302 ± 0.001. This corresponds to an absolute value of the product $Gm_{\mathbb{C}}$ (where $m_{\mathbb{C}}$ stands for the mass of the Moon and G for the gravitational constant) equal to $(4\,902.70 \pm 0.06) \times 10^9$ m^3 s^{-2} which compares with the value $Gm_{\oplus} = (398\,601 \pm 1) \times 10^9$ m^3 s^{-2} for our Earth.

From the value of the Earth : Moon mass-ratio quoted above we can calculate the mean distance from the Earth's centre to the centre of gravity of the Earth-Moon system. This turns out to be 4 671 km. This is, of course, less than the radius of the Earth, which means that the centre of gravity of the Earth-Moon system lies well inside the Earth's mantle. Also, since the value of the gravitational constant, G, is $(6.670 \pm 0.015) \times 10^{-11}$ N m^2 kg^{-2} we can show that the absolute masses of the Earth and Moon are 5.978×10^{24} kg and 7.353×10^{22} kg, respectively. The uncertainties in these masses are of the order of one part in a thousand, and are almost solely due to the uncertainty in the measured value of the gravitational constant.

The lunar *mass* of 7.35×10^{22} kg may loom large in comparison with all terrestrial standards but on cosmic scales it constitutes but a tiny speck. The *mean density* of the lunar globe, obtained by dividing the mass by the lunar volume of 2.199×10^{19} m^3, is 3.34×10^3 kg m^{-3}. This is just about equal to the density of rocks that form the upper part of the Earth's mantle and considerably less than the mean density of the terrestrial globe $(5.54 \times 10^3$ kg m$^{-3})$. The gravitational acceleration, $Gm_{\mathbb{C}}/r_{\mathbb{C}}^2$, on the lunar surface is, therefore, only some 1.62 m s^{-2}, less than one sixth of the value on Earth; and the velocity of escape, $(2\,Gm_{\mathbb{C}}/r_{\mathbb{C}})^{\frac{1}{2}}$, from the lunar gravitational field is only 2.38×10^3 m s^{-1}, whereas on Earth it is 11.2×10^3 m s^{-1}. The relatively low gravitational acceleration prevailing over the lunar surface means that much less effort is required there to lift weights or throw stones. However, it also means that a man's own body-weight works less effectively when he wishes to compress anything, as when driving a shovel into the ground by stepping upon it.

The smallness of the mass of the Moon and the relatively low value of the surface escape velocity have several important consequences. Perhaps the most important one is the complete absence of any atmosphere or protective gaseous mantle above the Moon's surface. Under temperatures prevailing on the lunar surface, ranging between some 130°C at noon in the lunar tropics and –200°C in the later part of the lunar night, any hydrogen or helium atmosphere would dissipate from the Moon in a matter of days. Water vapour would escape more slowly, but would still be lost quickly on a geological time scale, and at noon temperatures oxygen and nitrogen as well as carbon dioxide would completely escape in a relatively short time. In fact, the rate of dissipation into space of all but the heaviest gases, which are cosmically very scarce, is so high on the Moon that we should not expect to find any appreciable permanent atmosphere around it. All observational evidence available to us supports this conclusion.

If, therefore, the Moon possesses no detectable atmosphere because of its low gravitational field and relatively high daytime temperature, it cannot maintain any liquid on its surface. Near its poles, to be sure, depressions may exist which are never reached by direct sunlight or which, at best, are illuminated by sunlight scattered from neighbouring landscape. In such regions, condensed volatile substances may possibly be present in the form of some kind of ground frost, but should they ever evaporate, they would be irretrievably lost to the Moon in a very short time. Hence, no water or ice could be present almost anywhere on the Moon's exposed surface except possibly for very short intervals of time. The surface of the Moon is thus bone-dry, and has been so from time immemorial. No feature visible on the surface of the Moon could have been formed or modified by the effects of running water or by the effect of repeated freezing and melting of water entrapped in surface rocks. One of the most important terrestrial agents of erosion therefore is, and always has been, absent from the Moon, and cannot be invoked to explain any of the Moon's surface features which we shall discuss in the next section.

Before we do so, however, we must turn our attention to the interior of the Moon; for, as is

7. 2 Two photographs of the lunar crater Tycho near the Moon's South Pole—a crater close to 100 km across, whose ramparts rise 5 400 m above the level of its floor, though only 1 600 m above the level of the surrounding landscape. The top photograph was taken on 30 March 1966 with the 43-inch reflector of the Observatoire du Pic-du-Midi (Manchester Lunar Programme); the one below by Lunar Orbiter 5 on 15 August 1967 from an altitude of 105 km above the lunar surface.

true of every celestial body, be it a star, a planet or a satellite, the 'engine room' which controls the large-scale structure and cosmic evolution of that body is internal. The Moon's surface is controlled by internal thermal and stress processes but is also an 'impact counter' of external events which the Moon has experienced since the days of its formation.

If we wish to examine the essential properties of the Moon's interior, the primary clues are already in our hands, namely its observed mass and its size, combining as they do in a mean density of $3.34 \times 10^3 \, \mathrm{kg \, m^{-3}}$. What is the pressure prevailing inside this mass? Even in so small a celestial body as the Moon, existing under the influence of its own self-attraction, the internal pressure is essentially hydrostatic throughout most of the interior. In other words, the strength of the Moon material is unable to withstand its own weight anywhere except, possibly, in the very outer part of its crust. An application of the elementary theory of hydrostatics reveals that the internal pressure in a globe of the lunar mass and size cannot exceed some 50 000 atmospheres even at its centre—a pressure exceeded at a mere 150 km below the Earth's surface. Furthermore, this calculation for the pressure at the Moon's centre holds whatever the Moon consists of and whatever the internal temperature happens to be.

The internal temperature of the Moon is now known to be relatively low—much lower than is true of our Earth. The mean electrical conductivity of the lunar globe (as inferred from its interaction—or, rather, lack of it—with the 'solar wind' of charged particles emanating from the Sun) is that of common silicate rocks at a temperature of less than 1 000° K. Such a temperature is well below the melting point of silicate rocks; therefore the Moon should behave as a solid body throughout most of its mass.

That it actually does so is now well attested by several observed facts. An analysis of the lunar gravitational field (deduced from the motion of lunar artificial satellites since 1968) disclosed the presence, at shallow depths, of localized mass concentrations (*mascons* for short)—of total mass equal typically to $5–10 \times 10^{-5}$ of that of the Moon, and exerting a load of the order of 1–10 kg/cm² on their substrate 50–100 km below the visible surface. The locations of such mascons are usually found to coincide with the maria (in particular, *circular maria*—see Fig. 7. 1) of which Mare Imbrium is the largest; and it is very probable that their formation is due to the cosmic impacts (see next section) which produced the respective mare.

The excess loads exerted by the mascons represent departures from isostasy which would not be tolerated by the Earth's crust for any (astronomical) length of time. That the Moon has tolerated them (as we shall see) for more than 3 000 million years discloses that its crust is at least 1 000 times more rigid than the terrestrial one —consistent with the fact that the visible face of the Moon exhibits no evidence whatever of lateral shifts (folding); all mountain chains we see on the Moon are nothing but partly preserved ramparts of impact formations ranging from craters to maria (see next section).

The Moon is not only rigid enough in the crust to support the mascons and other, more global departures from isostasy. The seismic evidence obtained since 1969 by three successive *Apollo* missions has disclosed that its material behaves like rigid matter at much greater depths, and transmits transversal 'moonquake' waves from the depth of more than 800 km. Seismometers deposited on the Moon by *Apollo 11–15*, and operative almost continuously for the past three years, have disclosed the Moon to be seismically very much more quiet than the Earth — only about 10^{15} ergs of seismic energy (equivalent to that of about 200 tons of TNT) gets dissipated in the Moon through seismic waves per year (in contrast with 5×10^{24} ergs/year for the Earth).

Apart from general seismic background of very low noise level, occasional moonquakes have been registered by the Apollo seismographs—due partly to external (meteoritic impact) and partly to internal causes. One of the most interesting of the latter are tectonic moonquakes emanating repeatedly from the same epicentre (of lunar coordinates $\lambda = 28°W$, $\beta = 21°S$) at a depth of 800 ± 40 km below the surface of the Moon (i.e. almost half way between the centre and the surface). A number of additional centres of moonquakes have already been detected, and some of them found to give rise to whole swarms of such quakes in rapid succession. Their activity was, moreover, found to fluctuate with varying distance from the Earth in the period of exactly one month. The frequency of seismic events appears to be maximal when the Moon is closest to the Earth—a fact which suggests that bodily tides may trigger deep-seated tectonic instabilities of as yet unspecified nature. However, the magnitude

7.3 **The lunar craters Walter (upper half) and Regiomontanus (lower half) photographed on 18 May 1967 by Lunar Orbiter 4. A little cone inside Regiomontanus (Regiomontanus A), indicated by an arrow, is some 17 km across and looks, by analogy with terrestrial forms, like a volcanic cone.**

of such events is really tiny in comparison with current terrestrial standards. The energy of individual moonquakes is, on the average, less than that liberated by the explosion of one kilogram of TNT. Virtually thousands of events so small (with maximum Richter magnitude between 2–3) occur daily on the Earth, and can stand out on the Moon from background noise only because the Moon as a whole is seismically so quiet. The fact that these tiny moonquakes can be detected at all and their records transmitted through the intervening gap of space represents a veritable triumph of field seismology.

The seismic 'signature' of the observed moon-quakes is seen to be very different from what we observe in the Earth: for while the terrestrial seismic disturbances are damped out in a few minutes, lunar tremors—whatever their origin—persist for 60–100 minutes. Such long durations can be understood only if the seismic waves are scattered strongly in the mainly basaltic crust of the Moon—about 20 km deep—which must be highly fragmented (regolithic). At between 20–70 km in depth the seismic velocities are indicative of the presence of solid layers of pyroxene-rich eclogite; while the mantle below the crust may consist mainly of dunite and other similar silicates. The occurrence of moonquakes at greater depths (around 800 km) implies that the lunar interior there is rigid enough to support appreciable stress and to transmit transversal waves. Neither could be the case if the internal temperature at these depths were to approach the melting point of rocks.

LUNAR SURFACE AND ITS COMPOSITION

Now that we are acquainted with some of the fundamental physical properties of the lunar globe as a whole, let us look more closely at some of the diverse surface features of our satellite and try to understand the nature of the forces which have been responsible for producing them.

What is so arresting about the Moon's face, and what can we learn from it? The Moon is very old—not less than 4 600 million years of age—and has probably been a close companion of the Earth since the days of its formation. Moreover, the permanent absence of any air or water on the Moon makes it virtually certain that many of the features we see on its surface were created in the far distant past. It is possible that the Moon's oldest landmarks were produced when our solar system was being created. On the Earth or other nearby planets, all features of this antiquity have fallen prey to, and have been completely obliterated by, the joint erosive action of their atmospheres or oceans many aeons ago. However, as any changes on the lifeless Moon can proceed only at an exceedingly slow rate, its present wrinkled face still bears the scars and traces of the many events which have taken place in the inner precincts of our solar system since the days of its formation some 4 600 million years ago. If this is so, then the correct interpretation of these features holds a rich scientific prize indeed.

Even to the naked eye the Moon is beautiful, diversified with markings which have been associated with numerous popular myths. If we look at its wrinkled, pock-marked face through a telescope, the most cursory glance reveals that the lunar surface consists of essentially two different types of ground. One type, rough and broken, is comparatively light in colour, reflecting, in places, as much as 18% of incident sunlight; the other is darker, reflecting, on average, only 6–7% of incident sunlight, much smoother, and frequently so flat as to simulate the surface of a liquid. The first type of ground we shall generally call the 'continents'. They occupy large continuous areas, particularly on the far side of the Moon, and cover a little less than two-thirds of the entire visible face of our satellite. The darker flatlands, or 'maria' (seas) as they were misnamed by early observers before their true nature was properly understood, occupy the rest. The maria, whether they be small or large, are remarkably uniform in reflectivity and general appearance.

A closer look at the Moon with the aid of a telescope, or at photographs of the lunar surface, such as that reproduced as Figure 7. 1, reveals an almost bewildering array of formations and structures, no two of which are exactly alike. However, the dominant type of formation among them is the ring-like walled enclosures commonly called the 'craters'. They occur almost everywhere on the Moon, in continental regions as well as in the maria, in truly prodigious numbers, giving the lunar surface its appearance of a pock-marked face. The largest of these craters, mostly on the Moon's far side, have diameters of 300–400 km; and the number of craters with diameters of more than 1 km is in excess of 300 000 on the visible side of the Moon alone, and there are many more on its far side. Despite detailed surveys, the total number of craters with diameters less than 1 km is not known.

No two craters on the Moon are exactly alike. However, apart from individual distinguishing features, they have many characteristics in common. Their distribution over the lunar surface appears to be essentially random; the heights of their ramparts are, in general, very small in comparison with their dimensions; and their floors are depressed below the level of the surrounding landscape (Figs. 7. 2 and 7. 4).

With very few possible exceptions, the craters were produced by impacts on to the Moon's surface of other celestial bodies moving through

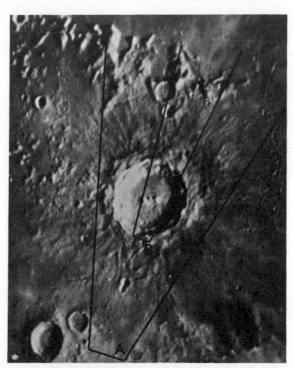

7.4 The lunar crater Copernicus, as photographed from the Earth (upper right), in comparison with views of the same structure as recorded by Lunar Orbiter 2 on 23 November 1966. The photograph on the upper left shows an oblique view of the crater taken from the position marked by A on the ground-based picture to its right; the lower half of the plate is a high-resolution view of the interior of the crater, from the vantage point B, at an altitude of 46 km above the lunar surface.

interplanetary space with which the Moon happened to collide. The interplanetary space through which the Earth and the Moon continue to circle the Sun is not empty—far from it. It contains a wide variety of objects of all weights and sizes, ranging from the ubiquitous gas of free electrons escaping from the Sun and extending well beyond the radius of the Earth's orbit through microscopic specks of dust and larger meteoritic debris (probably representing left-overs from the time of formation of the solar system) to major meteorites, asteroids, or comets whose orbits through space may intersect the path of the Moon so that they occasionally collide with it. The frequency with which the Moon, like the Earth, suffers direct hits by major meteorites, asteroids, or comets is known approximately from terrestrial evidence of such impacts in the recent past, and is unlikely to have been less in the more distant past. However, let us accept the low frequency of meteorite falls on Earth in the past few thousand years as a measure and extend it back in time. Then, even at this low rate of impact, it is entirely feasible that all the lunar craters, including the maria, are impact features. In this, the currently most acceptable theory, the lunar surface presents us with a *cumulative* scoreboard of celestial target

practice; for, once the features are formed, there are no erosive processes on the Moon to obliterate or modify them.

The strong probability that most lunar craters are of impact origin and go back in time to the first thousand million years of the Moon's history does not preclude the existence of some craters which may be of internal volcanic origin. Such craters probably do exist on the Moon (Fig. 7. 3) but they are limited in number and relatively smaller in size. There is indeed no evidence that volcanism on the Moon ever existed on a scale larger than on the Earth; if anything, the opposite may be true. There are no volcanic cones on the Moon that compare in size with terrestrial volcanoes such as Mount Fuji in Japan or Vesuvius in Italy.

To be sure, we see on the lunar surface other structures like the domes or wrinkle ridges (Figs. 7. 5 and 7. 6) which are undoubtedly of internal origin, but the specific mechanism of their formation remains unknown. On the other hand, there are no mountain chains on the Moon that could have been formed by folding nor, for that matter, are there any features to indicate lateral movement of the Moon's crust. The structures in the mountain chains of the Moon—the lunar Alps, Apennines and Carpathians surrounding Mare Imbrium (Fig. 7. 1) or the Cordilleras bordering on Mare Orientale (Fig. 7. 8)—originated, so far as we can tell, by the same impact processes that produced craters; there is no indication of folding. Thus, one of the most important terrestrial orogenic processes seems to be completely lacking on the Moon, and its absence suggests that the rigidity of the Moon's 'crust' is much greater than that of the terrestrial mantle.

Let us consider the physical and chemical composition of the lunar surface in the light of the samples of Moon rocks brought back to the Earth by successive *Apollo* missions of 1969–1971. By atomic contents, the most abundant constituents of the material are—as in the Earth's crust—

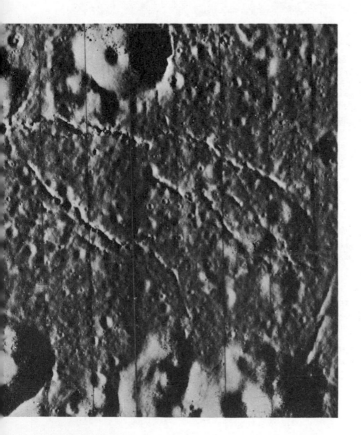

7. 5 A section of the floor of the crater Copernicus, as recorded by the high-resolution optics of Lunar Orbiter 5 on 16 August 1967 from an altitude of 103 km. It shows a striking example of a 'lunar dome' in the upper left-hand corner of the frame. Ground resolution is close to 2–3 metres on the lunar surface.

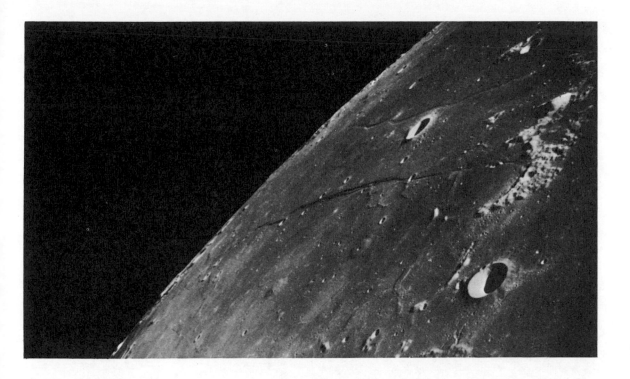

7. 6 Above: the 'Cauchy hyperbolae' rilles and a cleft flanking the crater Cauchy on each side, photographed from Apollo 8 on 24 December 1968. Below: a Lunar Orbiter 3 close-up of a rille running north-west and east from the crater Hyginus. The crater is 11 km wide and 800 m deep, nearly in the centre of the earthward side of the Moon.

7.7 The wrinkle ridges on the surface of Mare Tranquilitatis (with the 'ghost' crater Lamont near the centre of the field indicated by an arrow). The impact of the hard-landing Ranger 8 on 20 February 1965 and the first manned landing by Apollo 11 on 20 July 1969 occurred in the field covered by this Pic-du-Midi photograph, taken with the 24-inch refractor (Manchester Lunar Programme).

oxygen (about half by weight), silicon, and aluminium; followed by magnesium, titanium, and others in diminishing amounts. It is not until we come to less abundant elements that differences between the Moon and the Earth's crust become apparent. Thus titanium, chromium and visconium are more abundant on the Moon than on the Earth (or in the Sun); whereas nickel or the alkali metals (sodium, potassium, rubidium, etc.) or rare earths like europium are very much less abundant. The ratio of iron to nickel appears to be larger in the Moon than in any other sample of cosmic matter available for analysis; and carbon or nitrogen are conspicuously deficient on the Moon.

What kind of molecules and minerals do these elements constitute? About 40% of lunar surface material by weight appears to be silica (SiO_2); followed by FeO (around 20%), Al_2O_3 and CaO

(10%). Numerous minerals well known from the Earth—such as olivine, pyroxene, plagioclase, feldspar and ilmenite—have also been found on the Moon. Only three minerals not previously known on the Earth have been identified in lunar samples so far. The bulk density of the five-grain component of surface material proved to lie between 1.5 and 1.6 g/cm^3, but the densities of individual grains range from 3.1 to 3.5 g/cm^3. About half of the total volume of the material forming the topmost layer of the surface is, therefore, empty space; and this accounts for the relatively low bearing strength (about 3×10^3 N m^{-2}) encountered on the surface.

Many rocks brought back from the Moon suffered from evident effects of shock metamorphism produced, in all probability, by meteoritic impacts. All rocks are igneous (no trace of sedimentary material has been found anywhere on the

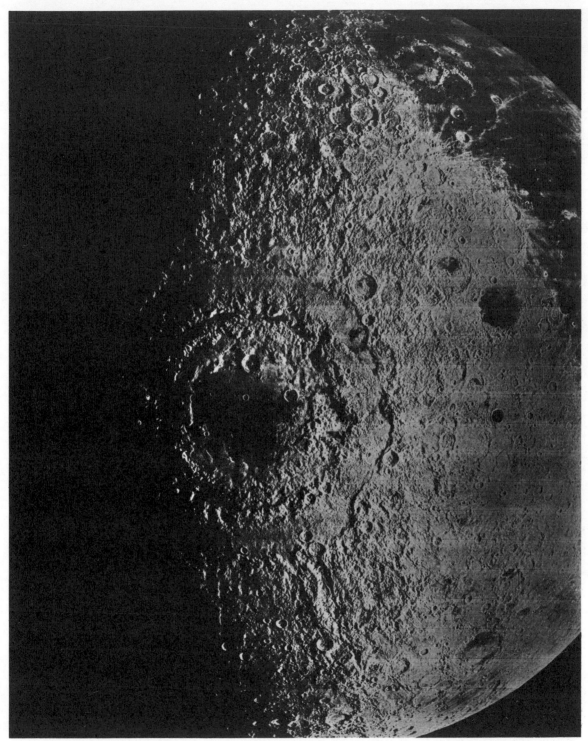

7. 8 The lunar Mare Orientale—a huge triple-walled circular basin almost 1 000 km across—as photographed by Lunar Orbiter 4 on 25 May, 1967 from 2 700 km above the surface.   This formation is at the extreme western limb of the visible lunar hemisphere.

Moon), and generally of basaltic composition. Their crystalline structure indicates, moreover, that lunar rocks solidified at temperatures between 1 000° and 1 200°C under highly reducing conditions, in which the partial pressure of free oxygen was less than 10^{-13} of an atmosphere; for all oxides present correspond to the lowest states of oxidation.

The sources of heat which produced melting and thermal differentiation were undoubtedly local rather than global in nature. Global melting of the Moon at any time in the past is inconsistent with the shape (and gravitational field) of our satellite. A completely molten Moon would not have had time to solidify in 4 600 million years; and if at least its surface has been all molten at any time, it would have solidified to a form approximating a state of hydrodynamical equilibrium much more closely than it actually does. More localized melting could be the result of impacts, and local pockets of lava may have been produced in this way; but 'volcanoes' as we understand the term in terrestrial geology probably played a much smaller role in shaping the face of the Moon and its composition than they did on the Earth.

The greatest single piece of information furnished by man-made spacecraft in recent years has been a determination, by chemical methods, of the age of the lunar rocks from different localities on the lunar surface—or, more specifically, of the time which has elapsed since these solidified. When radiometric methods were applied to determine this age of crystalline rocks which the *Apollo 11* astronauts brought back from Mare Tranquilitatis in July 1969, it proved to be close to 3 700 million years. The rocks from Oceanus Procellarum, brought back by *Apollo 12* in November of the same year, were dated as 100–200 million years younger; and Mare Crisium, visited by the unmanned *Luna 16* in September 1970, proved to be 3 400 million years of age. The semi-mountainous Fra Mauro region on the fringes of Oceanus Procellarum—the target of the *Apollo 14* mission in February 1971—was found to be 4 000 million years old; and the Apennine shore of Mare Imbrium—the landing place of *Apollo 15*—is 3 300 million years old. Therefore, the principal pattern of lunar maria as we see them to-day (cf. Fig. 7. 1) was formed between 3 300 and 4 000 million years ago, as a result of an apparently unrelated series of distinct events that occurred within this time.

However, perhaps the most interesting result which has emerged from radioactive dating of lunar rocks so far is the fact that, on each landing site, the smaller-size debris (the *fines*) were found to be of substantially greater age than the lumps of rock. The time which has elapsed since the solidifications of the fines proved, in fact, to cluster around 4 600 million years in all localities visited so far; and this age is virtually identical with that of the oldest known meteoritic material intercepted by the Earth. These fines provide an irrefutable testimony that solid matter existed on the lunar surface as far back as 4 600 million years ago and that it has not been re-melted since that time.

Nothing of comparable age has been found anywhere on the Earth, where the oldest rocks preserved in rare localities are not more than 3 800 million years old (all older strata having been 'consumed' by the interior). In other words, the first 1 000 million years of the Earth's history represent the 'dark aeon', of which we find no testimony engraved in the terrestrial stony strata. Since 1969 we have, however, unlocked a source of material which suddenly illuminated for us the earliest chapter in the history of the Earth-Moon system, and is enabling us to reconstruct an almost uninterrupted story of what has happened in the inner precincts of our solar system since the days of its formation.

In contrast to the Moon, the Earth exhibits a cosmic face of almost eternal youth, continuously rejuvenated by geological processes of erosion and denudation and probably by (quantitatively more important) continuous continental movement and consumption of its ocean floors. All these processes indicate an active mechanism, a heat engine, within the Earth. Very few parts of the Earth's surface, or of its ocean floors, are older than a few hundred million years. The Moon, on account of its small mass and heat capacity, can afford none of these cosmic cosmetics to make up her face. Her face truly mirrors the ages gone by and preserves a record of events that occurred long before our own terrestrial continents were formed and long before the first manifestation of life on Earth flickered in our shallow waters. As a record of the past, the Moon constitutes the most important 'fossil' in the solar system, and an interpretation of the hieroglyphs engraved by Nature on her stony face holds a fascinating story as yet only partly unravelled.

7.9   Apollo 15 mission: the lunar surface near the rim of Hadley's rille, August 1971.

TIDAL EVOLUTION OF THE EARTH-MOON SYSTEM

It is a comparatively simple matter to determine the dynamic characteristics of the Earth-Moon system as they are now—but how have they changed, if at all, over time? *A priori* evidence that such changes are probable comes from a simple qualitative consideration of the effects of tides. The most obvious property of an ocean tide, for example, is that the water moves relative to the solid Earth. This relative movement between two components of the same rotating Earth is thus likely to have some effect on the rotation. The effect may be, and indeed is, very small, but we have to remember that it has been operating over an extremely long period of time and that the cumulative effects may be appreciable. Let us attempt to reconstruct, then, not only the interplanetary environment in which the Earth and the Moon have been revolving around the Sun thus far, but also the secular influences they have exerted on each other and the changes in their respective motions which have occurred as a result of mutual gravitational forces.

If the Earth and Moon were each perfectly spherical—as they could be if each were absolutely rigid and impervious to tides—their internal structures and axial rotations would not be coupled with their motions in space. Gravitationally they would act as individual point masses revolving around a common centre of gravity. But if, as is true of all planetary or cosmic bodies, their masses are capable of yielding slightly to a mutual tidal interaction, the dynamic system they form will cease to conserve energy and angular momentum. Dissipative forces arising from viscosity, or imperfect elasticity, will tend both to redistribute the kinetic energy between the two constituent bodies and to reduce the total amount of energy by converting it into heat.

The oceanic tides raised on the Earth by the Moon are, of course, a well known phenomenon. Less well known, because their effects are not immediately visible, are the tidal motions produced in the solid Earth. On the Moon, where there are no oceans, the tides raised by the attraction of the Earth are produced entirely in the solid body. What, then, are the effects of these tides on the dynamics of the Earth-Moon system?

**7. 10 Footsteps on the Moon, 21 July 1969.
Note the soft nature of the lunar ground.**

Suppose first that the solid parts of the Earth and Moon were perfectly elastic and that the liquid part of the Earth were perfectly fluid. Under those conditions both solid and liquid components would respond instantaneously to changes in gravitational forces—and the maximum height of the tides would always occur in the direction of the attracting body. Figure 7. 11a shows, for example, a tide produced on the Earth by the Moon. Because of symmetry, the tide exerts no torque on the Moon and thus has no effect on its orbital motion. For the same reason, tides on the Moon would perturb neither the rotation nor orbital motion of the Earth.

However, if the solid parts of Earth and Moon are not perfectly elastic, and if the liquid part of the Earth has a finite viscosity, the maximum heights of the tides no longer lie along the direction of the attracting force. In the case of a tide produced on the Earth by the Moon, the rotation of the Earth carries the bulge forward; and the tide is high not when the Moon is directly overhead but at some time later (Fig. 7. 11b). The gravitational attraction on the bulge is assymmetrical to the line joining the centres of the

Earth and Moon. This gives rise to a torque on the Earth which tends to retard its rotation; and an equal and opposite torque on the Moon which tends to accelerate its orbital motion. The torques thus transfer angular momentum from the Earth to the Moon. The Earth loses some of its rotational energy, part of which is dissipated as heat and part of which goes into the orbital motion of the Moon.

These frictional effects are due to the relative rotation of the Earth with respect to the Moon. The reason why the tidal bulge on the Earth is carried forward in the first place is that the Earth's rotational velocity is greater than the Moon's orbital velocity. However, the Earth-Moon system would still lose energy even if the Earth were not rotating relative to the Moon. Suppose that the Earth were rotating with an angular velocity equal to the Moon's orbital velocity, and that the Moon moved in an eccentric orbit. In this case the tidal bulge on the Earth would always lie along the line joining the centres of the Earth and Moon, just as in the previous case of perfectly elastic and fluid bodies. But because of the eccentric orbit of the Moon, the tide on the Earth would not be high when the Moon is closer to the Earth, at perigee, but at some time later—and energy would again be lost from the combined system. A similar mechanism accounts for tides on the Moon.

The most important characteristic of tidal friction is that its effectiveness in transferring angular momentum between rotation and revolution depends strongly on the distance separating the interacting masses. When the Earth and Moon are separated by a distance r, the tides raised by either body on the other have a height proportional to r^{-3}. The tidal torque is thus proportional to r^{-6}. It is clear from this that when r is large the torque is small, and angular momentum will be transferred very slowly. Conversely, when r is small the transfer of angular momentum will be rapid.

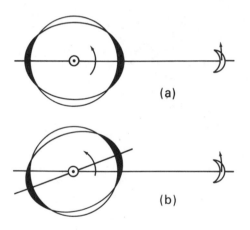

7. 11 (a) In the absence of friction, the tidal bulge would lie in line with the centres of the Earth and the Moon. (b) Friction causes the Earth's rotation to carry the bulge forward and so to delay the time of high tide.

The full mathematical treatment of the effects of tidal friction on the dynamics of the Earth-Moon system is complicated. What emerges from such a treatment, however, is that the effect of transferring angular momentum from the rotation of the Earth to the orbital motion of the Moon is to increase the distance r separating the two bodies. But as we have seen, as r increases the transfer of angular momentum becomes less rapid and thus the rate at which the rotation of the Earth is decreasing will slow down. Ultimately there will come a time, albeit many millions of years in the future, when no more angular momentum of rotation is lost by the Earth to the Moon because the torques have been reduced to zero, and the system should then be stable against further changes.

In the meantime, however, the rotation of the Earth is decreasing. But for how long has it been decreasing? One of the effects of a decreasing rotation is, of course, an increase in the length of the day. Astronomical observations over the past two centuries show that tidal friction is lengthening the day by 1.8×10^{-3} seconds per century, which is equivalent to a fractional change of 2.1×10^{-10} per annum. That this figure is also the average value for the last 370 million years has recently been confirmed from palaeontological evidence. The daily growth rings in corals show both monthly and yearly variations due to seasonal changes. It is thus possible to determine the number of days in a year by counting the number of growth rings in corals of various ages. Since tidal friction does not appreciably affect the length of the year, the ring counts lead directly to the length of the day. Recent corals show about 360 rings a year, whereas fossil corals from the Middle Devonian have about 400 a year. Thus about 370 million years ago the day was only 21.9 hours long.

Calculations based on the figures for the lengthening of the terrestrial day show that the semi-major axis of the Moon's elliptic orbit is currently increasing at the rate of 3.2 cm yr^{-1} and that the orbit is becoming even more eccentric. It is also possible to extrapolate these calculations backward in time in an attempt to determine what the orbit of the Moon looked like in the remote past. The results of such an extrapolation should, of course, be regarded with some reserve because they are inevitably based upon the effects of tidal friction as they are observed now. The internal structures of the Earth and Moon could have been very different from what they are to-day—for example, the Earth may not then have had its present metallic core.

Nevertheless, with these reservations in mind, it is easy to see that since the separation of the Earth and Moon is increasing, the Moon must have been much closer to the Earth in the past. This led G. H. Darwin to suggest, in 1880, that the Moon may have originated as part of the Earth. It is

7. 12 One of the largest Moon boulders (brecciated in structure) investigated by man 'in situ' so far. Apollo 16 photograph taken near the crater Descartes, 21 April, 1972.

7. 13 The Davy rille on the lunar surface, as photographed from the command module of the Apollo 14 mission in February 1971.

possible to show, however, that the Moon reached a minimum distance from the Earth, at which time it revolved about the Earth synchronously with the Earth's rotation—that is to say, the Moon remained above the same point on the Earth as it orbited. Darwin calculated that if the Moon were combined with the Earth the resulting body would rotate with about a four-hour period. He then suggested that the fundamental resonant frequency of the Earth would correspond to a period of about two hours, so that the period of solar tides (the Sun's attraction also produces tides on the Earth although these are usually considerably smaller than those produced by the Moon) would coincide with the free oscillation of the Earth. Because of the resonance produced, the height of the tide would then increase until a mass separated from the Earth. However, many years later Jeffreys knocked this theory on the head by showing that internal friction would prevent the tidal build-up suggested by Darwin.

Another theory of lunar origin supposes that the Moon was an independent body captured by the Earth during the early history of the solar system.

As originally proposed by Urey in 1963, the Moon came close enough to the Earth to be captured in a prograde orbit which then expanded slowly to its present size under the effects of tidal friction. An objection to this particular idea, however, is that for capture to have taken place the Earth would have had to be rotating with a period of 3.2 hours. This means that the Earth would have had to possess an initial angular momentum and density much higher than that characteristic of the present solar system.

In 1955 Gerstenkorn worked out a process by which the Moon may have been captured in a retrograde orbit. In attempting to unravel the evolution of the Earth-Moon system prior to the time of closest approach, Gerstenkorn showed that under conditions of reversed torque the Moon could recede from the Earth until a retrograde polar orbit is achieved at a distance of 4.7 Earth radii. The idea, then, is that the Moon is captured in such a retrograde orbit, spirals in towards the Earth until it reaches the point of closest approach and then spirals back out in a prograde orbit. However, Gerstenkorn, and later Alfvén, assumed for the purpose of calculation that the Moon's orbit was circular. It is possible to show that for an eccentric orbit the Moon's motion cannot be retrograde.

Gerstenkorn did note one further interesting fact, namely that the minimum distance of approach between the Moon and the Earth comes remarkably close to the terrestrial *Roche Limit*, R_L—the minimum distance from the Earth at which a satellite will hold itself together by its own gravitational field—given approximately by:

$$R_L = 2.44\,R_e \left(\frac{\rho_e}{\rho_m}\right)^{1/3}$$

where R_e is the mean radius of the Earth, and ρ_e, ρ_m the mean densities of the Earth and Moon, respectively. Putting $\rho_e = 5.52 \times 10^3$ kg m^{-3} and $\rho_m = 3.34 \times 10^3$ kg m^{-3}, we find from the above equation that $R_L = 2.88\,R_e$, which is practically identical with the distance of minimum approach of the Earth and Moon.

What would happen if the Moon were to enter the Roche zone? Would it suffer physical harm, and how gruesome would this experience be? This thought was recently developed by Alfvén who, whilst accepting the qualitative features of Gerstenkorn's retrograde orbit theory, argued that the initial angular momentum of the Moon might have been somewhat less than Gerstenkorn calculated. In this case the distance of minimum approach would have been well inside the Roche Limit. As a result, at least the surface of the Moon would have been torn into pieces and scattered in space. Some of the fragments would have fallen directly on to the Earth or have arrived after one or two orbits in the gravitational field of the Earth-Moon system. Such events would have speeded up the Earth's rotation and would possibly have added an extra ingredient to the outer crust.

Moreover, this process could have continued until the Earth acquired the same angular velocity, projected on the Moon's orbital axis, as the Moon. At that point the tidal couple would have vanished, the Moon would have ceased to approach the Earth and then would have begun to recede from it. Meanwhile, other lunar fragments not captured by the Earth may eventually have fallen back on the Moon's surface. As the Moon was receding from the Earth it might well have been bombarded by its own debris. It is possible that many of the Moon's impact craters were produced during such a brutal rendezvous with the Earth.

It may be that the Moon originally formed in close proximity to the Earth; but the theory that it originated outside the gravitational field dominated by the Earth and was captured by our planet may still be regarded as the more likely hypothesis. If there was capture, it must have been virtually simultaneous with the formation of the Earth, and at that time moon-like objects may have been more numerous in interplanetary space than they have become since then. But whatever uncertainty still surrounds the genesis of the Moon, its future is not in doubt: the continuous operation of tidal friction will make it recede from the Earth until its distance from us has 1.6 times the present value and its orbital period is approximately 60 days of our present time. The same cause will have slowed the rotation of the Earth to the point where a terrestrial day is as long as a month, and the Earth and the Moon will henceforward continue to show each other the same face. Several thousand million years are likely to elapse before this comes to pass; but it is improbable that the topography of the Moon will undergo any important changes in the meantime.

We now take it as certain that the solid crust of the Moon (constituting perhaps the major part of the lunar continents) was already formed as far back as 4 600 million years ago, and that for the main part its surface features were sculptured during the first few hundred million years of its existence; since then it has probably been disturbed only by incidental impacts. For reasons which we do not yet fully comprehend, the greatest events which disfigured the Moon— namely, the formation of the maria—occurred between 500 to 1 500 million years after the Moon was formed; and since that time things have become increasingly quiet.

FURTHER READING:

z. KOPAL 1968. Exploration of the Moon by Spacecraft. Contemporary Science Paperbacks, *Oliver & Boyd.*

z. KOPAL 1969. The Moon. *D. Reidel Publishing Co., Dardrecht.*

SIR BERNARD LOVELL *et al.* 1969. The New Space Encyclopaedia. *Artemis Press.*

J. A. WOOD 1970. The Lunar Soil. *Scientific American,* 223, 2.

8.0   The Arizona Meteor Crater: 1300 m across, 180 m deep, with the rim raised 40 m.

We invited Dr. Brian Mason of the Smithsonian Institution in Washington D.C., a leading authority on meteorites, to write the following article. Given below are his own introductory comments—no further introduction is needed.

WHY STUDY METEORITES?

This may seem a rhetorical question, since any natural object is worthy of scientific study, and meteorites, the only tangible samples of the universe beyond our planet prior to the receipt of the lunar rocks, are clearly of unique interest.   For a geologist, meteorites provide rock types existing in the solar system but differing from anything that occurs in the Earth's crust, and this material has just the sort of composition that petrology, geophysics and astronomy combine to indicate as likely to make up the interior of the Earth. For the geochemist the study of meteorites is of special significance, since it is widely believed that the average composition of meteoritic matter provides the best information on the relative abundances of the non-volatile elements. Tables of cosmic abundances of the elements are partly based on the interpretation of meteorite analyses.   Such tables are of fundamental importance not only to geochemists but also to nuclear physicists and astrophysicists concerned with processes of element formation.   The examination of radioactive nuclides, both original and produced by cosmic-ray bombardment, has provided a wealth of data bearing on the origin, age and history of meteorites, the solar system, and the universe as a whole.   As one scientist has neatly expressed it, a meteorite is the poor man's space probe.   It carries within it a history, albeit imperfectly preserved and difficult to decipher, of events in the universe over the past five thousand million years.

8 Meteorites

by BRIAN MASON

Meteorites are a gift from heaven: their composition and structure provide some of the most direct evidence we have concerning the evolution of the planets.

EVERY DAY THOUSANDS of meteors enter the Earth's atmosphere. Most of them are smaller than a grain of sand and burn up completely, leaving a luminous trail which we observe as a shooting star. These trails are easily photographed, and when two plates of the same meteor are taken from different stations the orbit can be calculated. Many granules are so small that they do not offer sufficient air resistance to become incandescent; they sink to the ground as *micro-meteorites*, and are extremely difficult to recover and to identify reliably. Very occasionally, a much larger piece of material survives a violent passage through the atmosphere and reaches the surface of the Earth, thereby becoming a *meteorite*. Its weight may range from a few kilogrammes to many thousands, and it may explode during entry or on impact.

PHENOMENA OF FALL

A meteorite fall is a rare event. It has been estimated that there are approximately 500 falls over the whole Earth annually. Since the oceans cover about 70% of the Earth's surface, the number falling on land, and thus accessible for recovery, should be about 150 per year. The actual number recovered is about ten. The low recovery rate is understandable when one considers how thinly populated is much of the Earth's surface.

The light and sound effects accompanying a meteorite fall can be most impressive. A bright fireball, with a long luminous tail, may light up large areas with an intensity approaching that of sunlight. The light is usually white, but may appear greenish, reddish, or yellow, and may vary from place to place along the path. The sound effects, which are usually heard a few seconds to one or two minutes after the light has extinguished (because sound travels less rapidly than light) are described as resembling artillery fire, thunder, or the passage of an express train. The fireball accompanying the meteorite fall near Belfast on the evening of 25 April 1969, with a track from the Dorset coast to Inishowen Head on the northern Irish coast, was observed from many points in the British Isles.

Since meteorite falls are rare, and unpredictable as to time and place, planned observations are extremely difficult. For only two meteorites, the Pribram (Czechoslovakia) fall of 7 April 1959, and the Lost City (Oklahoma) fall of 8 January 1970, do we have precise information on their trajectory, thanks to simultaneous photography of the fireball from several points. From the photographs it has been shown that these meteorites were travelling in elliptical orbits whose farthest points were between Mars and Jupiter, indicating an origin in the asteroid belt. It has long been conjectured that meteorites are fragments of asteroids which have been deflected into Earth-crossing orbits, but these are the first instances of positive verification.

Since the early days of meteorite research it has been standard practice to give each meteorite a name, so that they can be readily catalogued, compared, and studied. A meteorite is named from a locality near which it was found, usually the nearest town or village, or a topographical feature that can readily be identified on a large-scale map. The British Museum catalogue of meteorites, which is generally accepted as the authoritative guide to the names of meteorites, lists about 1 700 well-authenticated meteorites, each meteorite being a single fall. However, a single fall may comprise many individual pieces. Thousands of stones were picked up from the fall of the Allende meteorite in northern Mexico on 8 February 1969, and tens of thousands of nickel-iron fragments of the Canyon Diablo meteorite have been collected around Meteor Crater in Arizona.

Like terrestrial rocks, meteorites are usually classified on the basis of composition and structure. The basic division into irons, stony-irons, and stones is simple and straightforward; however, most of the stones are characterized by the presence of small ($\sim$ 1 mm diameter) spheroidal aggregates known as *chondrules*. Stones with chondrules are known as chondrites, those without as achondrites. The current fourfold grouping of meteorites into irons, stony-irons, chondrites, and achondrites, though sanctified by long acceptance and an apparent logic, may actually obscure significant genetic relationships. A genetic grouping would probably be into chondrites and non-chondrites, the latter being derived from the former by a variety of secondary processes.

About sixty minerals are known from meteorites, but many of these are rare accessories. The common and abundant minerals are listed in Table 1. Some contrasts to terrestrial mineralogy should be pointed out: nickel-iron is practically absent from terrestrial rocks; the common minerals in meteorites are largely magnesium-iron silicates, whereas in the Earth's crust the commonest minerals are quartz and aluminosilicates; the common meteorite minerals are anhydrous, whereas hydrated minerals are common and abundant on Earth. These features indicate that meteorites formed in a highly reducing environment, in which nickel and iron were largely in the metallic state. A small but remarkable class of meteorites, the carbonaceous chondrites, differ fundamentally; they consist mainly of serpentine, $(Mg,Fe)_6Si_4O_{10}(OH)_8$, their nickel is present largely in silicates and sulphides, and they contain considerable amounts of organic compounds of extraterrestrial origin. A notable feature of the overall mineralogy of meteorites is the absence of phases, such as pyrope garnet and jadeitic pyroxenes, indicative of high pressure (i.e., large parent bodies); the origin of the diamond in the Canyon Diablo iron has been plausibly ascribed to the shock of impact with the Earth, which formed Arizona's Meteor Crater, and the presence of diamond in the small group of ureilites appears to be due to extraterrestrial shock effects.

The classification of meteorites based on mineralogy and structure is set out in Table 2; photographs of some typical examples are given in Fig. 8.1–8.4. Analysis of the observed falls shows that the populations of the different classes vary widely (the figures for observed falls are used as being the best approach to actual extraterrestrial abundances; irons dominate meteorite finds, since they are resistant to weathering and are readily recognized as meteorites or at least as very unusual objects). Over 80% of meteorite falls are chondrites, and over 90% of these belong to two classes, frequently referred to jointly as the common or ordinary chondrites. The average composition of the common chondrites is about 40% olivine, 30% pyroxene, 5–20% nickel-iron, 10% plagioclase, and 6% troilite. A comparable terrestrial rock would be peridotite, but this lacks nickel-iron and troilite, and the chondritic structure. The achondrites are a diverse group which do not contain chondrules and which are usually much more coarsely crystalline than the chondrites. Many achondrites resemble terrestrial igneous rocks in composition and texture, and so have probably crystallized from a silicate melt. The stony-iron meteorites are made up of nickel-iron and silicates in approximately equal amounts. Two distinct groups, the pallasites and the mesosiderites, are recognized. The pallasites are made up of a continuous base of nickel-iron enclosing grains of olivine which often show good crystal forms. In the mesosiderites the metal phase is discontinuous and the silicates are mainly plagioclase feldspar and pyroxene. The irons consist essentially of a nickel-iron alloy (Ni usually between 4% and 20%, rarely more), generally with accessory troilite as globular inclusions. Those irons with Ni content between 7–14% usually show the Widmanstaetten structure, which is brought out by etching a polished surface with an alcoholic solution of HNO_3. This structure consists of lamellae of kamacite bordered by taenite. The lamellae are parallel to the octahedral faces of an originally homogeneous crystal of nickel-iron, and meteorites showing Widmanstaetten structure are therefore known as octahedrites. The structure is typical of exsolution in an alloy which has cooled very slowly from a high temperature; cooling rates for most octahedrites are 1°–10°C per million years, consistent with their having formed inside asteroid-sized bodies. Hexahedrites are irons with less than 7% Ni and consisting entirely of kamacite; ataxites usually contain 14% or more Ni and are made up largely of taenite.

Not only are the chondrites the most abundant meteorites, but many features indicate a primary origin for them and a derivate origin for the other meteorite groups. Of the different classes of chondrites, the carbonaceous chondrites, in particular a subclass known as Type I, show a remark-

olivine
pyroxene
feldspars

TABLE 1

The Common Minerals of Meteorites

Kamacite	α-(Fe, Ni)	(4–7% Ni)
Taenite	γ-(Fe, Ni)	(30–60% Ni)
Troilite	FeS	
Olivine	$(Mg, Fe)_2 SiO_4$	
Orthopyroxene[1]	$(Mg, Fe)SiO_3$	
Pigeonite	$(Ca, Mg, Fe)SiO_3$	(About 10 mole per cent $CaSiO_3$)
Diopside	$Ca(Mg, Fe)Si_2O_6$	
Plagioclase	$(Na, Ca)(Al, Si)_4O_8$	

1. Divided into enstatite, with 0–10 mole % $FeSiO_3$, bronzite, 10–20%, and hypersthene, > 20%; these minerals are orthorhombic, and have monoclinic polymorphs known as clinoenstatite, clinobronzite, and clinohypersthene.

TABLE 2. The Classification of Meteorites (figures in parentheses are the numbers of observed falls in each class)

Group	Class	Principal minerals
Chondrites	Enstatite (11)	Enstatite, nickel-iron
	Bronzite (227)	Olivine, bronzite, nickel-iron
	Hypersthene (303)	Olivine, hypersthene, nickel-iron
	Carbonaceous (31)	Serpentine, olivine
Achondrites[1]	Aubrites (8)	Enstatite
	Diogenites (8)	Hypersthene
	Chassignite (1)	Olivine
	Ureilites (3)	Olivine, clinobronzite, nickel-iron
	Angrite (1)	Augite
	Nakhlite (1)	Diopside, olivine
	Howardites (14)	Hypersthene, plagioclase
	Eucrites (26)	Pigeonite, plagioclase
Stony-irons	Pallasites (2)	Olivine, nickel-iron
	Siderophyre (1) (Find)	Orthopyroxene, nickel-iron
	Lodranite (1)	Orthopyroxene, olivine, nickel-iron
	Mesosiderites (6)	Pyroxene, plagioclase, nickel-iron
Irons	Hexahedrites (7)	Kamacite
	Octahedrites (32)	Kamacite, taenite
	Ni-rich ataxites (1)	Taenite

1. Sometimes subdivided into calcium-poor achondrites (aubrites, diogenites, chassignite, ureilites) and calcium-rich achondrites (angrite, nakhlite, howardites, eucrites).

8.1 Photomicrograph (transmitted light) of a thin section of the Clovis meteorite, a bronzite chondrite; individual chondrules, which range up to 3 mm in diameter, are made up of olivine and/or pyroxene. (Smithsonian Institution)

8.2 Photomicrograph (transmitted light) of a thin section of the Moore County meteorite, a eucrite consisting of pyroxene (grey) and plagioclase (white); average grain size is about 2 mm. The composition and structure of this meteorite resemble some terrestrial gabbros. (Smithsonian Institution)

able correspondence in elemental abundances with the Sun (Fig. 8. 5). If elemental abundances were the same in both, the points in Figure 8. 5 would lie on the 45° line. The close approach to this line for most elements is the basis for considering the chondrites, and specifically the Type I carbonaceous chondrites, as approximating in composition to the primordial non-volatile matter of the Solar System.

Many chemical analyses have been made of meteorites. The irons form a rather homogeneous group, differing from one to another essentially in nickel content, and an average composition is easily derived. The stony-irons and the achondrites are very diverse groups, and are comparatively rare, so averages of their compositions have little significance. The chondrites are abundant and show a remarkable homogeneity in chemical composition. Table 3 gives a selection of the available compositional data. The close correspondence between the average composition of iron meteorites with the average composition of

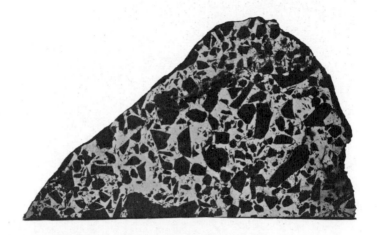

8. 3 Polished surface of the Huckitta pallasite, showing angular fragments of olivine (black) in a nickel-iron matrix (white); specimen is 26 cm long. (Smithsonian Institution)

8. 4 Polished and etched surface of the Canyon Diablo iron meteorite, showing typical Widmanstaetten structure and rounded inclusions of troilite (FeS) and graphite; specimen is 30 cm long. (Smithsonian Institution)

the metal from chondrites strongly suggests a common source. The iron meteorites probably represent metal segregated by the partial or complete melting of material of chondritic composition. Comparative data for the Earth's crust show that it is relatively more oxidized than the chondrites, enriched in Si, Al, Ti and the alkali metals, and depleted in Mg, Fe, and especially Ni, Co, Cr and S.

THE ORIGIN OF METEORITES

For many years the origin of meteorites was ascribed to the break-up of a hypothetical planet situated between Mars and Jupiter, in the region now occupied by the asteroids. However, it is now clear that the asteroids are not fragments of a disrupted planet, but represent solid material that failed to aggregate into a planetary body. It is

TABLE 3. THE CHEMICAL COMPOSITION OF METEORITIC MATTER (WEIGHT PER CENT)

	Metal (from irons)	Metal (from chondrites)	Silicate (from chondrites)	Average Chondrite	Earth's Crust
O			43.7	33.2	46.6
Fe	90.8	90.7	9.88	27.2	5.00
Si			22.5	17.1	27.7
Mg			18.8	14.3	2.09
S				1.93	0.026
Ni	8.59	8.80		1.64	0.075
Ca			1.67	1.27	3.63
Al			1.60	1.22	8.13
Na			0.84	0.64	2.83
Cr			0.51	0.39	0.01
Mn			0.33	0.25	0.09
P			0.14	0.11	0.11
Co	0.63	0.48		0.09	0.025
K			0.11	0.08	2.59
Ti			0.08	0.06	0.44

also clear that, however the meteorites originated, they have not come from a single parent body; for one thing, there are too many chemical and mineralogical discontinuities between different groups of meteorites.

As a result of the great resurgence of interest in meteorites as part of the extensive developments in space research, there is now a vast accumulation of data on these remarkable bodies. Chondrites, especially carbonaceous chondrites, are now recognized to be relatively well-preserved samples of the non-volatile matter of the Solar System. Compositional differences between different classes of chondrites indicate that the ancestral solar nebula probably underwent some chemical fractionation prior to the accumulation of the meteorite parent bodies. The extreme differences in mineralogy and oxidation state between the carbonaceous chondrites and the enstatite chondrites probably reflect differences in the environment of accumulation of their original material, and in the physico-chemical conditions within the meteorite parent bodies. The major events in the genesis of meteoritic material—the birth of the ancestral solar cloud, the formation of chondrules, the accumulation of meteorite parent bodies and the subsequent differentiation of some of them to give the material of the nonchondritic meteorites —all took place within a relatively short time interval, about 4 600 million years ago.

The wide variety of meteorites, and the contrast between chondrites and nonchondrites, may have their origin largely in different sizes of meteorite parent bodies. Some chondrites—the carbonaceous chondrites and a few of the common chondrites—have not been subjected to significant reheating since accumulation, and presumably are derived from relatively small parent bodies. Most chondrites show features indicating that they have been heated within their parent bodies to temperatures of 800°C and higher. Ultimately, if temperatures of 1 400°–1 500°C were reached inside the parent bodies, the material melted. The degree of reheating was probably controlled in large part by the size of the parent bodies; the larger the parent body the greater the heating-up of the interior. Heating-up was probably a comparatively rapid process, brought about not only by the greater abundance of the long-lived radioactive elements 4 600 million years ago but also by short-lived radioisotopes such as [26]Al, [129]I, [244]Pu, and others.

It is conceivable, of course, that each meteorite class represents a single parent body. However, it seems more likely that a large number of potential parent bodies formed, probably up to some hundreds of kilometres in radius (the largest asteroid, Ceres, has a radius of 385 km). The cooling rates for the pallasites indicate that they formed the core of a body with a maximum radius of about 300 km. Since the pallasites have the lowest cooling rate for any class of meteorites, this suggests an approximate upper limit for the size of meteorite parent bodies.

In this connection the populations of the different classes of meteorites may provide some relevant information. The abundance of the common chondrites indicates that most of the parent bodies must have had similar compositions. Parent bodies giving rise to the achondrites, irons, and stony-irons must be comparatively rare. This suggests that comparatively few of the parent bodies were sufficiently large to heat up to the melting stage. The low abundance of the carbonaceous chondrites, and their specific properties indicating a low-temperature origin for much of their material, suggest that the parent bodies for these meteorites accumulated farther from the centre of the ancestral solar nebula, where temperatures were low. Their parent bodies never grew large enough for extensive heating and recrystallization, perhaps because the proportion of solid matter to gas in that region of the nebula was considerably lower than where the other meteorite parent bodies accumulated. Within the region between Mars and Jupiter there is probably room to provide for the initial chemical and mineralogical differences and the divergent evolutionary paths of different parent bodies, whose results we observe today in meteorites.

TEKTITES

Tektites are small glassy objects found in limited areas in a few regions of the Earth's surface (unlike meteorites, which show a random distribution over the whole Earth), under conditions that preclude a volcanic origin. The principal areas are in Czechoslovakia, Texas, Ivory Coast, Australia, and southeast Asia. Tektites consist of a silica-rich glass (averaging about 75% SiO_2), superficially resembling obsidian, yet distinct from terrestrial obsidians in composition and texture. They have an unusual chemical composition, which consists of the conjunction of high silica and comparatively high alumina, potash, and

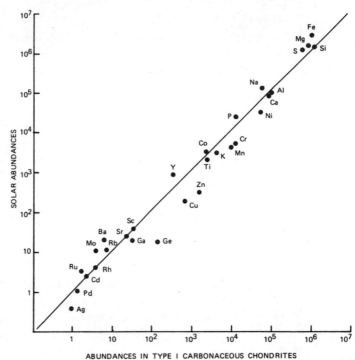

8. 5 Comparison of elemental abundances in carbonaceous chondrites (Type I) with those in the Sun. Abundances are relative to Si = 10^6 atoms in the Sun and in the meteorites.

lime with low magnesia and soda; this composition resembles a few granites and rhyolites, and some silica-rich sedimentary rocks. Unlike the meteorites described above, tektites have not been observed to fall, and their identification as meteorites is disputed; some authorities regard them as the product of the impact of comets or gigantic meteorites on the Earth. In view of their enigmatic origin and their aberrant composition they are not further considered here.

FURTHER READING:

F. HEIDE 1963. Meteorites. *University of Chicago Press.*

B. MASON 1962. Meteorites. *John Wiley & Sons.*

J. A. WOOD 1968. Meteorites and the Origin of Planets. *McGraw-Hill.*

A record of known meteorites, with a summary of the principal literature on each, is provided by

M. H. HEY 1966. Catalogue of Meteorites. *British Museum.*

The next six chapters are concerned with the history of our planet.   Observations and speculations are combined to produce models which explain past events simply.   Several of these chapters discuss how life on the Earth originated and developed, and how the advent of life radically changed the Earth's surface environment.

Organized life of the kind with which we are familiar to-day is preserved only in rocks younger than 570 million years.   But the fossil record stretches further back in time, for remains of primitive unicellular organisms and organic compounds have been found in rocks nearly 3500 million years old. From evidence already given in Chapters 2 and 8 it is probable that the Earth and its companions in the solar system are 4500 million years old.   So what happened during the missing one thousand million years?   Life must have originated in this period—but how?   No rocks formed during this period remain unaltered, but it is likely that some existed and were metamorphosed about 3500 million years ago (the oldest known rocks on the Earth). Therefore to construct a picture of the environment on the Earth in which life might have originated is extremely difficult, and any such result is largely conjectural.   Professor P. C. Sylvester-Bradley, of the Department of Geology at the University of Leicester, explains one hypothesis for the origin of life.   In doing so he draws on many fields of knowledge, including astronomy, biochemistry and volcanology.

9 An Evolutionary Model for the Origin of Life

by P. C. SYLVESTER-BRADLEY

Life arose as the result of continuous evolutionary processes which have also produced the solar system and all its parts. Organic compounds had been synthesized over a long period in a variety of environments, and concentration of these compounds by geological processes established reservoirs of prebiological food. It was within these areas of concentration that the spontaneous generation of life must have taken place.

IN BOTH GEOLOGY AND ASTRONOMY great use is made of the construction of hypothetical models. In the early stages of any investigation, a large number of alternative and highly speculative models are tested against observational facts. As knowledge grows, many of the models are abandoned, and those that survive are modified. In this way our hypothetical models gradually approach more nearly what we believe to be the truth. Nevertheless we are often still left with several alternatives, all of which can account for the observed facts in different ways. That is what has happened to our ideas about the origin of life. In this case the models which have been devised are still rather speculative. This is because they themselves depend on other models, some of which are astronomical, like the origin of the Earth and the solar system; others are geological, and involve the origin of the atmosphere and the geochemistry of carbon; yet others are biological and deal with methods of nutrition and reproduction.

The model presented in this chapter involves just such a chain of hypotheses. Our knowledge in all the fields is increasing very rapidly, and it is certain that the model will have to be constantly modified as each new discovery introduces new restrictions. Even in the present volume you will find that other authors have given in other chapters alternative explanations of how certain of the events postulated in this chapter could have occurred. This is as it should be. Much modern science depends on our skill in staging competitions between opposing ideas. Experimental and observational data provide the ammunition that is used in this war, in which one idea is pitted against another.

All forms of present-day life have many biochemical characteristics in common. They are characterized especially by macromolecules which are polymers; that is, each macromolecule consists of a large number of repeated basic units which are held together by chemical bonds. These macromolecules have characteristic shapes which are large enough to be photographed under an electron microscope.

Two different kinds of macromolecule are found in all forms of life; proteins and nucleic acids. Each kind exists in innumerable forms which are grouped together in various ways within the protoplasm of cells. Before life can have originated, it is likely that proteins and nucleic acids will have already been synthesized in a prebiological environment. There is also some evidence that prebiological cell-like aggregates were in existence; these have been called *pre-organisms* or *organized elements*. Biological cells are separated from the environment by a wall or membrane which possesses physical characteristics that bear on the function of the cell. Proteins, nucleic acids, cells and membranes are four *universal attributes* of life; cell-growth, cell-division and self-replication are *ubiquitous functions* of living cells. The search for the origin of life has to account for the acquisition of these four attributes and three functions.

The model that I am about to put forward postulates that the origin of life took place as an inevitable consequence of an evolutionary process. Prebiological evolution must differ in some ways from biological evolution; the processes that were involved in prebiological time are still operating at the present day, but now they are supplemented

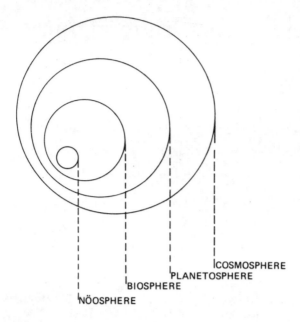

NÖOSPHERE
BIOSPHERE
PLANETOSPHERE
COSMOSPHERE

9. 1 The four 'spheres' of evolution. Each is characterised by time and space scales of decreasing magnitude from cosmic to psycho-social; each involves units of increasing magnitude; each achieves products of increasing complexity. The lower spheres contain within them the higher spheres and act as 'entropy sinks'; there is always a flow of entropy (and therefore a temperature differential) from the higher order to the lower.

on the Earth by biological processes. In order to understand the origin of life in these terms it is necessary to extend Darwin's theory of biological evolution, and to outline a general theory.

A GENERAL THEORY OF EVOLUTION

Evolution is a time-controlled process. It is brought about by a reaction which results in an increase of organization, and a decrease of random arrangement. In thermodynamic terms, there is a decrease of entropy. In keeping with the second law of thermodynamics, this decrease of entropy can only be brought about if there is a flow of energy through the evolving system. The evolving body must lie between an energy source and an energy sink. The energy sink is in almost all cases an external environment lying at a temperature below that of the evolving system, and immediately adjacent to it. The energy source may be more remote.

Classical thermodynamics frequently invokes an *adiabatic* system—one in which there is no exchange of energy or material with the environment. If such a system could exist, it would not evolve. There would be a gradual increase of entropy, and the system would in time break down and become uniform, randomly dispersed and unorganized. Such a system is a theoretical concept not known to exist in any part of the universe, though some cosmologists have maintained that the universe could be regarded as an entity with no energy-sink external to it; it would then be gradually suffering an increase of entropy. In that case the universe would be 'devolving' in a direction opposite to that of most of its parts, which are gaining in evolutionary organization. Sir Karl Popper maintains that the only model in which entropy could accumulate in such a way would imply a finite and non-expanding universe. It seems, in fact, that the adiabatic system is purely fictional. All natural systems are either in a steady state, or can be considered as *diabatic*, and are situated in an energy stream. Not *all* diabatic systems are necessarily evolving. In any reacting system, there is a change of free energy involving entropy production. If the entropy loss from the system is greater than the entropy production, the system will be in a state of evolution. If the production is greater than the loss, the entropy within the system will increase, organization will break down, and 'devolution' will take place.

These considerations enable us to define evolution in terms which lead to a general law:

An evolving system is one which during the course of time shows an increase in organization and a decrease in entropy. All evolving systems are diabatic. They receive energy from one or more sources, and lose energy to the external environment. An evolving system must necessarily maintain a higher temperature than its environment.

It is not easy to devise a quantitative method of assessing evolutionary progress. However, it is often possible to make a qualitative comparison, and to discover which of two systems is more evolved. A survey of all evolving systems reveals that there are four orders of evolution, each characterized by different scale-effects and different evolutionary processes.

The lower orders of evolution contain smaller, more highly evolved orders within their own spheres. Thus the *cosmosphere* involves physical processes and incorporates the smallest units (atoms and atomic particles) and gives rise to the largest products (galaxies and other cosmic aggregates).

The *planetosphere* involves chemical evolution as well as physical evolution. It incorporates larger units (inorganic molecules) and gives rise to smaller products than the cosmosphere (stars and planetary systems). Chemical evolution passes through stellar, planetary and geological phases. It is during the geological phase of planetary evolution that the differentiation of the planet occurs. Core, mantle, crust, hydrosphere and atmosphere become organized, and continents and oceans are formed.

The *biosphere* involves giant organic molecules; these are the genes which form the units of biological evolution. The products are species.

The nöosphere, (a concept first named and described by Teilhard de Chardin, but long championed by Sir Julian Huxley) involves ideas as units, and produces idea-systems by processes of psycho-social evolution.

These evolutionary spheres are contained one within another. A special interest in the study of the origin of life is that it is also a study of the transition between processes of chemical evolution and those of biological evolution.

In our model, all the processes which precede the origin of life must be limited by the application of physical and chemical laws. But as soon as life has originated, *biological* processes are added to those responsible for evolution up to that time. The study of the origin of life is, therefore, a study of the chemical and physical processes which affect carbon compounds in non-biological environments. All known extra-terrestrial environments and many geological environments are still non-biological. One current problem of exobiology (extra-terrestrial biology) is the examination of extra-terrestrial organic compounds in the hope that they will throw light on the origin of life on this planet. Likewise, the study of terrestrial juvenile carbon (carbon which has never been involved in a biological organism) has a special relevance to all theories on the origin of life.

CARBONACEOUS ENVIRONMENTS

Carbon is an element found in varying proportions in all stars and in all parts of every planet in the solar system. Measured by the number of atoms present, carbon is the fourth most abundant element in the solar system, the twelfth most

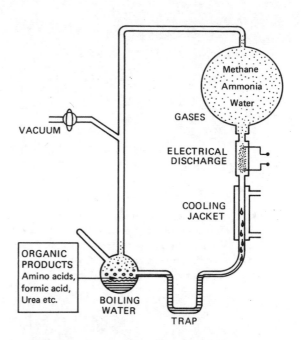

9.2 **Miller's experiment, in which amino acids and other organic compounds were synthesized from a mixture of gases. The organic products are collected in the trap for analysis.**

abundant in the Earth. But carbon compounds are not evenly distributed through these environments. Planetary evolution leads to different kinds of concentration in the various parts of each planet. The first two viable hypotheses which attempted to account for the origin of life considered the distribution of carbon. A. I. Oparin, nearly fifty years ago, postulated that an environment similar to that responsible for producing the organic compounds in carbonaceous meteorites (see Chapter 8) might have provided the materials from which life arose. A few years later J. B. S. Haldane suggested that organic compounds formed in the prebiological atmosphere would eventually find their way into the early oceans, which would form a hot, dilute soup of gradually increasing concentration. These two hypothetical environments for the origin of life are not at all the same. One suggests that the scene is set in the solid, crustal part of the planet; the other in the fluid environment of atmosphere and hydrosphere. But both have one feature in common. They postulate that the origin of life occurred in a non-oxidizing environment.

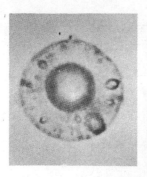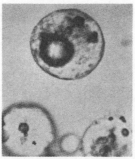

9.3 Left: a 'spurious' organized element from the Orgueil Meteorite, at first thought to be organic, but later found to be neither organic nor indigenous, but a particle of furnace ash (right) such as is common in the airborne dust of all big cities.

A great impetus to the study of the origin of life came in 1953 when S. L. Miller demonstrated experimentally that an energy source (such as an electric spark discharge, or ultraviolet radiation) applied to a mixture of gases and vapours including methane, ammonia and water vapour, synthesized a whole range of organic compounds. Miller's experiments have stimulated many others, and the accumulated results have convinced many people that the primitive atmosphere of the Earth must have been a *reducing* one in which carbon was present in the form of methane. The experiments have also been regarded as evidence that Haldane was right, and that Miller-type syntheses produced a gradually increasing concentration of organic compounds in the oceans. This primitive soup provided a nutrient 'broth' for the first living organisms which finally arose within it.

However, though no one doubts the validity of the experiments, the conclusions are now being seriously questioned. For example, P. H. Abelson contends that if methane had been dominant in the primitive atmosphere, this would have been recorded in sedimentary rocks of Precambrian age, which ought to contain an unusually large proportion of carbon. Moreover, L. G. Sillén claims that thermodynamic considerations show that the accumulation of organic compounds in the ocean could only have occurred at a concentration so dilute as to make the concept of a 'probiotic soup' an entire myth. In this chapter I am therefore going to explore other geological environments in which the concentration of organic compounds

derived from non-biological sources is greater. One of the most significant of these environments is that (noted by Oparin in his first paper) found in carbonaceous meteorites.

Meteorites do not, of course, represent a terrestrial environment. However, they certainly are products of what we have termed 'planetary evolution', and may, indeed, represent fragments of a planetary body. Carbon occurs in meteorites in several forms: as diamond, as graphite, combined with iron to form cohenite (Fe_3C), as carbonate, and in a diverse series of organic compounds. Although it is found in amounts up to 1% or so in most classes of meteorite, it is most abundant in those known as carbonaceous chondrites. These are particularly interesting in that they represent the least metamorphosed of all meteorites, and have retained more volatile compounds than any others. There are three kinds of carbonaceous chondrite (C_1, C_2 and C_3), and of these the C_1 have retained more carbon and more volatiles than the others. Carbonaceous chondrites have been considered the rarest of all kinds of meteorite, but this may be wrong, for although prior to 1965 only 28 had been identified, three have fallen since that time, and one of these weighed more than a ton.

Most of the carbon in C_1 carbonaceous chondrites is present in the form of organic compounds, but a small amount of inorganic carbon may occur as veinlets of carbonate, when it is associated with minerals which suggest deposition in the presence of liquid water. Organic carbon may be present in amounts of up to 5% in total weight, and about 80% of this organic material is insoluble and so difficult to extract, and even more difficult to analyse. It probably consists of high molecular-weight random polymers. The organic compounds which can be extracted by normal solvents are present in great variety. They include normal alkanes (paraffins) ranging from methane to dense paraffins with 31 carbon atoms, branched alkanes, cyclic alkanes, aromatics, benzoic acids, alcohols, fatty acids and sugars. Amino-acids, purines and porphyrins have all been recorded, and among the branched alkanes are the isoprenoids, pristane and phytane. If such an assemblage were found under normal terrestrial conditions it would be assumed that they were the product of biological activity, and this was suggested as a possibility by Bartholomew Nagy (now of the University of Arizona in Tucson) and his co-workers when they first discovered what a wide range of organic compound was present. When later Nagy discovered in one

C_1 carbonaceous chondrite (the Orgueil meteorite, which fell in France in 1864) a considerable variety of what looked like microfossils (but which Nagy cautiously named 'organized elements') the suggestion of a biological origin for the carbonaceous contents seemed much more reasonable. Nagy's findings were questioned vigorously by other workers, and more particularly by Edward Anders from Chicago. Anders showed that some of Nagy's organized elements were certainly terrestrial in origin, and that the Orgueil meteorite must therefore be regarded as contaminated. One of the contaminants was shown to be a distorted pollen grain of the American ragweed, and another a particle of fly-ash which is an abundant contributor to the dust in the atmosphere of most industrial cities. As a result many scientists lost interest in Nagy's discoveries, and thought that all the puzzling contents of carbonaceous chondrites could be explained away by invoking contamination. But this is very far from the case. Nagy's findings of organic compounds have now been confirmed by many independent laboratories, and although a few of the compounds present in very minute amounts are likely to be due to terrestrial contamination, the major part of the contents must certainly be indigenous, and this is agreed by all workers. But what is more puzzling, and more difficult to explain, is the confirmation that the majority of Nagy's organized elements are also indigenous. Those that have been demonstrated as contaminants are rare, aberrant forms. The two most common kinds are more or less spherical, or hexagonal. Martine Rossignol of Paris has confirmed by microprobe methods that the former are organic. (Microprobe methods use an electron beam to analyse extremely small areas, less than 10^{-7} cm^2, of solids.)

They occur abundantly, and there is no doubt that they are indigenous. The second kind form box-shaped hexagonal prisms. The wall of each hexagonal box bulges inwards, and looks as if it behaved plastically. Lois Nagy and Gerhard Kremp have undertaken detailed investigations. The size and the hexagonal shape suggest that they could be pseudomorphs of troilite crystals, but the wall-material may well be organic, and if so it would be difficult to account for the removal of the troilite from the interior. Kremp believes that they have probably resulted from a secondary distortion of organic shells that were once spherical.

It is important to realise that the evidence that some organized elements are both organic and indigenous to the meteorite is not in itself proof that they are the products of life processes. There are in fact three possibilities:

1. The organized elements are fossilized remains of life-forms, and the organic compounds are biogenic;

2. both the organized elements and the organic compounds are abiogenic;

3. the organized elements are fossil life-forms, but the organic compounds are mainly abiogenic.

Further, if the organized elements do represent fossil organisms this could be explained either by postulating that they record the previous existence of extra-terrestrial life, or by claiming that the carbonaceous chondrites once formed part of the Earth.

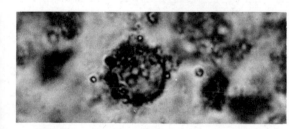

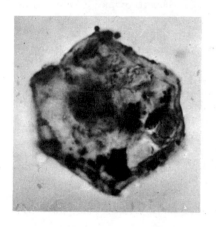

9.4 Two 'organized elements' from the Orgueil meteorite, almost certainly indigenous, probably organic, but unlikely to be biological. The spherical form is perhaps coating a 'microchondrule'; the hexagonal form is perhaps pseudomorphic after troilite.

Of these various possibilities the most credible would seem to be the second. This postulates that the organized elements are not life-forms, but are abiogenically-formed organic aggregates, in which case they are regarded as 'pre-organisms' or 'prebiological cellular aggregates'. This hypothesis also leaves us with the necessity of explaining the origin of all the organic material as abiogenic. If material of the same nature as that comprising the C_1 carbonaceous chondrites ever occurred on the primitive Earth, it would represent an aggregation of organic matter far exceeding in concentration anything that could be conceived to have formed as a dilute soup in the oceans. This is because most of the organic compounds in carbonaceous chondrites are insoluble in water; any such material discharged into the oceans would have floated, and would have been further concentrated by sedimentological processes along the shore-lines of the time. Ultimately, after incorporation in sedimentary rocks, it will have behaved like crude oil, and will have been concentrated in oil-traps.

THE PREBIOLOGICAL ORIGIN OF FOOD

Almost all present forms of life depend for food on the photosynthetic activity of green plants, which use the energy of sunlight to synthesize organic compounds from the carbon dioxide in the atmosphere. Photosynthesis depends on the action of enzymes which are themselves the products of life-processes. In prebiological times there were no enzymes and there could be no photosynthesis. The first life-forms must therefore have depended on food produced by abiogenic processes. If we are correct in postulating that this food had something of the nature of the organic compounds found in carbonaceous meteorites, then it would have resembled what would to-day be regarded more as fuel than as food. Yet even to-day there are some organisms that can use oil as food. Oilfields are all situated at varying depths below the surface of the Earth in sedimentary rocks. They may or may not be capped by a reservoir of gas, but below the gas comes a layer of oil, and below the oil a layer of brine. These three layers are developed within the pores of the rock that contains them. The brine contains a relatively sparse fauna and flora of organisms specially adapted to this hot, dark, subterranean environment. The main organisms present are bacteria, and these include sulphate-reducing bacteria and methane-forming bacteria.

Their combined action leads to the anaerobic decomposition of oil with the production of methane and nitrogen. The nature of the metabolic reactions involved is not yet understood, but there can be no doubt that the whole ecological assemblage in the brine depends for its ultimate food supply on the oil. That oil can be used as a food is also demonstrated by the exciting experiments that some oil companies have recently conducted. For example, BP have been able to grow yeasts on crude oil, and have so processed the crop as to provide commercial-type rations for pigs and poultry. These experiments may soon yield a new source of protein for our expanding human population. In any case it seems quite certain that before the evolution of photosynthesis the first organisms depended on their ability to decompose pre-existing stocks of abiogenically-produced organic compounds. In physiological terms, these organisms must have been anaerobic heterotrophs (i.e. they ingested food in an environment lacking oxygen). What we now need to examine is whether there is any evidence that carbon derived from the mantle can still be subjected to the processes postulated for the synthesis of prebiological food. What abiogenic processes affect juvenile carbon before it enters the biosphere?

JUVENILE CARBON

We speak of water as 'juvenile' when we want to distinguish water which has been derived directly from the mantle from that which has passed through the weather cycle. In the same way we can speak of juvenile carbon as that which has never entered the biological cycle. As with water, it is assumed that juvenile carbon is derived from the mantle, but I shall in this chapter be using the term in the rather special sense that carbon is juvenile even if it has had a complicated crustal history, providing this does not involve incorporation in any living organism. Using this terminology, I shall regard all prebiological carbon as juvenile, whether it occurs in core, mantle, crust, ocean or atmosphere. All compounds involving juvenile carbon must, of course, by this definition be abiogenic.

Although there is no direct evidence relating to either the abundance or nature of carbon in the mantle, E. Roedder has shown that fluid inclusions in the olivines and other minerals of basalts all contain minute quantities of liquid carbon dioxide. In so far as these basalts are derived from the mantle, it follows that the mantle is still making a contribution of juvenile carbon to the crust. Moreover, most theories of geological evolution maintain that the Earth has gradually differentiated to give its present layered structure. Once it consisted only of a homogeneous mantle. Gradually core, crust, hydrosphere and atmosphere originated as differentiation proceeded. This process of differentiation must have occurred at a much greater rate in pre-geological time than it does at the moment. The small amount of juvenile carbon contained in the fluid inclusions mentioned above is included in the basalts, and this is supplemented by carbon dioxide released as volcanic gas. To-day, basalt extrusion is largely confined to the oceanic ridges (see Chapters 16 and 21). Much of this same basalt returns to the mantle when the oceanic crust is forced back into it (see Chapter 19). But a proportion of this is regurgitated, together with other juvenile matter, in the form of andesite, and later further metamorphosed to give granite. During this process of regurgitation, a further supply of juvenile carbon escapes from the mantle. Most of this goes straight into the atmosphere as a component of volcanic gases; it then speedily enters the biological cycle, and thus loses its juvenile status. But this does not apply to all the carbon. Most granites are cut by late, low-temperature mineral veins, and quite often these contain carbon compounds as accessory minerals—both carbonates and organic compounds. The carbon involved seems (from isotope studies) to incorporate a mixture of juvenile and recycled carbon. The organic compounds may form sticky bitumens like those known in veins cutting the Mountsorrel granite in Leicestershire, or in some of the Cornish granite mines. Bitumens are also commonly associated with low-temperature hydrothermal minerals like the mercury ores, and juvenile carbon is mixed with recycled carbon in the carbon dioxide and methane given off by hot springs and fumaroles.

The contribution of juvenile carbon made to the crust is probably greater along the mobile belts

9.5 Oldoinyo Lengai, Tanzania — one of the active volcanoes of the East African rift valley, from which large quantities of carbon are emitted, either as CO_2 gas or in the form of sodium and calcium carbonate (lavas and ashes — 'carbonatites').

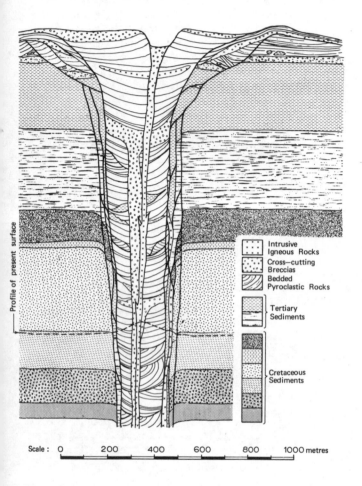

 (legend)

Intrusive Igneous Rocks

Cross—cutting Breccias

Bedded Pyroclastic Rocks

Tertiary Sediments

Cretaceous Sediments

Profile of present surface

Scale : 0   200   400   600   800   1000 metres

9.6 Idealized section through a diatreme, showing downward slumping and faulting of the vent walls, and successive infillings of volcanic debris (breccias and pyroclastics).

(such as 'the ring of fire' and ocean trench system around the Pacific) than it is along the ocean ridges. But it is even greater along a third fracture system —that involving the great intra-continental rift-fault systems typified by the rift-valleys of Africa. Here huge quantities of carbonate are intruded as 'carbonatite' (lava and ash largely composed of Na_2CO_3 and $CaCO_3$), active volcanoes belch out immense volumes of carbon dioxide, and *diatremes* are drilled through the crust and are responsible for such phenomena as diamond pipes and volcanic debris. Although most of the carbon is in oxidised form, bitumens do occur, and I. A. Petersil'ye has shown that in the Kola Peninsula of Russia and in Greenland both the hydrocarbons and bitumens involved are juvenile in origin. His most convincing evidence depends on the analysis of carbon isotopes.

CARBON ISOTOPES

There are three terrestrial naturally-occurring isotopes of carbon, with atomic weights of 12, 13 and 14 respectively. Of these, ^{12}C and ^{13}C are stable; ^{14}C is radioactive and has a half-life of about 5700 years. Lighter artificial isotopes are also known. Although the detection of the radio-active isotope ^{14}C is important as a means of dating archaeological material (back to about 70000 years before the present), it has no bearing on the juvenile status of carbon. The two stable isotopes always occur together, but ^{12}C is dominant; it is usually about 90 times more abundant than ^{13}C. The exact $^{12}C : ^{13}C$ ratio varies in carbon from different environments, and it is the cause of this variation that helps to determine whether a particular sample is juvenile or recycled. For example, the ratio for the carbon in the CO_2 of the atmosphere is 89.2 : 1, but it is rather less in marine limestones (88.6:1) and rather more in the carbon of organic compounds of the biosphere (91 : 1). In general it may be said that in all environments the carbon in reduced (organic) compounds is richer in ^{12}C than it is in oxidized (inorganic) compounds.

In order to understand this variation in isotopic ratio, it is necessary to postulate an initial condition in which the Earth received its carbon from an ancestral source. I have already suggested that the Earth was once more homogeneous than it is now, but in fact it seems unlikely that at such a time the carbon present was all of one kind and all possessing the same isotopic ratio, for extraterrestrial carbon shows a much greater range of isotopic ratio than does terrestrial carbon. For instance, it seems that in the Sun the $^{12}C : ^{13}C$ ratio is much higher than it is on the Earth. But in other stars this is not always the case. In the so-called 'carbon stars', the carbon:hydrogen ratio is greatly enhanced, and, in some of these, ^{13}C becomes dominant over ^{12}C. In meteorites, too, the variation is greater than it is in terrestrial rocks; inorganic compounds (carbonates) are much richer in ^{13}C, whereas organic compounds have about the same range as on the Earth.

The variation in isotopic ratio in all these different situations must arise in two ways. Firstly, carbon is formed in the interior of stars by more than one kind of nuclear process. Fractionation effects in these processes vary according to such environmental parameters as the mass of the star and the temperature of the reaction. Secondly, fractionation arises also during chemical reactions, for the two isotopes concerned have different atomic weights and therefore react at different rates. Every chemical reaction involving carbon must affect its isotopic ratio. These are known as kinetic effects. A different effect also occurs if equilibrium is approached, and fractionation can be produced by physical processes during diffusion. It therefore seems probable that in the primitive Earth juvenile carbon existed in several forms, each of which had evolved as the result of earlier chemical processes, and was therefore characterized by a different isotopic ratio. Four different kinds of juvenile carbon may well still occur in the mantle: diamond, graphite, carbon dioxide and methane. The isotopic ratio of each is likely to be different.

It is not easy to achieve very precise determinations of the absolute value of the $^{12}C : ^{13}C$ ratio by mass spectrometry. Much more accurate results can be obtained by measuring the difference in isotopic constitution between two samples. For this reason it is customary to compare isotopic constitutions not in the form of ratios, but in units which express the difference between the sample and a standard which (quite arbitrarily) is now almost universally accepted. This happens to be the inorganic, biogenic carbon of a belemnite (an extinct group of cephalopods) from the Cretaceous of America. It is known as the 'PDB standard'. The units are calculated in terms expressed (in parts per thousand) as δ (^{13}C) values, where

$$\delta \, (^{13}C) = 1000 \, \left(\frac{(^{13}C : ^{12}C) \text{ sample}}{(^{13}C : ^{12}C) \text{ standard}} - 1 \right)$$

If $\delta(^{13}C) = 0$, the sample has the same value as the standard. Consequently (as the standard is a marine fossil) marine limestones have a mean δ value approximating 0.

If the δ-value is positive, the sample has a higher $^{13}C:^{12}C$ ratio than the standard. Meteoritic carbonates have δ-values ranging from $+ 25‰$ to $+ 62‰$. They are enriched in ^{13}C.

If the δ-value is negative, the sample has a lower $^{13}C:^{12}C$ ratio than the standard. Most organic compounds have negative δ-values, averaging $- 13‰$. Atmospheric CO_2 is also negative ($- 7‰$) as is the carbon of diamonds (ranging from $- 4$ to $- 8‰$). All these environments have ratios in which ^{13}C is variously depleted when compared to the standard.

The way that biological processes can affect isotopic fractionation may at first seem rather simple, but is in fact very complicated and little understood. The main fractionation is believed to occur during the normal photosynthetic assimilation of CO_2 from the atmosphere. All plants show a depletion of ^{13}C as a result of this process; land plants have a δ (^{13}C) value ranging from $- 20‰$ to $- 30‰$, marine plants somewhat less ($- 7‰$ to $- 17‰$). Animals show δ-values which seem to depend on their food supply, and which therefore agree well with the plants in the same environment. But in detail this simple pattern is complicated by the discovery that an animal can display considerable variation within the different organs of its body. Most noticeable are the differences to be found in bone-carbon and shell-carbon, in both of which there is a considerable enrichment of ^{13}C. In land animals the bones have more ^{13}C than the food-plants, and the proteins have less.

Coal has much the same δ-value as wood, but crude oil differs somewhat from the biogenic carbon of sedimentary rocks, and tends to become more depleted in ^{13}C. Perhaps this is a result of a maturation process, or perhaps there is some fractionation during migration. It is possible that some crude oils contain an abiogenic component in which the δ-values are more negative than the biogenic component.

The isotope ratios of juvenile carbon vary more widely than biogenic carbon. Methane is particularly variable, but it is not often easy to tell juvenile from biogenic methane. That collected from hot springs and fumaroles probably contains a large juvenile element, and this may reach a ^{13}C depletion of as much as $- 63‰$. On the other hand some values are as low as those of atmospheric CO_2 (-7‰).

Inorganic juvenile carbon, such as that found in carbonatites and in hydrothermal mineral veins, have δ (^{13}C) values ranging from -4 to -8‰, and diamonds (almost certainly mantle-derived) fall into the same range.

Juvenile organic carbon seems to have a similar range to that of biological organic carbon. This is suggested by the range found in the organic carbon of meteorites (-18 to -28‰, much the same as in terrestrial plants) which is certainly abiogenic. This means that the isotope ratio cannot in most cases be used to discriminate between the juvenile and biological carbon of organic compounds. However, special cases occur in which anomalous ratios suggest peculiar conditions. For example, in the case dealt with by I. A. Petersil'ye which I quoted above, the abiogenic organic compounds (hydrocarbons and bitumens) are unusually rich in ^{13}C, with δ-values ranging from -3 to -13‰, which overlap the usual range of carbonates. Petersil'ye supposes that this result is due to a late-stage synthesis from mantle-derived carbon with a ratio comparable to that of magmatic diamond (say -8‰).

Other anomalous ratios may occur in the reverse direction, and can be explained in an analogous manner. Sometimes inorganic carbon has a ratio as depleted in ^{13}C as is normal in organic carbon. Ratios as low in ^{13}C as -56‰ have been quoted. These are explained by supposing they represent oxidized organic carbon.

CARBON IN EARLY PRECAMBRIAN ROCKS

The evidence of isotope ratios can therefore be used in only a few cases to distinguish between juvenile and recycled carbon, and no method has yet been devised for giving an estimate of how much carbon is juvenile and how much biogenic when samples of mixed origin are being examined. Nevertheless there can be no doubt that igneous rocks contain a proportion of juvenile carbon. Most of this must of course be of relatively recent origin, much more recent than the origin of life. This is because almost all igneous rocks are themselves more recent than the origin of life, and all the processes which we believe give rise to juvenile carbon are still active at the present time. There is nothing unique about them, and although it seems likely that in early Precambrian ('Katarchaean') time geologic processes were rather different from to-day's, it is not supposed that the rate of contribution of the mantle to the crust was any less than it is now. Quite the reverse. John Sutton believes that the mobile belts crossing continental crust were much more extensive. Almost the whole crust may have been mobile.

The oldest known rocks of the Earth are over 3000 million years old. They occur in two forms —high-crustal rocks of relatively low metamorphic grade, of which the least metamorphosed and best known form the Swaziland System of South Africa; and rocks from lower crustal levels that have suffered considerable metamorphism. B. F. Windley of Leicester believes that the latter are older and more primitive rocks than the former; he believes that they may represent remnants of a primitive continental crust, thinner than we have at the moment, characterized by intrusions of calcic anorthosite (plutonic rocks consisting almost wholly of the calcium feldspar *anorthite*). These anorthosites are petrologically similar to those found by the *Apollo 12* mission to the Moon and believed to form part of the lunar highlands. This is consistent with Windley's suggestion that they represent the first-formed continental crust. Both the Swaziland System and the lower level metamorphic complex contain carbonaceous rocks. The latter take the form of graphite schists, and have not yet been examined for organic contents. The former are being intensively investigated by several laboratories at the moment.

The two lower divisions of the Swaziland System have proved of particular interest, the Onverwacht Series at the bottom and the Fig Tree Series above. The former is dominantly volcanic (a thick sequence of ophiolites with subsidiary cherts, carbonates and pelites) whereas the Fig Tree Series is mainly sedimentary and dominantly fine-grained. *Both* series have yielded microfossils and both have been found after analysis to contain a wide variety of organic compounds, some extractable by organic solvents, the rest as insoluble kerogen.

The microfossils are very simple, and those in the Onverwacht Series show some resemblance to the 'pre-organisms' already noted as occurring in the Orgueil carbonaceous chondrite. Those in the Fig Tree Series are more varied and seem more highly developed. They contain both filamentous and globular structures, and the latter occur in more than one size-group. Chemical analysis also shows that the organic contents of the two series present some differences. A wide range of organic compounds is found in both, though most of them only in trace amounts. These include hydrocarbons and fatty acids. There is some difference in the organic chemistry of the two series, just as there is a difference in the fossils contained. For example, the ozonolysis products of the kerogens are aromatic in the Onverwacht, aliphatic in the Fig Tree. In short, it seems certain that life of a very primitive kind was already established in the Fig Tree series, and may have originated during Onverwacht times. If we are to discover any sedimentary rocks older than the Onverwacht, and therefore more likely to have formed in pre-biological times, we may have to rely on the graphitic schists cited by Windley, and these are still to be examined from this point of view. In any case, if life did originate on this planet, and did not migrate from some extra-terrestrial source, it must have done so in very early Katarchaean time, or even in pre-geological time. Moreover, if it did not originate from a very dilute oceanic soup, but from a more concentrated complex of high molecular-weight compounds like those which occur in carbonaceous chondrites and in terrestrial igneous rocks, perhaps we should look more closely at the sort of processes involved and the sort of environment implied.

CARBONACEOUS GASES AND FLUIDIZATION

Most of the carbon that escapes from the mantle does so as a component of volcanic gas. This is mainly in the form of carbon dioxide, though carbon monoxide, methane and higher molecular weight hydrocarbons are also included. Although carbon is the fourth most abundant element in the Sun, it has sunk to twelfth in abundance on the Earth. If, as we believe, the Earth and the Sun were formed at the same time from the same solar nebula, then clearly more carbon has escaped from the Earth than from the Sun. It has escaped because carbon compounds are volatile. It has escaped, in fact, in the form of gas. It is clear that the Earth underwent an outgassing period during its formation. The same outgassing mechanism is continued at the present time on a much smaller scale as part of the phenomenon of volcanism. Carbonaceous gases are always involved.

When gases escape through solid or liquid material they may do so either explosively or more gently. The nature of the physical reaction between gas, liquid and solid depends in part on the velocity of escape, in part on other parameters.

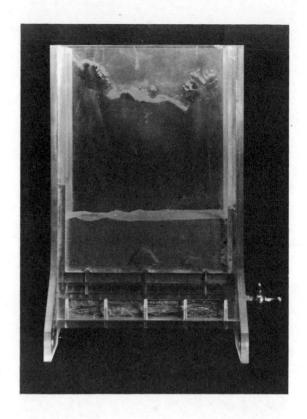

9.7 **Apparatus used to produce fluidized bed. Compressed air enters at the tap on the right and passes into the powdered rock via the porous floor of the container.**

If the solid material is particulate, it may be carried upwards in the stream of gas, and it then behaves as a fluid. Much use of this property is made in industrial chemistry, for a *fluidized* bed of solid particles possesses properties which lead to much more rapid catalytic reactions than occur with a fixed bed. Fluidization is a phenomenon that also occurs frequently during certain classes of volcanic eruption, and it must also have occurred during the outgassing episodes that accompanied the formation of the planetary bodies during the evolution of the solar system. A. A. Mills of the University of Leicester has concluded that fluidization must have played a part in the formation of the carbonaceous meteorites.

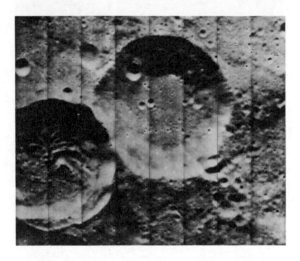

9.8 Above, artificial craters produced in fluidized bed apparatus by gradually turning off the compressed air supply. Note the great morphological similarity to lunar features (below).

In an attempt to reproduce in an experimental model the conditions that might have existed during an outgassing episode on the surface of a planet, Mills constructed a tank, part-filled with powdered rock, through which he passed gas under pressure. The gas fluidized the powder in the usual way, and an increase in pressure produced a 'boiling bed', a phenomenon well known to occur as a part of fluidization. When the pressure was turned off, the bed subsided, for during fluidization the volume of the fluidized powder

expands as the result of the gas it contains. One day Mills turned off his apparatus more slowly than usual, and was surprised to find that it did not subside in the usual way. The bed continued to 'boil' as the gas pressure was reduced, apparently exploiting channels through the gradually subsiding bed that had first been opened when the gas was escaping at full pressure. When the pressure was finally turned off, the whole surface of the bed was scarred with craters which bore an astonishing resemblance (on a small scale) to the cratered surface of the Moon or of Mars. It now seems highly probable that many of the lunar and martian craters are not the result of impact or explosion, but are fluidization scars formed quite gently as relics of outgassing episodes. Terrestrial craters are also produced by fluidization, and these, again, may be explosive or not. The well-known 'maars' of the Eifel district of Germany are believed by Arthur Holmes to be fluidization craters which have become flooded to form circular lakes.

In all these cases the gases involved include large quantities of water and carbon. Under extraterrestrial conditions, hydrogen would certainly have been dominant, and it is reasonable to suppose that on the Earth, during the course of geological time, the gas-mixture will have changed in composition from one that was highly reducing in nature (with carbon in the form of CH_4 or CO) to that at present characteristic of volcanoes (with carbon in the form of CO_2).

One of the phenomena that always accompanies fluidization is the building up of static electric charges. Mills has shown that this can result in glow discharges. In volcanic eruptions, lightning is always a spectacular phenomenon.

It seems likely that the carbonaceous contents of meteorites were synthesized during an outgassing episode involving either carbon monoxide or methane or both. Energy may have been provided by electric discharge. Catalysis may have been aided by the fluidization of particulate minerals. High temperatures are not necessary, but may have been involved. It is important to realise that carbonaceous meteorites consist of a mixture of high temperature and low temperature constituents and so must have had a complex history. George Mueller of the University of Miami has attempted to elucidate this history by a rather extensive petrological survey of carbon-rich meteorites. These included not only very representative samples of all three classes of carbonaceous chondrite, but also two other kinds of meteorite rich in carbon (the enstatite chondrites

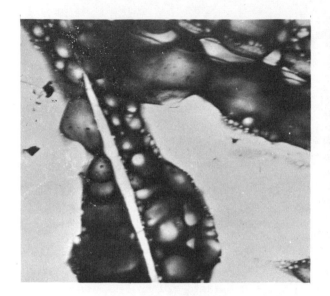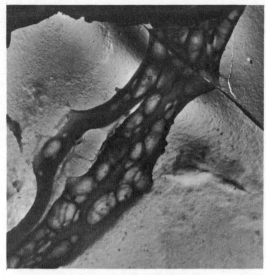

9. 9 'Precellular aggregates': left, microspheres (×3000) produced by A. E. Smith and his collaborators at Montreal, which are remarkably similar to the electron micrograph on the right showing material from the early Precambrian Bulawayo Formation in Rhodesia (×4000).

and the problematical diamond-rich ureilites, which are interpreted by Mueller as carbonaceous chondrites metamorphosed by impact shock). He also includes in his survey, for comparative purposes, a small number of ordinary chondrites. These are meteorites which contain spherical globules—*chondrules*—of minerals which look as if they may have crystallized from molten droplets (see Fig. 8. 1). He is able to show that the nature of the chondrules varies according to the percentage volatile content, and postulates that chondrules have evolved by condensation from a fine $(1 - 50\mu)$ incandescent microspray. The various stages in this series are preserved in the carbon-rich chondrites. The first stage, composed of *microchondrules* which preserve the form of the initial fine spray, is characteristic of Type I carbonaceous chondrites. In other carbonaceous chondrites stages are preserved in which the microchondrules coalesce and are sintered to produce chondrules, and these chondrules break up into fragments which are preserved in the ground mass. Mueller compares his hypothesis of chondrule formation with the way that rain droplets coalesce in a cloud. He concludes that the conditions which distinguish microchondrule from chondrule formation can be compared to those which lead to the formation of either a terrestrial fog or a rain-producing cloud. They are somewhat contrasted to those which lead to the formation of spray droplets of lava under volcanic conditions. From Mueller's hypothesis it would seem that the conditions which lead to the formation of carbonaceous chondrites on the one hand, or to ordinary chondrites on the other, are developed in two different regions of condensation. In one, the primary condensates were microchondrules which at the time of their accretion were already solid and therefore cool. With them were accumulated the volatile constituents, including water and carbon compounds. In the other region the microchondrules coalesce at the time of condensation, chondrule formation takes place, and although the chondrules themselves may be solid at the time of their accretion, volatiles are absent, and the temperature of the environment would be much higher than in the former case. This conclusion suggests that the meteorites were either formed in more than one parent body, or that a considerable degree of fractionation had already taken place in the protoplanet before it condensed into a solid body.

It is possible to test the validity of some of these ideas by experiment.

The various environmental parameters postulated for the origin of life can be simulated in the laboratory. Mention has already been made in an earlier paragraph of S. L. Miller's first successful experiment in 1953. Soon Miller's results were confirmed and extended by others. Indeed the variety of conditions and energy sources that have proved successful is almost embarrassing. Almost any mixture of gases that contains the requisite elements seems to yield organic compounds if subjected to almost any energy source, provided the mixture is reducing. These experiments have yielded an extremely wide range of organic compounds, including purines, pyrimidines, hydrocarbons, polypeptides and porphyrins.

It is clear from the experimental evidence that there were many pathways leading to organic synthesis during prebiological times. It seems probable that all of them have at one time or another been active, and that this activity has continued through the whole period during which the solar system has been evolving. Organic synthesis will have been active in the primitive solar nebula and will have continued as the planets became discrete bodies. It is certainly still continuing in certain environments at the present day. Its products will have become incorporated in the crust, oceans and atmosphere of the developing Earth, and there will have been subjected to the usual geological processes of concentration and dispersal.

The experiments which are of particular interest to the thesis of this chapter are those that bear on the formation of organic compounds in meteorites. Theoretical considerations stimulated R. Hayatsu, M. H. Studier, Edward Anders and their colleagues at Chicago to mount an experiment designed to test the possibility that these compounds were formed in the solar nebula at relatively high temperatures and low pressures, at conditions approaching equilibrium. The Chicago team used the Fischer-Tropsch process as a model for their experiment. This is an industrial process developed in Germany during the nineteen-twenties as a means of synthesizing petroleum from coal. Water-gas (hydrogen and carbon monoxide) derived from coal was passed over heated cobalt or iron catalysts. The carbon monoxide was hydrogenized to yield a complex mixture of hydrocarbons, with aliphatic alcohols, aldehydes and

ketones as by-products. Notably after the war a fluidized version of the Fischer-Tropsch process was developed in America on an industrial scale with a greatly increased yield. The Chicago experimenters used meteoritic iron as a catalyst, and in some experiments added ammonia to the reacting gases. The yield of nitrogenous organic compounds was dramatic, almost all those identified in carbonaceous chondrites being recognized. They included urea and a whole series of purines and pyrimidines, among them adenine, guanine, and cytosine, which are biologically important as nucleotide bases of DNA and RNA. John Oró of the University of Houston has repeated some of the Chicago work and has subjected the products he obtained to more rigorous analytical techniques. His work confirms the efficacy of the Fischer-Tropsch process as one method for accounting for many of the carbonaceous components of carbonaceous chondrites. Both the Chicago group and Oró were attempting to simulate partial equilibrium conditions as they might have existed in the solar nebula. However, other considerations make it unlikely that equilibrium was approached during the formation of the carbonaceous chondrites. T. Belsky and I. R. Kaplan, for example, have shown that the light hydrocarbon gas content and the carbon isotope ratios are incompatible with near-equilibrium conditions. It seems more probable that Fischer-Tropsch reactions were not initiated in the solar nebula, but during the outgassing phase of a parent planetary body in which conditions far from equilibrium were established.

It must not be supposed that the Fischer-Tropsch reaction is the only one likely to have yielded the sort of products characteristic of carbonaceous chondrites. Cyril Ponnamperuma and his colleagues in the exobiology division of NASA have long been experimenting with varying kinds of electric discharge through mixtures of gases designed to simulate possible stages in the evolution of the primitive atmosphere. They have produced a wide range of hydrocarbons and organic polymers. Electro-polymerization at low temperatures seems to be an important process accompanying such discharges. High-pressure high-temperature effects, particularly if followed by a rapid quench, seem to be equally important, and these are just the effects that are produced in a gas by the shock-wave (thunder) that accompanies spark discharge (lightning). Akiva Bar-Nun and his associates at Cornell University have concluded that thunder and the shock-waves that accompany the entry into the atmosphere of micro-meteorites may be the principal energy source for pre-biological synthesis on the Earth.

In a previous paragraph I concluded that carbonaceous meteorites have passed through a complex history. It now seems that their carbonaceous contents may not have been synthesized during just one phase of this history. More probably contributions have been made during a sequence of events involving several contrasting environments.

Carbonaceous chondrites are, of course, one of the products of the evolution of the solar system. But whether they ever formed part of any planet is somewhat controversial. Nevertheless it seems reasonable to suppose that the sort of processes which gave rise to their carbonaceous contents were also active at an early stage in the evolution of the primitive Earth. All the experimental evidence of prebiology points to the fact that the carbonaceous products were synthesized from gases. At the time of this synthesis, the meteorites must have had a gaseous atmosphere. If similar products were formed on the primitive Earth, it could only have been at a stage when either the whole atmosphere had the requisite composition, or pockets of it existed in volcanic areas. Before an attempt is made to understand the terrestrial history of carbon compounds, the evolutionary history of the Earth's atmosphere must be considered.

THE EVOLUTION OF THE ATMOSPHERE

If we are right in assuming that the Earth was formed by condensation processes from the material of the solar nebula, it must follow that all the material of the Earth, including our present atmosphere, must have ultimately been inherited from the solar nebula. We know a good deal about the composition of the solar nebula, and its gaseous contents differ very markedly from the present atmosphere of the Earth. The most noticeable difference is in the amount of hydrogen and helium, which are the commonest elements in the solar nebula but are much less common on the Earth. If the Earth once had an atmosphere of the same composition as the solar nebula, then it must have subsequently lost most of the hydrogen and helium. This is quite easy to explain, as these are the two lightest elements in the periodic table, and their velocity of escape is such that they could

not have been retained on the Earth even at temperatures much below present-day surface temperatures. But there are other discrepancies more difficult to explain. There are other gases of the solar nebula which are much depleted on the Earth. These include the two heavy inert gases krypton and xenon. The Earth should have been able to retain these gases in its atmosphere without difficulty—unless, of course, some catastrophic event stripped off the whole atmosphere, which is what is believed to have happened. In that case the whole primary atmosphere and hydrosphere will have disappeared. Only gradually, as volcanoes continued to outgas volatile products still trapped in the mantle, will a secondary atmosphere and ocean have replaced the primary envelope. The event which caused the loss of the primary atmosphere must have occurred fairly early in the history of the planet, presumably in pre-geological time (over 3600 million years ago). Otherwise we would expect to find some record of it in the rocks. It is possible that the event was connected with the acquisition of the Moon. W. H. McCrea has suggested that the Earth, the Moon and Mars may have all been formed at the same time, as the result of the break-up of a protoplanet. If McCrea is right, such an event must have occurred about 4600 million years ago, and will have been accompanied by the loss of the primary atmosphere from all three bodies.

The primary atmosphere will have been dominated by hydrogen, and will have been strongly reducing. The secondary atmosphere is certain to have been less strongly reducing. Much of the hydrogen present will have been combined in the form of water, ammonia or methane. Nevertheless it will have been very different from the present atmosphere, and free oxygen could only have been present in trace amounts. The sort of processes which gave rise to the organic contents of carbonaceous meteorites could equally well have taken place in the primary atmosphere or in the early secondary atmosphere.

Hydrogen will have continued to escape from the secondary atmosphere as geological time went on, and so there will have been a gradual change in composition. The most profound effect on the atmosphere, however, must have been the acquisition by newly-formed organisms of the power to extract carbon by photosynthesis. This will gradually have led to a change-over from reducing to oxidizing conditions (see Chapter 10).

We have seen that the abiogenic synthesis of organic compounds began with the very beginning of planetary evolution and has continued until the present moment. Some of the earliest products of such abiogenesis may conceivably still be retained within the mantle of the Earth, for it is likely that they survived the loss of the primary atmosphere, and their longevity is demonstrated by their retention in the carbonaceous chondrites. More probably such early products of the Earth, although inherited from the protoplanet in pregeological time, have since been absorbed as food by organisms evolved later.

Since the acquisition of a secondary atmosphere, abiogenesis has continued in three regions: in the atmosphere itself; in the oceans; and in crustal situations as a by-product of igneous phenomena. The products of atmospheric abiogenesis will have been transferred to the sea, there to await exploitation as food by the first organisms. Some of the igneous crustal organic material will also have found its way into the oceans, but a significant proportion of it (the hydrocarbons and bitumens, which are immiscible with water) will have floated and been transferred to other environments, e.g. to shore-lines and to subsurface crustal oil-pools. These environments all formed reservoirs of prebiological food. The most widespread source of food was in the ocean; the most concentrated source of food was in the crust.

Igneous phenomena often involve temperatures well above that of boiling water. For this reason some people have dismissed the possibility that volcanoes may contain sites for the origin of life. But this reasoning is unsound, for igneous phenomena are characterized by a temperature régime in which low temperature end-effects are associated with liquid water. S. W. Fox of the University of Miami has long championed the volcanic milieu as one particularly adapted to the abiological synthesis of the precellular structures he terms 'proteinoid microspheres'. He has shown that environments with a fluctuating temperature and water régime have particular significance, for if a solution of amino-acids is evaporated to dryness and subsequently quenched in water (as, for example, by rain) cell-like entities are automatically produced. We have already seen that volcanic eruptions are characterized by the emission of copious volumes of carbonaceous gas. Fluidiza-

tion phenomena often occur. Thunder, lightning and rain are invariably associated. Violent fluctuations of surface temperature accompany the eruption of lava. Bar-Nun's experiments show that thunder-shocks in an atmosphere of the right composition must invariably synthesize amino-acids. Fox experimentally simulated the surface environment of a volcano which is in eruption, and has subjected this to 'rain' containing dissolved amino-acids, which resulted in the formation of proteinoid microspheres.

Fox's experiments with proteinoid microspheres offer comparison with Oparin's experiments with coacervates. As shown, in synthesizing microspheres Fox uses as his starting point a mixture of dry amino-acids. These are heated in an oven to 170°C for several hours until co-polymerization converts the powder into a viscous amber-coloured liquid. This is then washed into a vessel with a 1% sodium-chloride solution. The resultant turbid liquid is rich in proteinoid microspheres. The microspheres, after mild buffer treatment, reveal boundary layers which may be double and which possess some of the properties of membranes. For example, they have the property of permitting molecules to pass through or to be retained selectively. Budding can occur under conditions of fluctuating temperature, and the buds, having formed, grow by heterotrophic accretion.

In these experiments, Fox has been careful to simulate an environment consistent with prebiological conditions. Oparin has for many years been experimenting with coacervates, which in shape and gross structure resemble Fox's microspheres, but which are formed from quite a different starting point. To form a coacervate, Oparin starts by mixing two proteins, or a protein and a polypeptide (e.g. albumen and gum arabic). The mixtures become turbid as the coacervate droplets separate out. These droplets seem to have very similar properties to those possessed by Fox's microspheres, but have a more heterogeneous internal organization. They can selectively adsorb substances (e.g. amino-acids) from the surrounding solution, and catalysts can be introduced in this way. Complex organic reactions can be set up within the droplet. Oparin has induced his droplets to convert glucose into starch, and starch into maltose, which is then returned to the external solution. He has even succeeded in feeding them with chlorophyll.

Fox has been able to show that proteinoid microspheres will—given the correct environment—come into existence spontaneously on innumerable occasions. If the correct foods are available, they can proliferate heterotrophically indefinitely, and enter into competition with each other. If, ultimately, nucleic acids become ingested by such microspheres, it is conceivable that a coding system of inheritance could become established, and this would mean that life had indeed arisen.

It is possible, then, that the origin of the cell antedates the origin of life. More cautiously, we can speak of 'precellular aggregates'. There is some evidence that these can occur as fossils. For example, A. E. Smith and a team of collaborators in Montreal have produced some microspheres which bear a strong resemblance to Precambrian microfossils from the Bulawayo Formation. We have already suggested in an earlier paragraph that the abundant organized elements in the Orgueil meteorite may well represent fossilized precellular aggregates.

It is clear that fluctuating conditions form an essential part of the environment necessary for the creation of precellular aggregates of the kind produced by Fox and Smith, and these are just the kind of conditions that are characteristic of a volcanic environment on the Earth, and which must also have been involved during the formation of carbonaceous chondrites. However, such environments may not have been particularly propitious for the final stage postulated for the origin of life, i.e. the establishment of the nucleic acid coding system. If precellular aggregates are to maintain their existence for a period long enough to allow competition and experiment, and the final invention of nucleic acid, we must envisage a continuing culture in a stable environment, rich in prebiological nutrients. We must invoke the transport of proteinoid microspheres from their place of origin in a volcanic or hydrothermal environment to a new 'breeding ground'. Two such suggest themselves: under clots of bituminous organic matter stranded on a muddy shoreline; and at the oil-water interface in a sub-surface abiogenic oil-pool.

SPONTANEOUS GENERATION

All theories for the origin of life agree that life must at one time have originated spontaneously. In the early 19th century it was still commonly supposed that life continued to originate in this way even at the present day. Such ideas were exploded by Louis Pasteur, who showed that spontaneous generation never occurred in an environment which had been properly sterilized. Nowadays it is usually assumed that although life must have arisen, perhaps repeatedly and not always successfully, from some non-biological sources in Precambrian time, this event can never have been repeated in later times. The main argument against the continued origin of life is that prebiological foodstuffs cannot be expected to accumulate as they would have done before life was evolved. No dilute prebiotic soup can continue to exist in an environment teeming with present-day life. It will be absorbed as soon as it arises.

In so far as this argument refers to the terrestrial environment, it seems unanswerable. Once life had become universally distributed over the face of the globe, it must have prevented the further generation of new life-forms. The only possibility of detecting even early stages in the process must lie in the exploration of environments quite isolated from the biosphere. Such situations are not likely to favour the full process of life-formation, but they may conceivably reach intermediate levels of organization such as those indicated by precellular structures. One conceivable situation of this kind is an environment in which water and liquid organic compounds are found trapped in fluid inclusions in hydrothermal minerals which form during the last stages of magma solidification. G. Mueller and K. A. Kvenvolden have been independently working on the nature of the organic contents of some of these inclusions; their results suggest that even here there has been contamination.

The spontaneous generation of life in extraterrestrial environments is, however, quite another matter, and it seems almost certain that life does exist on other planets in our galaxy. It may even exist on other planets within the solar system. And it may still be in the process of generation in more than one of them.

LIFE ON OTHER PLANETS

Meteorites give undeniable evidence that prebiological foodstuffs are produced in an extraterrestrial environment. The evidence of the organized elements shows that at least precellular organization had been reached in that environment as a stage towards spontaneous generation.

The materials brought back by the *Apollo* missions to the Moon offer no evidence whatsoever for the existence of lunar life, and very little evidence for the presence of organic compounds as complicated as those found in meteorites. This is a little surprising. Maybe later missions will discover that carbon compounds are distributed unevenly, as they must have been on the primitive Earth. Maybe concentrations of carbon do occur, but have not yet been discovered.

Carbon certainly occurs in fairly high concentration in the atmospheres of Mars, Venus and the larger planets Saturn and Jupiter. In the former two it is mainly present in the form of carbon dioxide; in the latter two, in the form of methane. We have seen that a reducing environment favours the synthesis of organic compounds. We might deduce, therefore, that extraterrestrial life is more likely to be found on Saturn and Jupiter than it is on Mars or Venus. There is a rather slender possibility that life may once have existed on Mars if we can assume that the present atmosphere is a relic of one which was once denser and richer in hydrogen and water. The same would apply (though with even more doubt) to Venus.

The possibility that life may exist on the two larger planets depends on the correctness of the hypothesis (championed, among others, by Carl Sagan) that, though the surface temperature of both planets is very low, there is a steady increase in temperature at lower levels in their atmospheres. This hypothesis supposes that there is a level below atmospheric surface warm enough to melt most of the components of the atmosphere (including water) which at higher levels only exist as ices. Experimental work by F. Woeller and Cyril Ponnamperuma of NASA has shown that a simulated Jovian atmosphere subjected to electric discharge synthesizes an organic complex including a translucent ruby-red polymer. They have suggested that such a product may account for the colour of a puzzling phenomenon to be detected in the atmosphere of the planet. This is the famous 'Great Red Spot', which may thus offer visible evidence for the existence of organic nutrients within the atmosphere of Jupiter. Perhaps life already exists in Jupiter somewhere beneath the Great Red Spot.

The suggestion that organic synthesis is almost certain to take place within the solar system wherever there is a combination of electrical energy and reduced carbonaceous gases opens up a further interesting possibility. Obviously prebiological food must be widely dispersed within environments that are not otherwise conducive to the origin of life. What is the possibility either that such foods may be transferred to a fertile environment, or that life may be transferred within the solar system to an environment already stocked with abiogenic foodstuffs?

PANSPERMIA

The hypothesis of 'panspermia', which postulates that life first came to the Earth from extraterrestrial sources, is an old one which has never found much favour, as it does not in any way explain how such organisms could have originated in the extraterrestrial environment. However, now that it is clear that abiological synthesis is likely to have occurred at a very early stage of planetary evolution, it becomes important to reexamine the possibility that life itself may have originated in one planet, and then have been transferred to others. Human activity itself has already infected the Moon with terrestrial life, and Mars and Venus will soon be similarly contaminated. Before long it is likely that plans will be made for growing crops on other planets. Did terrestrial life originate on some other planet? This is not a fanciful idea, as the Earth has demonstrably been infected by meteorites containing both exobiological food and what look like fossil extraterrestrial organisms.

None of these questions is easy to answer. A few years ago they could have been dismissed as idle speculations. As scientific hypotheses they would not have been respectable, for no one could conceive a method of testing them. But that is no longer true. Our advancing techniques of probing into space are demanding new ideas. They have furnished us with a whole battery of new devices which can refine speculation. What we need to transform speculation into theory is ingenuity; we need to use our new techniques for devising new tests. Theories on the origin of life have had to ignore the boundaries that once separated disciplines. Almost every discipline in science has become involved. New tests of new ideas are being applied at such a rate that it is quite difficult to keep abreast of the subject. Indeed, most of the ideas put forward in this chapter are likely to be superseded by others within only a short time.

9. 10 Oblique aerial photograph of the Pulvermaar (literally 'powder lake') in the Eifel district of Germany — a crater which some believe to have been produced by fluidization. The water-filled part is over 700 metres wide and 74 metres deep, and the Devonian rock rises for a further 60 metres. The lake has no inlets or outlets.

FURTHER READING:

P. C. SYLVESTER-BRADLEY 1971. Environmental Parameters for the Origin of Life. *Proc. Geol. Ass.*, v. 82.

J. KEOSIAN 1965. The Origin of Life. *Chapman and Hall.*

S. W. FOX (ed.) 1965. The Evolution of Pre-biological Systems. *Academic Press, New York.*

R. BUVET, C. PONNAMPERUMA AND E. H. SCHOF-FENIELS (ed.) 1971. Molecular Evolution. *North Holland Publishing Co., Amsterdam.*

M. G. RUTTEN 1971. The Origin of Life by Natural Causes. *Elsevier Publishing Co., Amsterdam.*

C. PONNAMPERUMA (ed.) 1972. Exobiology. *North Holland Publishing Co., Amsterdam.*

Sometimes the publication of a new theory results in a re-appraisal of generally accepted ideas in other fields of knowledge. The theory outlined in this chapter (first published in 1965) is one such case. In it, the late Lloyd Berkner and Lauriston Marshall, of the South West Center for Advanced Studies, Dallas, Texas, consider the effects of ultraviolet radiation on the Earth's primitive atmosphere, and on organic compounds, including DNA, which are accepted as being the necessary precursors to the development of life. From this they conclude that the content of oxygen in the Earth's atmosphere both influenced, and was caused by, the progressive development of life. The theory proved attractive to many geologists, who proceeded to revise the geological record along the lines suggested by Berkner and Marshall's theory for the build-up of oxygen in the atmosphere.

Since the publication of the theory, the timetable of events during the build-up of the oxygen content of the Earth's atmosphere has been modified, but the sequence has remained much the same. The most notable change is the view that free oxygen reached one per cent of the present level much earlier than suggested in this Chapter, namely about two thousand million years ago instead of about six hundred million years ago. Reference to the Berkner and Marshall theory is made in Chapters 11, 12 and 26, showing how it provided a valuable stimulus to investigations concerning the evolution of our atmosphere.

10 Oxygen and Evolution

by Lloyd V. Berkner and Lauriston C. Marshall

When the close connection is traced between the build-up of oxygen in the Earth's atmosphere on the one hand, and the environment and physiology of living organisms on the other, a model emerges that accounts for periods of explosive evolution. It may also explain subsequent catastrophes, such as that which overtook the great reptiles.

THE POSSIBILITY of direct exploration of the planets with space vehicles has directed attention to the constitutions of planetary atmospheres. Moreover it is only since the International Geophysical Year that we have a reasonably detailed picture of the structure of the Earth's atmosphere. The facts at hand suggest that analysis of data obtainable on the Earth, combined with new data from the planets, may be exceedingly useful in developing a broad theoretical basis for the rise and stability of planetary atmospheres, and related development of life.

Our present study concerns the history of the Earth's atmosphere—not so much what it is now, but how it got here, why it is, and what may have been the intermediate steps. As will be seen, these questions have significant bearing on the initiation and evolution of life on Earth.

One of the most interesting facts about the Earth's atmosphere is the relatively large content of oxygen—about 21 per cent—a condition unique among the planets of our Sun. It presents an interesting paradox. Life could not have been generated in the beginning with the present quantity of oxygen in the atmosphere—yet we cannot live and breathe without it. Most modern forms of life, even many plants, are highly dependent upon oxygen, and in most cases a rather precise amount of it, for their existence. This means that oxygen in the atmosphere must have been relatively stable in its concentration for a considerable time. Yet the very origin of life on Earth would have been frustrated if oxygen had been abundant at that time. The amino acids and other vital but vulnerable materials needed for incorporation in the incipient organisms would have been promptly oxidized.

Until recently, scholars have not been able to agree fully on how a stable atmosphere came into being with the present amount of oxygen. On the one hand there is a very useful principle propounded about 150 years ago by Hutton which has found wide acceptance—it is called *uniformitarianism*. This explains geological and evolutionary variety by the slow changes we observe under the conditions encountered today. On this view, all developments have occurred at almost imperceptible rates compared with human events and concepts of time without the need to postulate at any time significant changes in the environment. But, as we shall see, a modern interpretation of the geological record can show sudden bursts of evolutionary activity that are hard to explain unless we allow for decisive changes in the environment.

Thus, we observe that multi-cellular forms of life in the sea came into being only at the opening of the Palaeozoic—that is to say, 600 million years ago. Plant life on dry land has existed only about 420 million years (preceded for a few million years by some fossil spores) and animal life on land perhaps about as long. The only evidence of any life before the Palaeozoic consists of certain bacteria, and other very primitive organisms. Yet such algal material is found by Hoering and by Abelson to be trapped in the oldest rocks in curious formations known as bioherms, some dating back nearly 3000 million years. The Gunflint shales of the Great Lakes, laid down nearly 2000 million years ago, teem with the fossils of single-celled organisms, and the deposition of certain ores may have required bacterial action in their chemical deposition. But there is no evidence that life existing in the waters progressed during these aeons of time, to even a modest degree of sophistication, until about 600 million years ago.

The Earth was probably formed from small objects —planetesimals or meteorites—which did not bring with them any primordial atmosphere. The birth of its atmosphere was a secondary phenomenon, arising after the Earth's formation from the escape of gases from its rocks, usually by localized heating, melting and volcanic action. As the primaeval atmosphere appeared, it would be dominated by hydrogen, water vapour and carbon dioxide, and subsequently modified by gradual escape of the lighter gases (hydrogen and helium) and by the absorption and precipitation of carbon dioxide in carbonates as the oceans became widespread.

At present there are about 500 volcanoes classified as active around the world; about 400 are situated in the Pacific ring. Wilson estimates that slightly less than a cubic mile a year of solids is added to the continents at the current rate of activity. Multiplying this number by the life-span of the Earth one has a figure of about 3 000 million cubic miles, which is very close to the present estimates of the total volume of all the continents.

Accompanying this solid effluent, the volume of volcano gases is considerable. The largest single constituent is water vapour, which can be quite as high as 97 per cent by volume. This is mixed with varying amounts of nitrogen, carbon dioxide, hydrogen, sulphur dioxide and chlorine, plus much smaller quantities of hydrogen sulphide, carbon monoxide, methane, ammonia and others. All these volcanic vapours and gases have been bound in the rocks, to be released, sometimes explosively, in the throat of the volcanoes.

The water vapour and gases released over 3 000 or 4 000 million years are thought to be more than sufficient to account for the volume of the oceans, and for the nitrogen and other constituents of the present atmosphere, excepting only oxygen. But this exception is particularly important. There is no free oxygen produced by volcanic activity. Volcanic temperatures and the presence of iron, sulphur and other materials in chemically reduced state lead to violent chemical reactions that ensure that all oxygen comes out in combined form.

Thus oxygen was virtually absent from the primitive atmosphere. This view is confirmed in other ways. Ancient sediments formed by erosion in those early days are known to be only partially oxidized, suggesting an atmosphere low in oxygen. More recently, Holland has concluded that the early atmosphere must have been chemically reducing in character (that is, dominated by hydrogen). The origin of life itself appears to forbid the existence of much oxygen at the beginning, as we observed earlier in posing the paradox.

Since the evidence suggests an absence of oxygen in the primitive atmosphere, the immediate question is how to account for a build-up of oxygen and a resolution of the paradox. The free oxygen must be derived chiefly from the break-up of water molecules into hydrogen and oxygen. It can occur via one of two routes: by direct photo-dissociation of water vapour exposed to ultraviolet light (primarily in the wave-band 1 500 to 2 100 angstroms), and by the indirect process that occurs during photosynthesis in green plants. On a primitive planet without life, only the first alternative is available. How much oxygen will this photo-dissociation yield?

We have made a study of the ultraviolet content of the radiation from the Sun. Observations of rockets and satellites have shown the amounts of ultraviolet light from the Sun that are reaching the atmosphere at the effective wavelengths. It comes from the visible photosphere of the Sun, and has been shown by Wilson, in his study of stars similar to the Sun, to be sensibly constant over geological time.

Of the products of photo-dissociation, the hydrogen is light enough to escape rather quickly from the atmosphere (the exact rate is still under debate). Nor would the nascent oxygen formed by photo-dissociation stay around long in the primitive atmosphere, where it was produced at the surface. It would react very quickly with partially oxidized rocks and materials. In the atmosphere, some of the oxygen forms ozone— reactive molecules containing three atoms of oxygen. Ozone itself reacts at a high rate with surface rocks. Thus, in a primitive atmosphere, oxygen is used up at a definable rate in surface oxidation; and possibly also by reconstitution of water vapour at certain levels in the atmosphere.

Simple calculation demonstrates that ultraviolet energy could release oxygen by photo-dissociation of water vapour which is more than 100 times that necessary to account for the oxidation of all the sedimentary rocks in the continents and on the floor of the oceans over geological time. But when we ask whether photo-dissociation can account for the present high level of oxygen in the atmosphere itself, the answer is no—not by a factor of more than a thousand.

Dr Harold Urey has pointed out that the mechanism becomes self-regulatory because the water vapour is shaded from the solar ultraviolet by absorption by the very oxygen that is so produced. We call this limiting process the 'Urey effect'. The detailed calculations allow but a slight build-up of oxygen in the atmosphere.

Water vapour is mainly limited to the lower atmosphere, since it is readily precipitated as rain from the colder regions above. As the water vapour is broken down by ultraviolet light to provide a supply of free oxygen to the atmosphere, this same oxygen is distributed above the principal water vapour levels. It turns out that this oxygen absorbs the very wavelengths that are needed for the continued breakdown of the water vapour. When oxygen reaches a certain very small concentration it simply cuts off its own further production.

The ultimate concentration of oxygen in the primitive atmosphere is determined by the rate of loss which, at equilibrium, must be balanced by the supply. This balance can be calculated and identified on Figure 10. 1. We find that the concentration of oxygen in the primitive atmosphere did not exceed about 0.1 per cent of the present oxygen concentration. We suspect that the same processes control the present oxygen content of the atmosphere of Mars as seen by *Mariner IV*.

In our present atmosphere we are shielded from lethal ultraviolet by very small amounts of ozone, principally at a height of about 50 km. The total quantity of ozone amounts only to about 0.05 to 0.07 ppm of the atmosphere below 50 km, yet it is sufficient to protect life from lethal sunburn.

But in the primitive atmosphere, when oxygen was less than one-thousandth of its present levels, ozone was produced very close to the surface. The protective ozone layer was very thin—not enough to protect incipient life at the surface from the lethal ultraviolet glare of the Sun at the surface. Our calculations, illustrated in Figure 10. 2, show that the combined effect of absorption of ultraviolet by the oxygen and ozone in the primitive atmosphere, plus a layer of water, provides conditions favourable to the synthesis of living molecules and promotion of the simplest forms of life processes in water at depths of about 10 to 13 m —but not on, or even near the surface. Carl

Sagan has independently reached a substantially similar conclusion.

Thus we can imagine organic compounds synthesized through the energy of ultraviolet light at the surface and gently convected downward away from ultraviolet damage into regions where more complex organisms could be synthesized in the presence of visible non-lethal radiation. This primitive ecology calls for shallow pools with a depth of 10 m or more, sufficient to shadow the deadly ultraviolet, but not so deep as to cut off too much of the visible light. The model also calls for convection currents in the water, too slight to sweep the primitive organisms up from the bottom, yet sufficient to carry organic nutrients downward from the surface. At this early stage, life in the oceans seems very improbable—the waves and currents of the oceans would circulate the early organisms to the surface or disperse them downward into darkness. Therefore, they could exist only in lakes and protected shallow seas as bottom-dwelling organisms at a depth of 10 to 13 metres.

In looking at Figure 10. 2, we notice a very curious coincidence. In the primitive atmosphere the greatest protection occurs around 2 600 angstroms. The location of this wavelength-zone of maximum protection is a fact of our solar system, depending only on the distribution with wavelength of ultraviolet radiation from the Sun, and the absorption characteristics of ozone and water. Now this is the very region where present-day nucleic acids (2 630 angstroms) and proteins (2 750 angstroms) are most sensitive to damage. This might suggest a selective advantage to nucleic acids and proteins synthesized with such absorption characteristics. We have not calculated how much another Sun with quite a different radiation pattern might shift this protective zone, nor do we know whether other nucleic acids or proteins with different absorption characteristics can exist, but this point ought to be kept in mind in projecting life to other solar systems.

We are still faced with the problem of accounting for the further build-up of oxygen to present levels capable of supporting higher life forms, and of protecting them from fatal radiation in oceans and on the land. The rise of oxygen can only have occurred through a process of photosynthesis whereby oxygen is released from water during the reduction of carbon dioxide into carbohydrate. The crucial point is that photosynthesis employs visible light, whose intensity is not curtailed by the presence of oxygen or ozone in the atmosphere. The Urey effect does not limit its production.

The first cell life possibly incorporating photosynthesis appears to have been identified at about 2 700 million years ago. The subsequent oxygen balance would be determined by:

+ *generating mechanisms*

    (1) photo-dissociation (Urey effect limits)

    (2) photosynthesis (Urey effect does not limit)

— *removal mechanisms*

    (1) oxidation of surface materials

    (2) biological processes, including decay and respiration

    (3) dissolving of oxygen in water

    (4) some re-association of hydrogen and oxygen in the atmosphere

At first, the effect of photosynthesis would not seriously disturb the primitive balance at 0.1 per cent of present atmospheric level. Eventually, however, as the areas covered by photosynthetic release of oxygen enlarged, atmospheric accumulation of significant amounts of oxygen would begin. For this to happen, oxygen must be supplied faster than it is dissociated and lost through its active forms, permitting us to estimate that the pools must have covered an equivalent of somewhat more than 1 per cent of present land areas in order to provide enough photosynthetic activity and consequent oxygen-production to overpower the Urey effect.

From this, one can form a graphic picture of how continental areas must have looked. The lands were barren, with no vegetation, but possessing bodies of water in which the photosynthetic activity took place at their shallow bottoms. Anyone who has flown over Northern Canada and Labrador can readily visualize how much of the Earth's surface must have appeared.

The mechanisms available so far have provided the early organisms with only limited amounts of energy. As oxygen began to increase slowly, to a point about ten times that which existed at the early stages, or 1 per cent of the present oxygen level, the possibility of energy-supply to organisms through *respiration* begins to appear. This mechanism provides 30 to 50 times the energy per molecule made available to organisms through more elementary chemical processes of fermentation. In many primitive organisms, the changeover from fermentation to respiration occurs when oxygen reaches about 1 per cent of its present concentration in the atmosphere. Pasteur pointed out this effect nearly a century ago during his study of the spoilage of wines, so it is known as the 'Pasteur effect'.

Respiration creates altogether new evolutionary opportunities—for a circulatory system to convey the oxygen, a digestive system to employ it, a nervous system to control the process—in short, for complex multi-celled organisms with advanced mechanisms of energy capture, control and utilization. Respiration has always been recognized as a major evolutionary development. What has been overlooked is that the physical opportunity for respiration is the key that unlocks the whole catalogue of advanced biological function and development. Only as oxygen concentrations permit respiration is there created an evolutionary demand for complex biological function to maximize its energy capture, with the consequent employment of that energy for the advantage of the organisms.

The increase in oxygen in the atmosphere to about 1 per cent of its present level is automatically accompanied by a calculable increase in the total ozone content. When we consider the absorption of ultraviolet in oxygen, ozone and water (Figure 10. 2) we find that the lethal rays of ultraviolet are cut off at depths of only 30 cm or so of water.

The level of oxygen in the atmosphere at 1 per cent of the present amount is referred to as the *first critical level*. Life is still confined to the water but now can survive at very shallow depths. The spread of life to the oceans is permitted since the danger is limited only to the very surface of the waters and more advanced organisms can now evolve which have some control over their situation in that environment. Indeed, as one reads Hardy's charming account of the unexplained diurnal depth-control of pelagic organisms, found so generally in the seas, it is suggestive of a protective response inherited from these very times.

So we ask, was there a sudden evolutionary spurt at any time in geological history that would correspond to this vast new evolutionary opportunity that was provided when oxygen reached 1 per cent of its present concentration? The answer

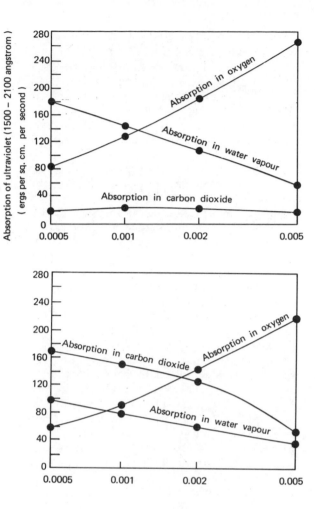

10. 1 The horizontal scale of all graphs on this page shows oxygen concentrations as fractions of present atmospheric level; above, on the assumption that the prevailing levels of water vapour and carbon dioxide were the same as the present levels; below, on the assumption that carbon dioxide was ten times more abundant than at present. Oxygen was produced by photo-dissociation of water vapour by absorbed ultraviolet but, at a certain level, oxygen's absorption of ultraviolet impeded its own production. In both diagrams the absorption in oxygen begins to exceed that in water vapour at about 0.001 of the present atmospheric level of oxygen.

10. 2 The depth of penetration into water of solar ultraviolet of different wavelengths, after penetrating an atmosphere containing increasing amounts of oxygen. (Penetration is the depth at which the intensity falls to less than $1\ erg\ cm^{-2}\ s^{-1}$ per 50 angstrom waveband.) Note that the depth of penetration becomes zero over most of the range at about 0.01 of the present oxygen level.

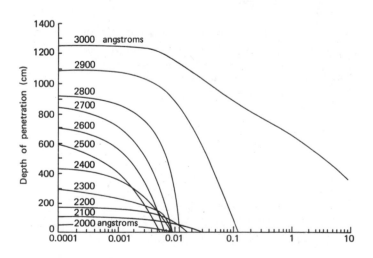

is yes—just one—at the beginning of the Palaeo-zoic era, 600 million years ago. Before this time there is no evidence of organisms more advanced than the most primitive of very ancient times. Within a few million years, with the opening of the Cambrian, there is widespread evidence for more advanced life forms. In a period of the order of 20 million years life branched out in such a variety of forms in the oceans, rivers and lakes as to lay the foundation for all modern forms. At least 1 200 new species of different life forms appeared during the ensuing Cambrian period in North American geological horizons alone. Some of these were quite large. One sees, in fossil beds of the Cambrian, skeletons of trilobites and large shell-fish comparable to modern scallops or abalone, some as much as one metre in diameter.

Following this evolutionary explosion, marine photosynthesis increased the total to a much higher rate, further accelerating the build-up of oxygen in the atmosphere. This was accompanied by an increase in the depth of the ozone layer, the sole shield in the Earth's atmosphere against lethal ultraviolet in the band 2 100 to 3 000 angstroms. At the oxygen levels of the early Palaeozoic, the lethal glare of solar ultraviolet would still wither any organism that tried to raise its head above the water. Continually life reached for the open air to maximize its energy for photosynthesis, but continually it was prevented from doing so by the burning radiation. As oxygen built up, living processes could take place at shallower and shallower depths until (as we see from Figure 10. 2) oxygen had reached a level of about 10 per cent of the present atmospheric concentration. Then life could finally exist at or near the surface of the water.

The 10 per cent point we have defined in our study as the *second critical level*. Now, for the hardier forms of life at least, the thickening depth of atmospheric ozone opens the opportunity to evolve ashore. Newly evolving organisms can spread in great profusion over the land surface, resulting in a further acceleration in the processes of photosynthesis and respiration. Can we identify this era in time?

The geological evidence shows such an explosion of life on land occurred about the time of the late Silurian—that is, 400 to 420 million years ago. Before that, in the mid-Silurian, there are spores—the first evidence that plants were growing above the level of the waters in protected localities. Then, very quickly, by the late Silurian, many classes of both plants and animals had moved ashore, and before long (early Devonian, 380 million years ago) great forests had appeared. As dense plant life spread over most of the land surface, it added to its own protection from ultraviolet light by the rapid build-up of oxygen by its photosynthetic action. With the accompanying increase in the thickness of the protective ozone, more sensitive forms of life began to appear on land. By the end of the Devonian period—350 million years ago—many forms of life pre-existing in the water had projected their evolutionary counterparts on land, including the amphibians and insects.

The oxygen content of the atmosphere may, in fact, have built up to a level *exceeding* our present atmospheric content by the time of the Carboniferous. This represents an overswing, with photosynthesis generating large quantities of oxygen and consuming comparable amounts of carbon dioxide. We can speculate that such an overswing would eventually correct itself for, with the reduction of carbon dioxide, the Earth would cool. The carbon dioxide absorbs the infrared radiation returned toward space by the Earth, and keeps that heat from being lost from the Earth, thus providing a 'greenhouse' effect. Diminution of this effect due to reduction of carbon dioxide during the Carboniferous might have led to the extensive ice ages of the Permian period, over 200 million years ago. Photosynthesis would fall sharply as organisms were constrained by the changed climate, leading to an abrupt, catastrophic reduction in oxygen concentration. One would like to know whether those great reptiles of the Cretaceous attempted to acquire a greater lung capacity before they disappeared suddenly from both land and sea.

This reasoning suggests that for the past 300 million years or so both oxygen and carbon dioxide may have fluctuated about the present levels in a series of saw-toothed oscillations, dropping suddenly to perhaps the 10 per cent level, and again rebuilding slowly. Perhaps a more critical study of the geological evidence will provide more precise estimates of these fluctuations.

A summary of the rise of atmospheric oxygen derived from these studies appears in Figure 12. 3, which relates the oxygen concentration to the geological column.

Let us, in conclusion, consider the implications of these events in historic geology and palaeontology coinciding with the rise of oxygen, particularly as related to rates of evolutionary modification and to planetary atmospheres generally.

Evolution during any geological period must be interpreted as a complex interaction between the level of oxygen generated by living organisms and the way in which that level produces new opportunities for evolution. Our thesis is that, from time to time, critical stages in the level of oxygen are reached, affording vast opportunities that are immediately seized by physiological responses of organisms. Scholars in the past have assumed that very long periods of time are required to account for the tremendous evolutionary rise from simple microscopic organisms to the large and highly organized forms of life which appeared suddenly about 600 million years ago. That no fossils of such developmental precursors have been found after a century of almost microscopic search of relatively undisturbed Precambrian sediments has been considered as a major 'geological puzzle.'

According to the present model, no such precursors over long periods should be expected; indeed, they are *forbidden* until the new opportunity arises from the increase in oxygen.

Thus the geological record must be read exactly as found, with no interpolations or suppositions. Preceding the Cambrian, levels of oxygen were insufficient to permit a smooth progression in evolution. The abrupt opening of the whole oceans to life, 600 million years ago, as a consequence of sufficient shadowing of the lethal ultraviolet, at *first critical level*, or 1 per cent level, seems to coincide with the implementation of respiration. The opening of the dry land areas to land-dwelling plants and animals at oxygen levels which shadowed the lethal ultraviolet at the *second critical level* (10 per cent) represents a similar climatological jump about 180 million years later. This, too, was followed by an immediate evolu-

tionary response by many groups of plants and animals in extending their counterparts ashore.

Conventional evolutionary thinking is strained by the speed and complexity of these great evolutionary leaps. Evidently, evolution can proceed as rapidly as combination, selection and adaptation permit. If our interpretation is correct, a considerable step can be taken in quantifying evolutionary rates.

With respect to planets generally, we are forced to conclude that atmospheric oxygen in considerable quantities (at least 1 per cent of the present terrestrial atmospheric abundance) would be *prima facie* evidence of advanced forms of living organisms. Lesser quantities of oxygen restrict organisms to primitive unicellular form, and then even such simple life forms can exist only provided other essential environmental requirements can be satisfied—such as an appropriate temperature range and availability of sufficient quantities of a suitable solvent. Mars falls into the latter category.

FURTHER READING:

A fuller treatment of the subject of this article can be found in the following papers by the authors:

'On the Origin and Rise of Oxygen Concentration in the Earth's Atmosphere', *Journal of Atmospheric Science*, Vol. 22, pp. 225–261 (1965).

'History of Major Atmospheric Components', *Proceedings of the National Academy of Science*, Vol. 53, pp. 1169–1226 (1965).

From the previous two chapters it is evident that the evolution of an atmosphere and of life on the Earth are intimately linked. In this Chapter, Professor Preston Cloud of the University of California, Santa Barbara, discusses some other events and processes which may have occurred during the earlier part of the Earth's history. He takes up the story from Chapters 3 and 7, and gives a geologist's viewpoint (written in 1969) of the Berkner and Marshall theory. In Chapter 26, he discusses the oxygen cycle together with Professor Aharon Gibor.

11 The Primitive Earth

by PRESTON CLOUD JR.

The Earth's atmosphere and hydrosphere probably arose as the result of processes which occurred after the formation of the solar system. If the Moon was captured by the Earth, the accompanying catastrophic events may have caused the initial outgassing; subsequently both living organisms and iron ores appear to have played leading rôles.

IT IS THE PREVALENT view nowadays that the Earth and solar system originated by the gravitational contraction of a cold cloud of dust and gas. A close study of the ensuing evolution of the Earth's atmosphere and oceans reveals, however, that neither can be directly attributed to this primary process. To read this ancient and extended span of Earth history the geologist must make use of several highly diverse lines of evidence, and consider the interactions between living organisms, and geochemical and geophysical systems. When he tackles the problem in this way a coherent picture of events on the primitive Earth begins to emerge.

One of the most surprising chemical features of the terrestrial atmosphere is its great depletion in the noble gases as contrasted with their cosmic abundances. This fact alone seems to require that our atmosphere be of secondary origin. Either the Earth originated without an atmosphere, or it lost such an atmosphere in a later thermal episode. Nevertheless, the continuous existence of sedimentary rocks dating back for more than 3 000 million years (three aeons), demonstrates the continuity of our atmosphere and hydrosphere over a period of at least that length.

At the same time, the existence of detrital grains of readily oxidized minerals in stream deposits only two aeons old means that the atmosphere contained no free oxygen as recently as that. The puzzle is further complicated by the presence of primary ferric oxide in marine deposits of the same age, which means that there *was* a source of oxygen within the large water bodies from which these deposits precipitated. Ferric-oxide-coated detrital sediments ('red beds') of non-marine origin as old as 1.8 aeons, moreover, indicate that

by then free oxygen was accumulating in the atmosphere; and the subsequent evolution of life and sediments hints at later steps in atmospheric evolution. From such fragile threads we can weave a strong, if coarse-textured, tapestry of events on the primitive Earth.

HOW AND WHEN THE ATMOSPHERE BEGAN

In fact, Earth scientists proposed an internal source for the terrestrial atmosphere long before anyone recognized the depletion of the noble gases. And they suggested that it arose from outgassing and weathering following the accumulation of the primordial rocks. But the compelling force of the evidence was not widely appreciated until 1951, when the American geologist W. W. Rubey published an incisive assessment of possible sources for the atmosphere and hydrosphere. Now the concept of accumulation of both from juvenile sources is generally accepted, and discussion focuses on the composition and time of origin of the primitive atmosphere and the amounts and changes in it of oxygen, nitrogen, carbon dioxide, and hydrogen—especially free oxygen, for which no primary source is available.

As to the time of origin of the atmosphere, an outside limit is the age of the Earth. Radiometric data show that the materials of which the Earth is comprised originated about 4.55 aeons ago, at the time of origin of the solar system as a whole, but they do not tell us exactly when these materials aggregated to form the Earth. A minimal age for that event is, of course, given by the most ancient terrestrial minerals dated. This age appears to be around 3.5 to 3.6 aeons. That, at least, is the age of the oldest radiogenic dates obtained on four

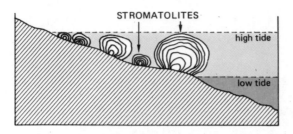

STROMATOLITES

high tide

low tide

11. 1   The relation of domal stromatolites to the intertidal zone.

intensively studied continents, and it seems that some very significant historical event must have taken place at about this time.   It also highlights the yet unsolved problem of what was happening during the first aeon or so of Earth history, assuming that accumulation of the planetary materials was completed not long after the big homogenization event 4.55 aeons ago.

The point of interest here, however, is that, inasmuch as we do not have rocks older than about 3.6 aeons, there are few constraints on conjecture about the nature and origin of an atmosphere or hydrosphere that might have antedated that time.

SOME CONSTRAINTS ON CONJECTURE

The oldest rocks that tell us something definitive about the early hydrosphere and atmosphere are sedimentary rocks in southern Africa believed to be somewhat more than 3.2 aeons old.   They could not have originated in the absence of atmospheric weathering and a substantial hydrosphere. They also limit conjecture about the nature of the atmosphere beneath which they accumulated.   In addition, the easily oxidized minerals uraninite and pyrite of detrital origin in various younger rocks combine with other evidence to indicate that the atmosphere between about 3.2 and 1.8 aeons ago could have contained little or no free oxygen. Other geochemical evidence implies that, contrary to the popular methane-ammonia model of the primitive atmosphere, there could have been little ammonia or methane in the atmosphere from about 3.2 aeons onward.   The early atmospheric

gases instead would have been those that are trapped in igneous rocks, or are juvenile components of volcanic and hot-spring gases—H_2O, CO_2, CO, N_2, HCl, H, S, and a few other trace gases.

In order for an atmosphere and hydrosphere to begin, the outer Earth would have to undergo melting sufficient to release its 'volatiles'.   Such melting might be produced by the combined effect of radiogenic heating, tidal friction, and the conversion of gravitational energy to heat.   Again, it might be the product of lunar capture, abetted by these same factors.

If, among the older geologic records, some provided clues to tidal amplitudes, we could test whether or not it was likely that the Moon was in orbit at any particular time—say as far back as 3.6 aeons ago.   Such geologic evidence does exist, although not older than about 3.2 aeons.   It includes dome-shaped sedimentary structures of algal origin known as stromatolites, plus other sediments indicating intertidal deposition.

Domal stromatolites that rise conspicuously above the surface on which they grow are found only between tides, where, among recent forms, they reach a maximum known relief of 0.7 metre. The West Australian geologist Brian Logan found that height (or amplitude of stromatolite layering) was determined by tidal range and position in the intertidal zone as in Figure 11. 1.   Fossil stromatolites can similarly give a clue to tidal ranges in the geologic past.

Now it happens that stromatolites of pre-Palaeozoic age often have much greater amplitudes than younger ones (2.5 to 6 times as great) and cover large areas.

The implication one may draw from this and other data is that tidal amplitudes too great to be accounted for by the solar component alone existed at least two aeons ago; that tides generally were probably greater then than now; and that the Moon was, therefore, already in orbit and closer to the Earth between 2.0 and 0.6 aeons ago than it is now (see Chapter 7).

The evidence of the more than 3.2-aeon-old Swaziland System of south-eastern Africa is also consistent with the Moon being already in orbit around the Earth that long ago.   Rocks at the top of this system have the characteristics of extensive intertidal deposits, implying tidal amplitudes too great for the solar component alone.

If the Moon was in orbit 3.2 or more aeons ago, and if it did not originate at the same time as the Earth, perhaps its origin had something to do with

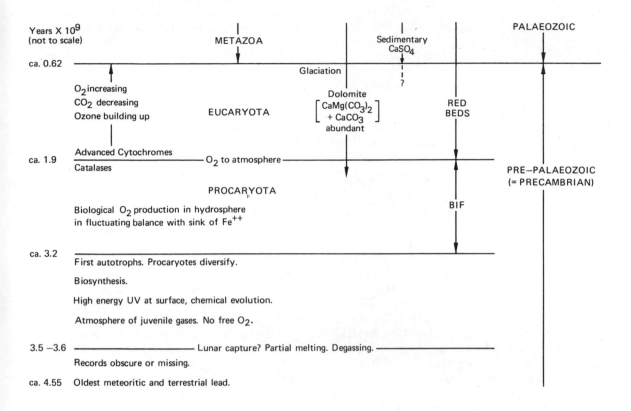

Years X 10^9
(not to scale)

ca. 0.62	
	O_2 increasing
	CO_2 decreasing
	Ozone building up
	Advanced Cytochromes
ca. 1.9	Catalases
	Biological O_2 production in hydrosphere in fluctuating balance with sink of Fe^{++}
ca. 3.2	First autotrophs. Procaryotes diversify.
	Biosynthesis.
	High energy UV at surface, chemical evolution.
	Atmosphere of juvenile gases. No free O_2.
3.5 –3.6	Lunar capture? Partial melting. Degassing.
	Records obscure or missing.
ca. 4.55	Oldest meteoritic and terrestrial lead.

METAZOA · Sedimentary $CaSO_4$ · PALAEOZOIC

Glaciation

EUCARYOTA · Dolomite $\left[\begin{array}{c} CaMg(CO_3)_2 \\ + CaCO_3 \end{array}\right]$ abundant · RED BEDS

O_2 to atmosphere

PROCARYOTA · PRE–PALAEOZOIC (= PRECAMBRIAN)

BIF

11. 2   Biospheric, lithospheric and atmospheric evolution on the primitive Earth.

the 3.5 to 3.6-aeon thermal event implied by the oldest known metamorphic and granitic rocks and by lead-isotope data (Fig. 11. 2).

If the Moon was captured, tidal friction sufficient to induce subcrustal melting would have been likely. Melting, in turn, would have promoted out-gassing and accretion of atmosphere and hydrosphere, together with a general resetting of geologic clocks. Any pre-existing terrestrial atmosphere and hydrosphere would have been lost at that time, and a new or first atmosphere and hydrosphere started. It is also conceivable that the postulated lunar capture and accompanying thermal episode could have given rise to a temporary lunar atmosphere and hydrosphere.

CHEMISTRY OF ROCKS AND FOSSILS

Thus our atmosphere and hydrosphere may have evolved, with additions, from one that began with pervasive melting about 3.5 to 3.6 aeons ago (Fig. 11. 2). To account for microfossils more than 3.2 aeons old, life must have originated soon after. Such life would have been anaerobic in the absence of oxygen, and dependent on external food sources. However, it could not have continued, giving rise to the observed evolutionary record without the appearance of an organism that could manufacture its own substance—an autotroph, probably a photosynthetic autotroph. And photosynthesis is the most likely process by which free oxygen might be generated in quantities sufficient to produce an oxygenous atmosphere.

In the absence of suitable oxygen-mediating enzymes, however, free oxygen is lethal to all forms of life! When oxygen-releasing photosynthesizers arose, therefore, they would have faced the problem of disposing of oxygen in such a way as not to burn themselves up. Unless oxygen-mediating enzymes preceded the origin of oxygen-releasing photosynthesizers, such organisms would have been dependent on an associated oxygen acceptor in the physical environment. This is where the ancient haematitic iron ores and red beds enter the story.

Ferrous ions whose oxidation produced the world-wide haematitic banded iron formation (BIF) between about 3.2 and 1.8 aeons ago may have been the oxygen acceptor. Nothing like the BIF—a feature of the old rocks on every continent—of any thickness or regional extent is found in younger rocks. The geochemical problem posed by the BIF is how to transport the iron in solution, under oxidizing conditions, or to precipitate it under anoxidizing conditions. This problem is resolved by the concept of a balanced relationship between organisms and BIF. The iron would be transported in solution in the ferrous state and precipitated as ferric iron on combining with biological oxygen. The banding suggests a fluctuating balance between oxygen-producing biotas and supply of ferrous ion.

The Earth's first biologically generated oxygen thus was locked in chemical sediments, and free oxygen did not appear in the atmosphere except in small quantities from photolytic dissociation of H_2O and CO_2, rapidly scavenged by the reduced substances then abundant in the atmosphere and at the surface of the Earth.

When efficient oxygen-mediating enzymes did arise, this balance would have collapsed. Primitive algae equipped with such enzymes could spread as widely through the hydrosphere as light penetration and ultraviolet shielding mechanisms would permit. They would have swept the ocean free of ferrous ions, and oxygen would accumulate in excess and begin to escape to the atmosphere. The last great episode of BIF, about 1.8 to 2.0 aeons ago, may mark such an event.

What would have happened when oxygen began to build up in the atmosphere? At that time, in the absence of an ozone screen, solar high-energy ultraviolet light would have been able to reach the surface of the Earth. Some of the molecular oxygen (O_2) would have been converted to atomic oxygen (O) and ozone (O_3). Iron would have been retained in the weathering profile of the Earth in the ferric state. Because of the great chemical activity of O and O_3, even a low rate of transfer of O_2 to the atmosphere would cause extensive oxidation of surface materials. Red beds should appear in abundance in the geological column at that time. The oldest thick and extensive red beds now known are about 1.8 to 2.0 aeons old—a little younger than, or overlapping slightly with, the youngest BIF. This date may mark the time in atmospheric evolution when free oxygen began to accumulate.

The appearance of atmospheric oxygen would also set the stage for the emergence of a new type of organism. Palaeontological evidence implies that until then all organisms consisted of procaryotic cells (those lacking a nuclear wall and being incapable of mitotic cell division and genetic exchange as a normal accompaniment to reproduction). The presence of free oxygen, even in small quantities, was presumably followed by the evolution of the eucaryotic cell, with nuclear wall, well-defined chromosomes, mitotic cell division, and the capacity for sexual reproduction and genetic recombination as the usual mode of replication.

How fast did oxygen accumulate in the atmosphere once it started, and what were its biological consequences? At first the green plant photosynthesizers would still be confined to protected sites in sedimentary mats, or where they would not be circulated into surface waters, until such time as an efficient ozone screen built up to exclude DNA-inactivating radiation in the neighbourhood of 2 600 angstroms. The late Lloyd Berkner and his colleague L. C. Marshall found that this happens at about one per cent present atmospheric level (PAL) of oxygen (see Chapter 10). Both they and (earlier) the Canadian biologist J. R. Nursall suggested that the appearance of differentiated multicellular animal life (Metazoa) was a consequence of the achievement of atmospheric oxygen concentrations sufficient to support a metazoan level of oxidative metabolism.

The appearance of the Metazoa in the geologic record obviously has two necessary, if not sufficient, preconditions. One is the origin of the eucaryotic cell, of which all Metazoans are constituted. The other is a sufficient level of free oxygen—although perhaps closer to three than to one per cent PAL.

Now the oldest rocks in which eucaryotic fossils are known are about 1.2 to 1.4 aeons old, although eucaryotes may have made their debut before this. The precondition of the eucaryotic cell, therefore, was satisfied well before the dawn of the Palaeozoic about 570 million years ago; and I have elsewhere documented the conclusion that there are as yet no records of unequivocal Metazoa in rocks of undoubted pre-Palaeozoic age. This suggests that sufficient free oxygen may have been triggering the event.

The apparent abruptness of early Metazoan evolution may be partially explained by a polyphyletic origin—a wave of multicellular forms derived from different possible Metazoan ancestors almost simultaneously. Since, moreover, all ecologic niches that could ever be occupied by Metazoa were then unoccupied, adaptive radiation probably contributed to diversification of the Metazoan root stocks. This biological revolution may have taken place over an interval of say 100 million years—roughly equivalent to the time indicated for chemical evolution leading to the origin of life itself, and somewhat more than that required for the Cenozoic diversifications of the mammals following extinction of the dinosaurs.

What evidence, other than the geologically rapid evolution of the earliest Metazoa at this time, might suggest that the dawn of the Palaeozoic approximately coincided with the appearance of a level of free oxygen adequate for Metazoan metabolism?

When ozone reached a level sufficient to cut off the DNA-inactivating ultraviolet radiation at about one per cent PAL of oxygen, it would open up the surface waters of the entire hydrosphere to occupation by photosynthesizing phytoplankton. That, then, could generate a large increase in the amount of oxygen in the atmosphere. A sudden big increase in atmospheric oxygen might correlate with several features in lithospheric evolution. The increase in oxygen would presumably have been paralleled by a decrease in the carbon-dioxide blanket serving to reflect heat radiation. The result could have been a temperature decrease sufficient (other conditions being suitable) to account for the widespread late pre-Palaeozoic glacial deposits recognized by many geologists. Such an oxygen increase would also be consistent with the observed abrupt increase in abundance of sedimentary calcium sulphate in basal Palaeozoic rocks and an episode of oxidative enrichment of the BIF in late pre-Palaeozoic or earliest Palaeozoic time.

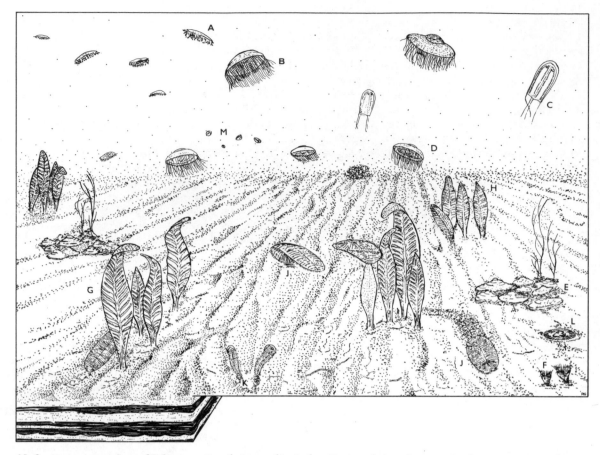

12. 0 Reconstruction of Ediacara, South Australia, in late Precambrian times.   Coelenterata: A (siphonophore), Eoporpita; B, Ediacara; C, Kimberella; D, Brachina; E, Cyclomedusa; F, Conomedusites; G, H (pennatulids), Arborea, Rangea.   Annelida: J, Dickinsonia; K, Spriggina. Echinodermata: L, Tribrachidium.   Arthropoda: M, Parvancorina.   The densities of distribution are speculative.

At the present time it is impossible to separate sedimentary from biological processes.   The very existence of organisms influences the rate and type of natural processes occurring around us, which in turn govern the character of their products such as sedimentary rocks.   Organisms trap and stabilise sedimentary grains, physically sort material during their feeding activities, accelerate the chemical and mechanical breakdown of rocks, and aid rock formation by contributing their hard parts.   A description of some sedimentary environments is given in Chapter 13, and the evolution of life is discussed in Chapter 14.   Here, Dr. Roland Goldring of the Department of Geology, University of Reading, describes some examples of where these realms meet, and how the environments they influence have changed through time.

12 Evolution in Environments

by ROLAND GOLDRING

Concern is mounting over Man's increasing influence upon his environment, and its repercussions. This chapter illustrates how other living creatures have had major effects on their surroundings.

THE RECORD OF LIFE for the past 600 million years is extraordinarily rich, especially for animals and plants which had mineralised parts, and particularly so if we consider the chances involved in the burial, physical and chemical modification and subsequent survival of organic remains over such a vast period of time. We would of course like to know much more about the soft-bodied forms which must at all times have formed the bulk of organic matter. Of these, such as jelly-fish and worms, we have only occasional glimpses due to unusual preservation. We do have, however, the evidence from activity traces (*trace fossils*) where soft-bodied as well as skeletoned organisms have left impressions of movement on and within sediments, and *chemical fossils*, organic molecules and isotopic fractionation effects associated with actual body fossils or disseminated in the rock.

Sylvester-Bradley in Chapter 9 discusses the origin of organic matter and life, and describes primitive life forms over 3 000 million years old. Metazoans and metaphytes have been present for only one-fifth of this period, and have been exerting an influence on Earth's environments which can to-day be most clearly seen in Man's activities. At the present time, the extent to which Man is rapidly changing the natural environments on the surface of the Earth is all too familiar. We can state with some certainty that Man's activities over the past few hundred years, stemming from his advanced state of psychological evolution, mark the beginnings of a major revolution in the pattern and type of environments present on the Earth's surface. However, many of the major events in the evolution of life during the past 3 500 million years have triggered off revolutionary changes in Earth's environments just as important; so that the Earth's natural environments may be considered as having evolved, though not at the rate of the present revolution.

The acquisition of photosynthetic ability was probably the most important event after that of the actual appearance of life, since all animals are ultimately dependent on plants. As has been discussed in Chapter 10, it seems likely that most of the free oxygen was gradually acquired as a direct result of photosynthesis. When the atmospheric oxygen reached 1% of its present level, aerobic respiration became possible, and the 10% level has been shown by Rhoads and Morse to correspond to the level of dissolved oxygen at which calcareous skeletons can be secreted and retained. This must have occurred close to the Cambrian-Precambrian boundary about 600 million years ago.

Before discussing further the effect which organisms have on their environment it may be useful to look briefly at some of the early life communities. The later fossil record is covered in Chapter 14.

Our evidence of procaryote and eucaryote life in the Precambrian is mainly from primary silicified stromatolites. Stromatolites are as frequent and are as much a feature of Precambrian rocks as they are of younger rocks. The oldest known are from limestones in the 2 600 million years old Bulawayan Group of Rhodesia (Fig. 12. 1). The organic origin of the structures is deduced mainly from their gross morphology. Unfortunately, no micro-organisms have been detected, because the limestones have been recrystallised. Associated cherts, possibly secondary, have also failed to yield micro-organisms. However, isotope analyses of the carbonate and organic carbon indicate enrichment of ^{12}C commensurate with photosynthetic isotopic fractionation.

In comparison, the silicified stromatolites of the 900 million years old Bitter Spring Formation of Central Australia have yielded a flora as diverse as can be found on algal crusts in shallow warm-water areas, such as the margin of the Persian Gulf, to-day. Filamentous, non-mitosing blue-green algae, observed in thin sections of the Bitter Springs chert, are shown in Figure 12. 2. The chert also yields algal forms that are planktonic. The organisms can be readily freed from the chert by hydrofluoric acid maceration.

The form *Eotetrahedron* (Fig. 12. 2c) appears to represent a tetrahedral association of algal spores suggesting derivation by meiotic cleavage of a eucaryote cell. Mitosis and meiosis mark two further evolutionary steps towards a metazoan, and the Australian cherts contain the oldest evidence of these. Over half the algal taxa recognised are similar at the generic or specific level to modern algae. Indeed, the overall organization and species composition is comparable to modern laminated algal communities. We can conclude that these cyanophytes had reached a peak of diversity by Bitter Springs time and have subsequently remained evolutionarily conservative.

Apart from some rather doubtful structures, one of the earliest pieces of evidence for a definite metazoan is a short track (Fig. 12. 3) in sediments 2 000 m below the top of the Precambrian in South Australia. The bilaterally symmetrical ridged track suggests that a molluscan-like animal was responsible. At this stage in Earth history, trace fossils are more likely to be preserved than body fossils; though in sediments of such age there is often considerable difficulty in establishing an organic origin at all. Indeed, it is not surprising that so many *pseudofossils* have been described from Precambrian rocks. Abundance, chemical analysis and symmetry cannot be used to establish an organic origin for a structure, the general form of which may be strongly suggestive of a metazoan. The best procedure at present is to show that the structure cannot reasonably be interpreted as having an inorganic origin. For this, the whole 'fossil' must be collected—part and counterpart. (For instance, sectioning the jellyfish-like *Aspidella terranovica* from the Precambrian at St. John's, Newfoundland, generally shows it to be a water- or gas-escape structure.)

The first body fossil metazoans occur at no great distance beneath Cambrian rocks. At Ediacara in South Australia and at several sections in the adjoining Flinders Ranges, a rich fauna of 34 species (including the trace fossils) has been found. The semi-desert climate, low dips and extensive outcrop have aided exploitation of the locality, now a Reserve. Elements of the same fauna have been found in South Africa, England, the Ukraine and Siberia and, except for the English locality, are at a similar stratigraphic level. The fauna is entirely unmineralised except for spicules in the pennatulids (sea-pens) (Fig. 14. 8).

A reconstruction of the Ediacara fauna is shown in Figure 12. 0. Figure 12. 4 shows an illustration of one of the most common fossils, *Dickinsonia*, a flatfish-like polychaete of which four species are known. This annelid had extensive fusion of the anterior body segments like the modern *Spinther*. Some specimens show evidence of partial decay with edges that had become frayed before final burial. As Professor Glaessner and Dr. Mary Wade have shown, it is absolutely essential with all these fossils to understand fully the preservation before making any palaeobiological analysis. The integument of *Dickinsonia* and the other annelid *Spriggina* were relatively strong, and most specimens are represented by external moulds. In contrast, the much less resistant jellyfish show a great variety of preservation, particularly as casts and composite moulds. Jellyfish dominate the fossilised fauna, and include floating forms and mobile forms like *Kimberella*, which shows some resemblance to the living, voracious sea-wasps. Others like *Eoporpita* had adopted a sessile mode of life at the water-air interface, aided by a chondrophore-type float. This is a niche which we would expect to have been occupied early in metazoan history, since the shallow surface waters would have been better oxygenated. In contrast, the stress of the intertidal zone would have made this habitat relatively difficult to occupy. Two species of arthropod and a possible echinoderm have also been recognised. Sponges, a group which would be expected because of their simple organisation, are only doubtfully noted. Six types of trace fossil are known, none of which can be attributed to any of the body fossils. Their mode of behaviour is of a relatively simple type, though one shows a grazing pattern. Most of the body fossils are not in a life position (whereas the trace fossils of course are) but they were obviously not transported far and were probably introduced and smothered by thin sand layers deposited during storms.

The first metazoans marked the real beginning of food webs and it is not unreasonable to attempt to reconstruct a food web (Fig. 12. 6) for the Ediacara fauna. What is remarkable, and so different from modern systems, is that virtually all the fauna were primary consumers: either substrate feeders, or suspension feeders on a microplankton that must have been overwhelmingly

12. 1 Stromatolithic limestone in the Bulawayan Group, Huntsman Limestone Quarries, Rhodesia.

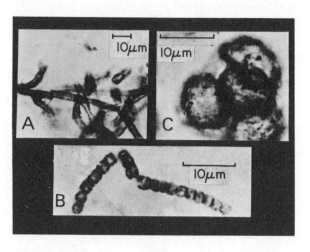

12. 2 Micro-organisms from the Bitter Springs Formation: A, fungal-like filaments; B, cyanophyte filament; C, eucaryote, tetrahedral tetrad.

12. 3 Track (slightly reduced) of late Precambrian metazoan from Brachina Formation, Bunyeroo Gorge, South Australia.

12. 4 Latex casts from dorsal moulds of two specimens of Dickinsonia costata, a very common annelid, and of Spriggina floundersi, a particularly convincing annelid, from Ediacara (× 1.5).

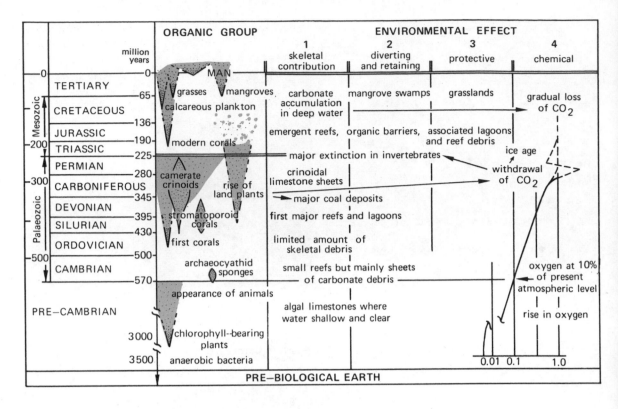

12.5　The effect of some groups of organisms on their environments.

phytoplankton. There is no evidence for an intermediate zooplankton. It seems unlikely that predators were limited to some of the medusae, and it is reasonable to suggest that animals such as nemerteans were also present (*phantom fossils*).

The Ediacara fauna scarcely affected the chemical balance of the Earth, and only slightly modified the look of some sediments. Infaunal animals often rework (bioturbate) the sediment, and finer grains may be re-suspended or brought to the substrate surface during excremental processes. Some Phanerozoic sediments should look quite different from those deposited in Precambrian times in a similar situation and under the same physical conditions.

The major control exercised by organisms on the Earth's environments is clearly chemical. Beside the release of oxygen into the atmosphere (above) the volume of free carbon dioxide has varied over biological time. The prolific increase in land plants during the Carboniferous (some 300–350 million years ago) and the formation of major coal deposits must have led to an immense removal of carbon dioxide from the

atmosphere. The increase in oceanic alkalinity that followed may have contributed to the major faunal extinctions about 225 million years ago.

There are three other ways in which organisms significantly affect sedimentary environments.

First, during the Phanerozoic, organisms contributed a vast bulk of skeletal material to sedimentary rocks. This has varied with time according to the type of organisms present (Fig. 12.5). The Chalk of north-west Europe is often 99% calcium carbonate of organic origin. In the Isle of Wight it is 600 m thick. The equivalent amount of sediment in Ordovician times, before the evolution of the minute calcareous plants whose skeletons make up so much of the Chalk, would have been only a few decimetres of claystone. In the Lower Cambrian, fossil archaeocyathids (Fig. 12.7) with a stout calcitic skeleton account for considerable thicknesses of limestones in Australia and Siberia. Their porous skeletons probably functioned like those of sponges. The group had become extinct by the Middle Cambrian, and their ecological niche does not appear to have been immediately filled.

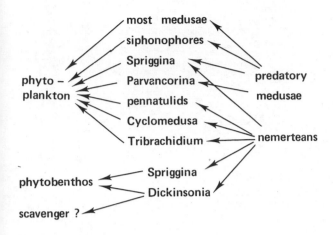

most medusae
siphonophores
Spriggina
phyto–
plankton
Parvancorina
predatory
medusae
pennatulids
Cyclomedusa
Tribrachidium
nemerteans
phytobenthos
Spriggina
Dickinsonia
scavenger ?

12. 6   Food web of Ediacaran sandy substrate, shallow marine environment.

12. 7   Block of limestone (× 1.5) with skeletal remains of fossil archaeocyathids, from Lower Cambrian, Ajax Mine, Flinders Range, South Australia, with reconstruction of group.

Second, organisms influence the erosion rate, as is all too obvious when the vegetative covering to the soil is removed. In pre-grass times (pre-Tertiary) the bad-land effect must have been much more widespread, with erosion proceeding at a greater rate than at present. Conversely, some of the boring organisms, notably certain molluscs and the sponge *Clione*, are effective erosional agents.

Third, organisms disturb the normal distribution of sediments by diverting and retaining materials. Lagoons are an interesting feature of tropical coastlines, but they owe their presence to the organically constructed reef. Coral reefs of modern type are a relatively recent evolutionary invention, and it is doubtful whether organic bioherms of the Palaeozoic rose above wave base. Also, the salt-tolerant mangroves have probably changed the whole appearance of post-Mesozoic tropical deltas, by trapping sediment between their prop-roots and thus controlling the movement of tidal waters. At all latitudes various organisms, particularly algae and polychaete worms, are able to create sediment baffles leading to the deposition and stabilization of sediment, which would otherwise be hydrodynamically mobile.

All these effects must act very slowly and with many fluctuations, so that only by careful analysis of each factor involved in a particular situation can it be possible to forecast large-scale environmental changes. In predicting for the future it is wise to take a backward glance at the evolution of environments over the past 1 000 or so million years, and to analyse the environmental effects of past biological revolutions.

FURTHER READING:

E. J. KORMONDY 1969. Concepts of Ecology. *Prentice-Hall.*
L. F. LAPORTE 1968. Ancient Environments. *Prentice-Hall.*

13. 1   A restoration of a Carboniferous 'coal' forest from scientific data.

The apparent permanence of our natural surroundings is relative.   Viewed from a life-span of three score years and ten there is a comforting constancy in the landscape around us, in the alternation of day and night, winter and summer, and the progression of life through the seasons.   The local valleys and hills remain much the same as when we were children, except where urban and industrial development have intervened to change both scenery and fauna.   But are such scenes static?   Consider subtle changes which may not immediately be obvious—rivers carrying sand and boulders from hills to sea, or rainfall washing away soil.   If in one lifetime a few tons of rock are removed, the effect is hardly discernible to a human observer, but over a period of tens of thousands or millions of years the effect on a land-scape will be considerable. Given *time*, mountain chains will be reduced to monotonous plains, and marine areas become dry land.

Although the human observer is limited in time, observation of present-day processes such as the movement of sand in rivers is the basis for the interpre-tation of sedimentary rocks.   These rocks are the products of past processes which sculptured ancient landscapes, and therefore they contain clues to the nature of past conditions on the Earth's surface.   It is this kind of detective work which forms the subject of this chapter.   In it, Professor E. K. Walton of the University of St. Andrews describes some of the clues which can be used to piece together a description of past environments.   Such results form the basis for constructing a detailed history of the Earth.

13 Looking back through Time

by E. K. WALTON

The environments in which ancient rocks were formed can be inferred from comparison with sediments accumulating to-day. This allows the geographies of the past to be reconstructed.

EROSIONAL PROCESSES belong to part of the cycle of geological change which was discovered and propounded by the Edinburgh geologist, Dr. James Hutton, towards the end of the 18th century. As described in Chapter 1, erosion and transportation are followed by deposition. The deposits are hardened and compacted into sedimentary rocks such as sandstone, limestone and coal. At a later time these sedimentary rocks may be uplifted, a marine area becomes land and the rocks are subjected to renewed erosion, or the region may be inundated by the sea to become again a site of accumulation of sediments.

The significance of these processes, for our present purpose, is twofold. First comes the recognition that the geological history of any area is made up of a series of changing geographies, changing relief, periods of marine occupation, periods of emergence and so on. Second, the history of the region, and therefore the changing geographies, are recorded in the sediments which have accumulated at different periods in that area.

There are obvious gaps in the record whose importance is difficult to assess. Periods of uplift and erosion are represented only by unconformities, and times when no deposition occurred are represented simply by planes between beds. Nevertheless the evidence is considerable and can be interpreted on the basis of one principle—that of *Uniformitarianism*. According to this principle, the processes leading to the production of sedimentary sequences throughout geological time have remained essentially the same. We can observe sediments accumulating under a specific set of conditions at the present time. Under these conditions the sediment has certain characteristic

features such as composition and arrangement of the grains. Ancient sediments with the same compositions and with the same structural features are assumed to have formed under the same conditions.

Comparisons with to-day's conditions allow us to define the environments in which older rocks were formed. They thus enable us to deduce the changing geographies through geological time, and these palaeogeographical reconstructions make it possible for us to indulge in intellectual time-travelling. They give us an insight into the world in which the giant dinosaurs existed; regressing further in time we can map out the limits of the Carboniferous sea and swamps and explore the forests (Fig. 13. 1) inhabited by the primitive reptiles and enormous dragon-flies. Beyond that we come to the period when the land was hardly clothed by plant-life and the early vertebrates were just beginning to evolve from their invertebrate ancestors. Finally, in the most remote times of the Precambrian, we can indirectly watch innumerable cycles of erosion and mountain-building and explore the epoch when life was just beginning to find itself.

Scenes like that depicted in Figure 13. 1 are fairly familiar. Not so well known are the palaeogeographical maps which have been produced, such as that for the Carboniferous period in the British area (Fig. 13. 2). These maps resemble those of to-day in that they depict the distribution of land and sea, pick out areas of highland and lowland, coastal regions with their beaches, deltas and lagoons and areas where the sea was very much deeper. They are based on the following studies:

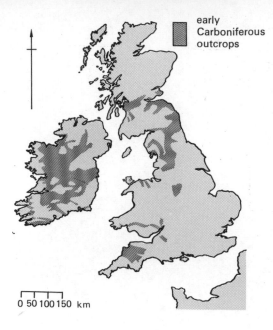

early
Carboniferous
outcrops

0 50 100 150 km

L A N D

LAND

Culm facies:
deep water shales and sandstones

shallow water limestones

shallow water sand

shales and sandstones
(shallower than culm facies)

Lagoon: oil shales, coal swamps

Yoredale facies

V volcanic activity

land

(i) Recognition of the age of rocks from the fossils they contain, and from a knowledge of the spatial distribution of these rocks.

(ii) Analysis of the features of the rocks to determine the environment in which they formed; for instance, which area was land, which river or which marine. We also need to know the direction in which rivers flowed and the nature of the marine currents.

(iii) Finally, integration of all this information to produce a palaeogeographical map.

Mapping of the rocks in Britain was begun at the turn of the 18th century, and William Smith in 1835 produced the first detailed map showing the distribution of rock strata in England. He made the discovery that each rock may be characterised by a particular assemblage of fossils. This allows both the tracing of the rock when exposures are intermittent, and the determination of the age of the rock, since the fossil forms are characteristic of certain periods in Earth history. By refinements which have been made over the last century and a half the fossil content of a rock can be used to determine its age within relatively narrow limits (see Chapter 2).

The present distribution of rocks formed in any period is readily available in Britain from the geological maps published by the Institute of Geological Sciences (available from HMSO). Information from the maps can be supplemented from work published by independent workers in geological journals. Sites of sedimentary basins can usually be recognised immediately from the geological maps, but interpretation is not entirely straightforward.

The integration of all available information to produce the palaeogeographical map (point (iii) above) we shall discuss later. Point (ii) now requires to be examined in some detail. Two procedures are involved; first, we must discuss features of sedimentary rocks (rock analysis) and secondly we must proceed to an analysis of present-day environments.

13. 2 **Upper left: areas where Carboniferous rocks are exposed at the surface in the British Isles.
Below: simplified palaeogeographical map of the British Isles in early Carboniferous times.**

Palaeogeographic maps are reconstructed mainly from the evidence of sedimentary rocks discussed in this chapter, though there may be some instances (e.g. pillow structures in lavas) when igneous and metamorphic rocks yield important information.

The presence of a feature in a sedimentary rock is a function of its origin and post-depositional history. Our problem is to peel off the 'outer' effects of tectonic disturbances (i.e. those associated with folding and faulting), and diagenetic effects (those associated with burial and the lithification of the soft sediment after deposition) to reveal the core which resulted from the environment of accumulation (Fig. 13. 3). The sediment will be determined chiefly by this environment of accumulation but it will also bear, perhaps less obviously, signs of the nature of transportation, signs of the type of weathering which predominated in the source area, and indications of the original geology of the source area, i.e. some evidence regarding the composition of the parent rocks.

We shall now discuss sediments in terms of their *mineral composition*, *texture*, *structure* and of their *lateral* and *vertical variation*. Each one of these attributes gives information about the formation of the sediment, and we can usually make deductions about parent rocks, weathering, transportation and deposition, along with the effects of post-depositional changes.

Composition

Sedimentary rocks are made up of detrital fragments derived from pre-existing rocks, or of minerals derived by organic activity, or of minerals formed by chemical precipitation.

Constituents of sedimentary rocks can be split into four categories, as described in Chapter 1: *terrigenous*, *allochemical*, *orthochemical* and *authigenic*.

In terrigenous constituents, many of the detrital fragments are derived directly from the parent rocks, others are secondary stable detrital minerals in the sense that they have been formed by the action of weathering on minerals of the parent rocks. A large proportion of the organically formed minerals are those of organic skeletons; another proportion, rather more difficult to estimate, is produced by the indirect activity of organisms in assisting or causing the precipitation of minerals. It is for example certain that bacteria

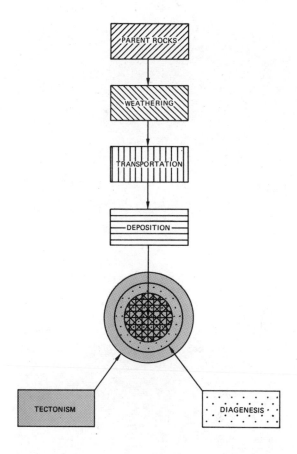

13. 3 Illustration of processes which have determined the nature of a sedimentary rock (represented by concentric circles). Diagenesis refers to processes operative after deposition, such as compaction and cementation. Tectonism refers to folding and faulting of rocks, which affects their overall distribution but can also change their texture and cause some changes in mineral composition.

play an important role in the production of H_2S in sediments, which leads to the formation of pyrites (FeS_2).

Conglomerates and sandstones vary in composition according to their content of fragments of pre-existing rocks and minerals. If weathering has been largely mechanical and the sediments have accumulated without much sorting, i.e. the sediments are of mixed composition and are termed *immature* (Fig. 13. 4), it is not difficult to infer the nature of the source rocks. The rock fragments in the conglomerates, such as granites, basalts and schists, give a direct indication of derivation. In general it is possible only to point

13. 4 Mature and Immature Sediments.

Above: immature coarse-grained sediment (breccia), large angular blocks of various lithologies showing very poor sorting. Old Red Sandstone (Devonian) of Kerrara near Oban, Scotland.

Below: mature conglomerate, with well-rounded pebbles largely composed of quartzite. Old Red Sandstone (Devonian), N. Ireland.

to a similarity in composition between rock fragments and older, presently exposed, possible sources. Some sandstones are similarly immature. These will be rather poorer in rock fragments, but the collection of minerals of different type will allow source rocks to be identified.

As conditions of weathering and transportation become more intense and prolonged, more and more *mature* sediments will be formed. One end member in this sequence is a sandstone composed almost exclusively of quartz. Attempts have been made to analyse the quartz grains to see whether they have come from metamorphic or igneous sources, but with limited success. Rather more satisfactory from the point of view of recognising parent rocks is the presence of a small proportion (often less than 1%) of what are known as 'heavy minerals'. These have a greater specific gravity than the common minerals quartz and feldspar (hence their name), and are separated by breaking up the rocks and floating off the lighter minerals in a heavy liquid (bromoform, S.G. 2.89 is commonly used). Assemblages of heavy minerals may be characteristic of certain groups of parent rocks.

Weathering and vigorous transportation tends to remove the unstable minerals like feldspar and pyroxenes. Frequently, their removal is due to chemical breakdown at the weathering site; occasionally, as in aeolian conditions, transportation is so vigorous (grain impacts are so numerous) that all grains with cleavage are reduced to very small flakes and removed as dust from the sand deposits. Mature sandstones may be the result of either weathering or transportation processes; we then have to look at other features to decide which of these two processes has been the main cause. For instance, as we shall see later, aeolian deposits produce a series of diagnostic structures. We have already noted that intensive weathering produces more mature sandstones. Also produced are stable secondary products—the *clay minerals*. These are separated from the coarser sand grains during transportation and segregated as muddy sediments. In principle the nature of the clay mineral is determined by the intensity of weathering. The so-called 'degradation series' (Fig. 13. 5) relates the igneous-rock minerals to their weathering products. It will be seen that assemblages of clay minerals correspond to different intensities of chemical changes during weathering. At late stages, kaolinite becomes predominant. Finally, in present-day tropical conditions of high rainfall and good drainage, chemical weathering culminates in lateritic soils composed of hydrated iron oxides, aluminium oxides (bauxite) and silica. Such lateritic soils appearing in the rock column are indicators of extreme weathering. Intermediate products such as kaolinite or chlorite can be taken to indicate progressively weaker weathering, because of higher relief, or a colder drier climate, or both.

However, application of these conclusions is made rather difficult because the clay minerals tend to respond to post-depositional changes. Thus weathering at source may result in the formation of the mineral kaolinite, but deposition of this mineral in a marine basin may cause it to change to the mineral illite. Therefore the presence of illite cannot be simply interpreted as a product of weathering less intense than that which produces kaolinite. During subsequent burial and even slight tectonic movements, similar changes in the clay minerals can also occur. Other minerals may not be so susceptible to post-depositional changes, but alterations to original composition do occur. Original grains may be replaced to varying degrees by silica or calcite deposited by percolating solutions. Feldspars are particularly susceptible to such infiltration; in some cases deep burial or slight metamorphism may be responsible for the formation of clay minerals from feldspar and ferro-magnesian minerals.

Other rocks formed by organic activity and/or inorganic precipitation are usually very distinctive in composition, for example coals, reef limestones, and evaporites. Their value in looking into the past lies in the insight they give as to conditions—usually climatic and topographic—within a basin (see below). The presence of fossils, whether as a dominant component as in reefs or just as a minor constituent, is often much more illuminating than a great amount of other chemical or physical information about a rock. Some groups, like the corals and the brachiopods, are found only in marine conditions to-day and, although fossil forms and present-day forms may not be precisely the same, there seems no reason to suggest very different habitats. Furthermore, many organisms have a restricted distribution in the sea with regard to both climate and to depth. At present, the reef corals are restricted to warm, shallow, tropical waters to which light penetrates. Muddy waters are anathema to them, so their occurrence in ancient rocks gives a clue not only to probable climate, but to depth of water and the absence of any river mouths in the near vicinity. Other forms of life indicate marginal, swampy conditions, such as the luxuriant ferns and trees of the Carboniferous, which flourished in flat-lying swampy conditions on the margin of a lake or sea. Yet other groups are exclusively terrestrial in habitat—such as the Arachnida (spiders, scorpions etc.) and most types of mammals. If it can be shown that their remains have not suffered trans-

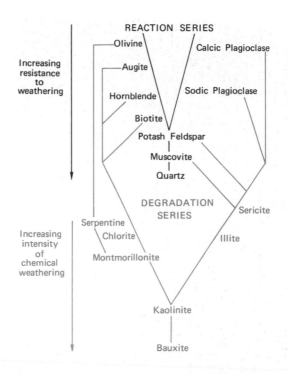

13. 5 The 'Degradation Series'—igneous rock minerals in order of increasing resistance to weathering, and their weathering products (the clay minerals) as they appear under conditions of increased chemical weathering.

portation, then a terrestrial environment is suggested when they are found in rocks.

In assessing the environmental significance of fossils account must be taken of their state of preservation. A fossil man may for example be found in marine sands, but this does not mean that he was a marine dweller, but that his body was transported to its present site. Signs of transportation are given by the state of preservation of the fossil. Some fossil assemblages (paradoxically referred to as 'life assemblages') are made up of shells in a position they occupied during life, and so they have been preserved without being moved from their habitat. Like the coral reefs their interpretation is straightforward. More often shells, and less frequently bones, are scattered over the sediment surface or through the rocks by transporting currents. Shells are frequently disaggregated, fragmented, and have been transported to varying degrees. In one sense this is a

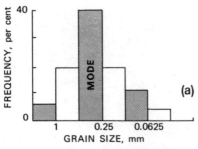

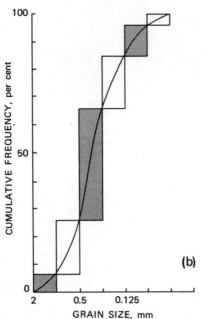

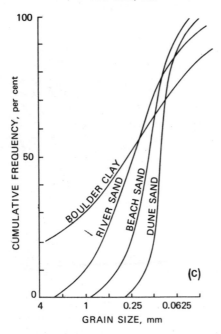

disadvantage in that there may be uncertainty regarding the exact derivation of the shells, but it gives positive evidence of the high energy level of the environment. There are, too, some obvious limits to the transportation of shells. Thus a mixture of marine and terrestrial forms is hardly likely to occur elsewhere than in the sea. Similarly, shallow-water forms may be transported to greater depths in the sea, but it is unlikely that deep-water organisms could ever be fossilised in shallow-water sediments.

Texture

Texture refers to the relationships between the grains of a rock. It is determined by the size, shape and arrangement (packing) of the grains. Some sedimentary rocks, the evaporites for example, have been crystallised from solution and their textures show how the grains grow together into a mosaic. Any early crystals have a well-formed outline with distinct crystal faces, while minerals forming later merely fill interstices in the rock. In these two features the evaporites are strictly comparable with the igneous rocks which have crystallised from a magma (Chapter 1). The textures associated with the fragmental sedimentary rocks are distinctive in that they are aggregates of variously shaped grains of differing size.

The grain-size distribution of the fragmental rocks is usually determined by analysing the individual grains once they have been disaggregated. This is accomplished by fracturing with a pestle or, depending on the cement of the rock, by the use of various acids (e.g. acetic or dilute hydrochloric). The fractions of the grains belonging to different size classes are then determined by sieving or by measuring the rates of fall in a sedimentation tube.

13. 6 Size analysis of fragmental rocks. (a) Histogram showing percentage frequency of size grades. (b) Cumulative frequency curve showing construction from the histogram data. (c) Cumulative frequency curves of sediments from different environments of deposition. Wind-blown sand is very well sorted; boulder clay, poorly sorted.

13. 7 Deposition at the mouth of the river Lachen, where the current loses velocity.

The second method relies on the fact that the rate of fall of a particle depends on its diameter. However, if disaggregation is likely to damage the grains to a great extent, optical measurements are made on extremely thin slices of rock. The resulting size distributions (Fig. 13. 6) allow the determination of a mean or modal size and a range. The range of sizes (or spread) is referred to as the *sorting* of the sediment (the civil engineer calls this *grading* but, as will be seen below, grading is used by geologists in another sense).

Statistically-minded readers will recognise the size-distribution curves as examples of frequency distributions of a log-normal type. A number of 'sorting coefficients' have been proposed, but there is a growing tendency to use the second moment of the distribution, i.e. the standard deviation. An indication of sorting is given by the slope of the cumulative frequency curve (Fig. 13. 6c): the steeper the curve, the better sorted is the sediment.

The mean (or mode) of the distribution is a measure of the competency of the transporting medium. Thus boulders (defined as particles with a diameter exceeding 264 mm) are to be expected in the high energy environment of a shore exposed to heavy waves, or in the valley of a swiftly flowing river as it disgorges from the mountains: sheltered bays or lagoons, on the other hand, may receive only deposits of clay size. Sorting in a sediment depends on the nature of the transportation and the deposition. In general, the

longer the phase of transportation the better may be the sorting. Sands which have undergone transportation over many miles of desert are particularly well sorted. Actual distance from source may not be great so long as movement is long-continued. Beach sands for example may not travel enormous distances from their parent rocks, but their continued to and fro movement under the action of tides and waves results in the separation of grades of different size. Poorly sorted sediments are formed by rapid transportation and deposition. If a sediment is dumped at the foot of a canyon, as it is in alluvial fans, then sorting is hardly possible. The stream simply drops, at once, all the particles which it is carrying. If deposition follows slow deceleration of the current, large particles are deposited before the smaller sizes and relatively well sorted deposits result. The peculiar conditions of glacial transportation produce singular results. The so-called '*boulder clay*', formed from the mixed debris scraped up by the glacier as it moves over the land surface and carried in its lower portion as moraine, is perhaps the best-known example of a poorly sorted deposit.

Grain shape and roundness

A distinction can be made between the shape of an individual grain, a property referring to the three-dimensional character of the grain, and the roundness, a property which refers to the state of the corners of the grain in a two-dimensional projection or in section.

The shape can be expressed in a number of ways. For our purpose it will be enough to note that three axes at right angles to one another can be picked out in each grain. Depending on the ratios of these axes the grain can then be described as spherical, discoidal, rod-shaped or tabular (Fig. 13. 8). The shape of the fragment depends very largely on its original development in its parent rock, and in the case of the larger rock fragments the spacing of the joints in the rock is of fundamental control. In some environments, however, the grains tend to take on a characteristic shape because of the mode of transportation, and the resulting abrasion of each grain is significant. In desert sands, for example, the quartz grains are described as 'millet-seed' sands, a reference to the almost perfect, spherical shape. This shape is the result of long-continued abrasion as the sand is driven over the desert floor by the wind. Other environments favour the concentration of grains of a particular shape, i.e. a shape sorting occurs.

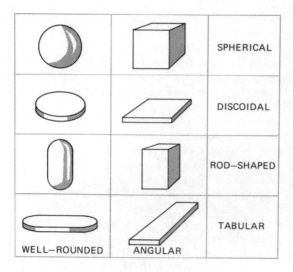

13. 8 **Extremes of roundness and angularity in the four basic grain shapes.**

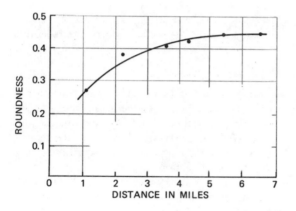

13. 9 **Increase in pebble roundness with distance down a valley; perfect roundness = 1.**

For example, Dr. B. J. Bluck of the University of Glasgow has shown that concentrations of discoid pebbles and cobbles may occur on the upper portions of beaches while the spheroidal particles have been concentrated in a lower, seaward, zone. The separation of the shapes is determined by their differing resistance to movement.

Roundness reflects directly the rigour of transport. The well rounded (as well as spheroidal) grains of the desert have been chipped by other flying grains during miles of transport; the angular fragments in some river deposits have suffered a minimum of chipping during a short journey from source to site of deposition. The controlling factors involved in the rounding process are not only distance of travel, but speed of movement and the difference in density of the grains and the enclosing medium.

Rounding of grains is a complex process involving a number of variables. If, however, grains of a particular size in a single environment are considered, then the roundness can be an indication of the distance of travel from their source (Fig. 13. 9).

Although there is every indication that in the larger grains roundness is directly proportional to distance of travel, in the finer sand and silt sizes there may be a decrease in the roundness in a downstream direction. This arises because the finer sizes are derived during transportation downstream by the chipping of larger fragments; the smaller sizes outrun the larger, and so rounding in water is a very slow process for the smaller sizes. Indeed one authority suggests that well rounded grains can only be formed under aeolian transportation.

Post-depositional changes can affect both grain-size and shape. Percolating solutions can replace some of the quartz grains or parts of the grains, leaving increased cementing material in the form, for example, of calcite, and silica can be deposited as overgrowths around detrital quartz grains. In the latter case the original shape can sometimes be seen because of impurities left around the margin, but where these are absent it is impossible to determine what the grain was like before deposition.

Sedimentary Structures

Sedimentary structures are those features in sedimentary rocks relating to a particular organisation of the grains or mineral components. The most characteristic structure in sediments is their stratification. This is a separation of the components of the succession into layers more or less parallel to the original surface of accumulation. The layers or beds are a response to changes in the environment of deposition such as variation in current strength. They are usually formed over long periods of time, and reach thicknesses of over

1 cm. Laminations are defined as layers less than 1 cm in thickness; they have usually formed during a relatively short time and occupy a small area.

Beds of rock form a convenient basis for the definition of other structures which are called surface or *external* features, and *internal* features which can be seen within the bed. The divisions are not completely unrelated, in that some of the features which are primarily expressed as external may be due to processes also affecting the internal part of the bed. Ripples (as described below) are of this nature. A further division can be made into structures which formed during the deposition of the bed (*primary* structures) and those which have formed afterwards (*secondary* structures).

Internal Features

Some beds show a lamination parallel to the bedding planes. The lamination may be made up of grains of differing composition and/or differing grain size. In fragmental rocks such laminations may have formed by separation of grains of different size and composition during transport and deposition. In other rocks, such as those formed by chemical precipitation, the laminae have been formed by changes in temperature of the water and other factors affecting the concentration of dissolved salts.

Cross-lamination (*current bedding* or *cross-bedding*) is also common in the clastic rocks. This commonly arises through the load being carried along by the current as a moving carpet of grains. One way in which these inclined laminae may form is through the forward growth into deeper water of a sand body or 'bar'. The sand grains are rolled along the floor to the edge of the sand and avalanche down into the deeper water.

A common response of the floor to the shearing action of the current is the creation of a series of undulations or ripples of varying size. The abundant ripples of beaches are about 10 cm long from crest to crest, and about 1 cm high from crest to trough. They move forward under the current by stoss-side erosion and lee-side avalanching (Fig. 13. 10). Internally, this movement is recorded in small laminae inclined in the direction of current movement and at an angle appropriate to the size and shape of the constituent grains. This *micro-cross-lamination* has its counterpart in larger structures formed during the development of much larger ripples (mega-ripples). This structure may be preserved in sandstones in units of varying size, but for water-borne deposits is

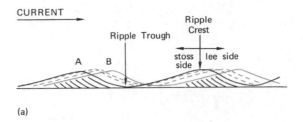

(a)

(b)

13. 10 (a) As ripple migrates down current from A to B with no net gain of sediment, sediment is removed from the stoss side and deposited on the lee side, producing cross lamination within the ripple; (b) migration of a train of ripples under conditions of net sediment gain produces small scale cross lamination, because ripples migrate laterally and upwards.

often up to about a metre in thickness. Under aeolian conditions small ripples occur, but such deposits are dominated by very large units (up to 50 m or so in thickness). These units record the growth of large dunes of which the crescentic-shaped *barchan* dunes are a common representative (Fig. 13. 16). These large, cross-laminated units along with the excellent sorting and the highly spherical grains provide diagnostic criteria for the identification of aeolian sediments of any age.

Not all inclined lamination is due to avalanching in front of a growing structure. After deposition of unconsolidated sediments, sliding may occur on the sea floor and a series of folds develop which may be buried by later sediments; or sinking sand may ruck up flat laminations in the underlying sediment (Fig. 13. 11a). Alternatively, deformation may be contemporaneous with deposition in the small-scale structure known as convolute bedding (Fig. 13. 11c). The flow of water over the crests of ripples is thought to produce a slight

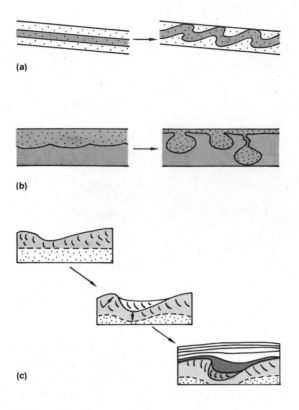

13. 11 Deformation structures produced during or soon after deposition.

(a) Slumping produced by strata 'sliding' down a gradient.

(b) Load structures produced by the relative movement of layers of differing plasticity; in this case sand has foundered down into mud.

(c) Convolute bedding; for further description see text.

suction effect, which together with concomitant deposition of further sediment in the troughs of the ripples causes the crests to grow upwards and the troughs to sink further. Of these structures the first—sliding and slumping—gives an indication of the direction of slope of the floor, and the last augments observations of undeformed ripples on the depositing currents. Deformation may also occur when one sediment type sinks into or is intruded by an underlying, softer plastic layer, such as a mud (Fig. 13. 11b).

When large-scale cross-lamination is absent, clastic rocks may show a gradation in grain size from the bottom to the top of the bed. This *graded bedding* may occur in massive beds, or there may be a gradation through a series of parallel laminae (Fig. 13. 12). It is found in a large group of sands and sandstones, including present-day deep-sea sands forming on the floors of the oceans, as well as in the great thicknesses accumulated in the deep waters of ancient geosynclinal troughs. The grading is thought to originate from large clouds of sediment (*turbidity currents*) which are thrown into suspensions on the edge of the continental shelf, flow down the continental slope to the oceanic plains and gradually lose their velocity and thereby their ability to transport sediment. From these turbidity or *density* currents, the coarser grains are deposited first and, as the weaker tail of the current passes by, the finer sediment gradually falls out. At these later stages of flow there is a drag effect along the floor, and ripple marking and parallel lamination may develop. But the bed from coarse grains below through laminated portion shows an overall grading (Fig. 13. 12b).

External Features

The surface between adjacent beds may be marked by a number of different structures. As in the preservation of fossils these can be seen as either originals (the mark) or as infillings (or moulds). Ripples often appear on the upper surfaces of beds and show a variety of patterns, from the straight-crested types to the crescentic or linguoid forms (Fig. 13. 16). Some structures are more often preserved as moulds on the undersides of sandstones in contact with shales. These so-called *sole-markings* are made by a sand filling in original sculptures in an underlying mud. There is a great variety of such markings, formed either by the scouring vortex action of currents, or by pebbles or shells being dragged along by the flow, or even by burrowing organisms (Fig. 13. 13).

Other surface features are good indicators of emergence from the water. When a lake floor or a flood plain dries out, the mud shrinks and forms a series of surface cracks roughly hexagonal in plan (Fig. 13. 14). These cracks penetrate into the mud and form traps for sand brought down in a subsequent flood. The pattern of mud cracks may then be preserved in the top of the underlying shale and the bottom of the overlying sandstone.

Smaller and often unnoticed structures are formed by rain drops falling on soft sediment (Fig. 13. 14). The surface is covered by a series of small pits which in rare cases may be preserved by subsequent, gentle sedimentation.

13. 13a   Tool marks, produced by objects bouncing along a muddy bottom, the resulting grooves being filled with sand.   Current from left to right.

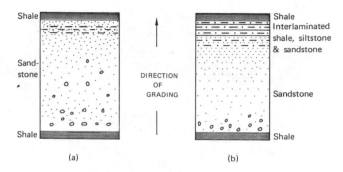

B

A

13. 12   Graded bedding: (a) simple grading; (b) grading in a composite bed; (c) graded bedding in Ordovician of New York State, U.S.A.   Sandstone (between A and B) grades from light-coloured (coarse-grained) base to dark (fine-grained), laminated (in part, cross-laminated) top.

13. 13b   Organic trail preserved as a sole marking.

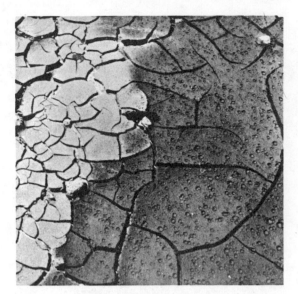

13. 14 Cracks in recent mud. If sand is deposited over this bed, then the pattern may be preserved as a mould on the base of the sandstone. In this example the rainprints may also be preserved as moulds.

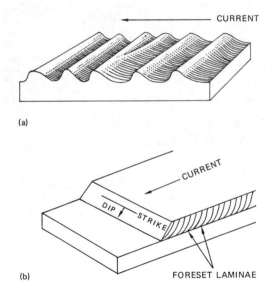

13. 15   Current directions inferred from (a) measurement of straight ripple crests, (b) measurement of dip and strike of foreset laminae in a cross-laminated set. The dip lies in the same vertical plane as the current direction.

Determining Current Directions

An essential part of the description of past environments is to determine the direction of transporting and depositing currents. In a general way the overall movement of a sediment is known when the position of the source rocks has been located. But directional studies may yield wider conclusions; for example, the study of wind systems associated with arid deposits of different ages gives evidence of palaeolatitudes and so has a bearing on continental drift. Fortunately a number of current-indicators are available. The ripples mentioned in connection with beach sands are not only characterized by a straight-crested form. They are also almost invariably developed with their crests running parallel to the shore. This is not coincidence. The movement of water under tidal and wave action is essentially up and down the beach, and ripples-crests are formed at right angles to the flow in response to the shearing action of the current. The trend of the formative current is readily determined from the crest-line, and the sense of the movement is indicated by the steep side of the ripple. Another way of determining both directions and sense is to measure the *dip* and *strike* of either the steep face of the ripple or of any of the foreset laminae within that ripple (Fig. 13. 15).

Where ripples or dunes have a crescentic form, as in the case of the linguoid ripples formed under water and the large cresent-shaped barchan dune formed in deserts by the wind, the current direction lies along the line bisecting the structure (Fig. 13. 16). Measurement of the dip of foreset laminae produces a spread of readings symmetrically arranged around the current direction.

Additional variation arises in different environments. The uni-directional flow of rivers is complicated by the development of meanders and the resultant complex flow pattern of water around them. Under marine conditions the combined effect of tidal and wave motion is to produce structures with laminae dipping in opposite directions as the currents are regularly reversed. The picture is further complicated by currents which transport sand along the shore. The *spread* and the *modes* found in current diagrams are in themselves an indication of the environment, and with careful interpretation yield valuable information.

Even where structures are not developed it may be possible to find current directions. In conglomeratic beds, for example, flat pebbles being

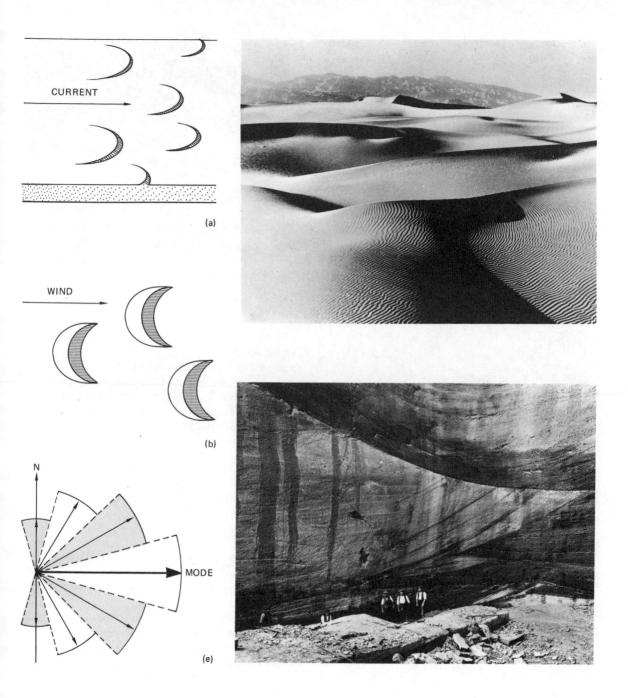

13.16 LINGUOID RIPPLES AND BARCHAN DUNES. (a) Block diagram of linguoid ripples (which may only be a few centimetres high); (b) plan view of barchan dunes (may be 30 m at highest point in the middle of the lee side of the dune); (c) a group of barchans, Death Valley, California, with aeolian ripples; (d) fossil dune bedding, of Permo-Trias age, in the Mauchline Sandstone, Scotland; (e) rose diagram representing readings of dip directions of the cross laminae of either linguoid ripples or barchan dunes (stylised, the length of each arrow representing the number of readings within each 30° arc).

13. 17 Limestone pebbles showing imbricate arrangement dipping up current (which flowed from left to right). New Red Sandstone (of Permian age), Torbay, Devon, England.

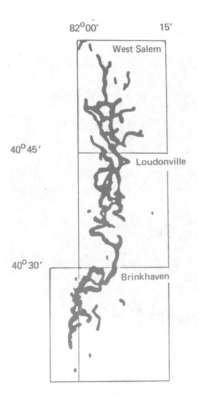

13. 18(a) Channel deposits in a meandering river of Lower Carboniferous age in Ohio.

trundled along the floor very often come to rest in a stable position within the current. This position is one in which the pebble is stacked, tile-like, with its neighbours and dips in an upstream direction. The pebbles are said to show *imbrication* (Fig. 13. 17). In sands and sandstones experiment and field observation have shown that the ellipsoidal sand grains tend to lie with their long axes parallel to the current. Furthermore, where the grain has a blunt and a sharp end, like an egg, it is usually rotated so that it lies with the blunt end upstream. Analysis of the preferred orientation of sand grains can therefore give a reliable indication of the orientation and sense of movement of the current.

On a larger scale, mapping of thicknesses and grain-size of beds may bring out directions of sediment movement—even the shape of old valleys. In some cases exposures and bore-hole records are sufficient to reconstruct a large part of an ancient river course (Fig. 13. 18a). Otherwise observations at individual outcrops on sandstones can be integrated to give a picture of the drainage pattern of a basin.(Fig. 13. 19).

In rock successions which have suffered gentle folding and which undulate only a few degrees about the horizontal, readings can be taken on current structures and no correction is necessary. Where rocks dip steeply, a rotation of the structures has taken place and current directions indicated by the structures have been altered. The effect of rotation during folding can be taken off by suitable calculations—as it were 'unrolling' the folds so that the beds are returned to the horizontal.

Vertical and Lateral Variation

The types of rock which occur in a succession are good indicators of the environment of deposition. The superposition of rock types may appear random, but in many sequences there is a tendency for certain lithologies to repeat themselves in a particular order. Such sequences are described as *cyclic*, and the units which are repeated as cycles or *cyclothems* (Fig. 13. 20 a & b). The vertical arrangement of lithologies is a reflection of the lateral variation in the environment of accumulation (Fig. 13. 20 b & c). If shallow-water sands pass out into shales which, in turn, pass into limestones in a marine area there will be a tendency, as the shore line migrates and depths vary, for neighbouring lithologies to occur above or below one another.

The extent of individual beds is also an indication of the environment. In small fault-bound basins in a mountain range the sediments will vary rapidly away from the mountain front in any

13. 18(b) Meander belt of the River Add, showing abandoned meander channels in dark grey. Compare this pattern with that in Figure 13. 18(a).

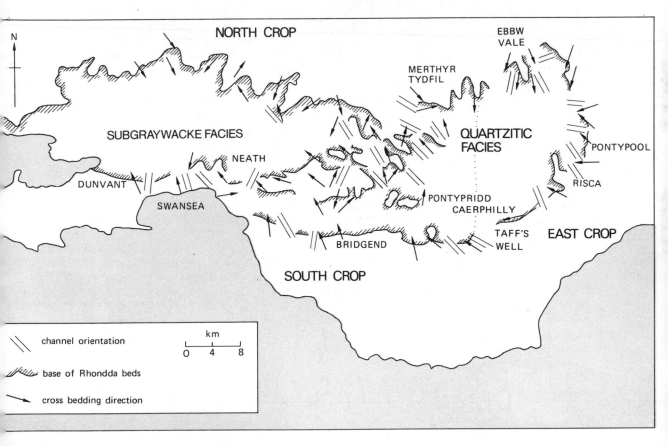

13. 19 Drainage directions in the Upper Carboniferous of South Wales (Rhondda Beds) based on the orientation of observed channels and cross-lamination (= cross-bedding).

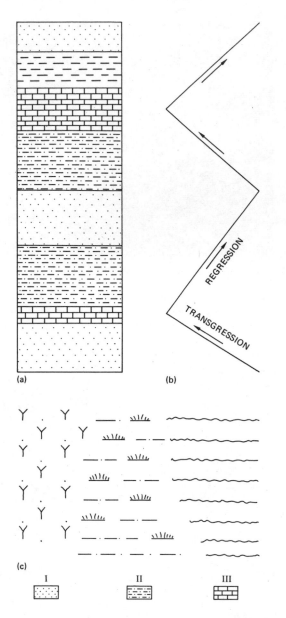

13. 20 Cyclic sedimentation. (a) Hypothetical succession showing repetition of sandstone, siltstone and limestone beds. (c) Stylised diagram showing three zones representing: (I) fluvial conditions with accumulation mainly of sandstones; (II) marginal complex with deposition mainly of siltstone; (III) marine area with limestone sedimentation predominant. The succession represented in (a) could form by the lateral migration of the three zones in (c). Migration of the marine conditions over the land is referred to as transgression, and the reverse movement, regression (b).

basin, and they will change particularly rapidly from the mouths of canyons which debouch large amounts of debris on to the floor of the basin. Contrast this with a deep ocean floor, lacking in relief, and far from any strong point-sources of debris; sedimentary units here will be constant over very large areas.

EXPERIMENTAL STUDIES

As we have seen, sedimentary features can be documented from rock successions, and we shall see below how these may be interpreted in the light of knowledge of recent environments. Field investigations may indicate the mechanism of formation of, say, cross-laminated structures or the chemical conditions necessary for the precipitation of certain minerals. Frequently, however, field studies allow only limited conclusions with regard to the genesis of the sedimentary features. Natural conditions are usually very complicated due to the great number of variables involved, and there are often physical difficulties in observing processes continuously over long periods of time. Experimentation under controlled laboratory conditions can therefore augment field observations and bring new insights into the processes operating in particular environments.

For example, interpretation of the laminated structures in fluvial sands has been helped enormously by experiments in flow channels (usually termed flume tanks). The structures formed under different depths and rates of flow have been determined. This has made it possible to use the size and form of laminated, and especially cross-laminated units to make inferences concerning conditions (for example, water depth and velocity) in ancient river deposits.

On the chemical side, experimentation can obviously help in determining the conditions necessary for the precipitation of different mineral phases. Conditions can be expressed in terms of acidity or alkalinity of solutions (expressed as the pH, the concentration of hydrogen ions) and in the oxidising potential (Eh, expressed in volts). Eh/pH diagrams can be used to portray the stability field of minerals. For example, the formation of different iron minerals is prescribed by Eh and pH, and their relations as empirically determined are shown in Figure 13. 21. For pyrite (FeS_2) to be precipitated, reducing conditions are necessary although the pH can vary on either side of neutral (pH = 7). Haematite (Fe_2O_3) needs oxidising conditions and alkaline or rather weakly acid conditions. The situation in natural conditions is complicated. Variation in the amounts of other components in solution affects the positioning of the stability field. For example, increased concentration of total iron allows precipitation of siderite in more acid solutions.

Perhaps this will suffice to illustrate the principle that the presence of certain mineral phases can be used to indicate the chemical conditions obtaining in the basin of sedimentation. The application of experimental results, chemical or otherwise, is limited due to the inevitable simplification involved in the laboratory work and to scale effects. Nevertheless the undoubted success of experimental work to date means that it will play a growing part in sedimentary studies of the future.

COMPUTER STUDIES

Actualistic studies can also be augmented by use of the computer. There are two main ways in which the computer can help. First it can be used for processing data. Techniques in this case vary in sophistication. On the one hand the computer can be used to find simple statistics such as sorting from grain-size distribution data. On the other, it can be used to study the distribution of data in three (or more) dimensions. Consider the study of sedimentary features on an areal basis. A simple parameter such as bed thickness of sandstones can be plotted at a number of observation points on a map (Fig. 13. 22a). Traditionally this has been contoured simply from inspection; in other words, the contours are drawn in by eye, using simple proportion between any two observation points. This often leads to complicated maps which may be difficult to interpret (Fig. 13. 22b).

Computer programmes are available which analyse the data and produce a surface (represented by the contours, Fig. 13. 22c) which is the best fit that can be made to the data with such a simple plane surface. Other more complicated surfaces can be computed if desired, but frequently this linear surface is sufficient. Taking this *trend surface* as representing the thickness of beds, we may conclude that the general trend is for a thickening of the sediments towards the NNE probably caused by greater net subsidence of the basin in that direction. Such a conclusion would prob-

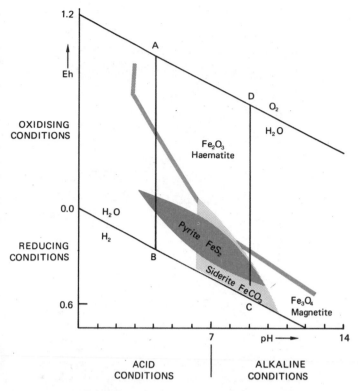

13. 21 Eh-pH diagram showing the conditions necessary for the precipitation of different iron minerals. The range of naturally occurring solutions is given by the parallelogram ABCD. The colour line defines the stability field of Fe_2O_3.

ably have been reached from the hand-contoured map, but the pattern in these raw data is not so clear. Now if we take the figures (X observed) at each observation point (L in Fig. 13. 22) and subtract the corresponding figures for the same localities derived from the computed surface (X_1 comp. for example at L_1) we obtain a difference map (Fig. 13. 22d). At each point on this map is plotted the difference or residual between the computed and observed values ($X_{res.} = X_{obs.} - X_{comp.}$). In this example we have supposed that there is a belt of positive residuals running somewhat irregularly from SSW to NNE. This means that the sediments in this zone are somewhat thicker than might be expected, and if other sedimentary features (such as fossils, cross-bedding etc.) give no contrary indication it might be concluded that this belt represents the site of a former river course. The general principle is that the trend surfaces which are computed allow us to pick out the general features within the area, and

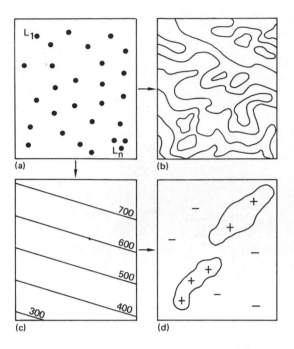

(a)   (b)

(c)   (d)

13. 22 Trend-surface analysis.   (a) Measurement of sedimentary feature, say bed thickness (X obs.) are recorded on the map at localities L_1......L_n, (b) The same data, processed by hand contouring. (c) A trend surface (a simple plane represented by the contours 700-300) is computed. From it, values of X at localities L are also worked out, giving X comp.   (d) Residual map based on differences between observed and computed values.   (X_1, residual at locality $L_1 = X_1$ obs. $- X_1$ comp.).   In this example only the fields of positive and negative residuals have been plotted.

the residual maps allow the pin-pointing of more local features which might be of significance from a palaeogeographical point of view.

The second use for the computer is only just developing but it will undoubtedly become very important: it is the production of simulation models.   In this technique a theoretical model is set up, based on processes known to operate in, say, shallow-water marine conditions.   Taking a variety of conditions of depth, wave action, temperature and so on, the computer can be programmed to predict the type of sediment and the fauna included in that sediment.   These results can be portrayed on a map which corresponds to a *facies* map.   The important difference lies in the fact

that the natural facies distribution is the result of the interaction of many processes which are only vaguely known, while the simulation map has been computed from *specified conditions*.   In so far as the sediment, fauna and their distribution correspond to known, real facies distributions we can deduce that the values for, say, temperature or wind-wave action prescribed in the model also apply to the natural situation.   Time may also be included in the model, so that the distribution of facies (rock types) can be investigated at successive geological stages.   And since we have seen already that lateral variation of facies produces the vertical changes in rocks, we can use the model to build up a geological succession.   Again, if our computed succession bears close comparison with a natural rock succession, then our prescribed conditions in the model might, with caution, be supposed to have caused the accumulation of the rock sequence.

The map shown in Figure 13. 23 is part of a typical set produced by the computer following a programme simulating shallow-water sedimentation written by John Harbaugh of Stanford University, California.

ENVIRONMENTAL ANALYSIS

The Beach

If we study the sediments in one of the most pleasant and accessible of present-day environments we can see how the sedimentary features form an assemblage which may be diagnostic for that environment.   Consider a sandy beach.

A beach accumulates in a well-defined area relative to sea level (Fig. 13. 24).   It is formed in a position marginal to the land, providing there is a supply of clastic fragments and the energy level of the waves and tides is suitable.   The foreshore is periodically covered by the sea during the tidal cycle, the backshore is normally exposed, and behind the backshore is a cliff of older material, or the beach may be backed by a line of dunes formed of wind-blown sand derived from the beach.

There are some differences between the features formed in different parts of the beach, but we shall treat the area as a whole.   From our experience of British and other European beaches we tend to think of their composition as being almost exclusively quartz.   They are repeatedly worked over by the sea and are therefore well sorted and mature.   If a rocky headland is near there may be some admixture of rock fragments.   Some

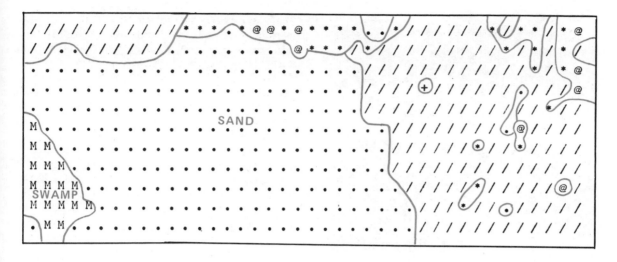

13. 23 Computer-produced map of different sediment types (facies) with different faunas.

Dots — Sediment with sand remains
M — Sediment with swamp remains
@ — Sediment with sponge remains

Asterisks — Sediment with crinoid remains
Oblique dashes — Sediment with algal remains

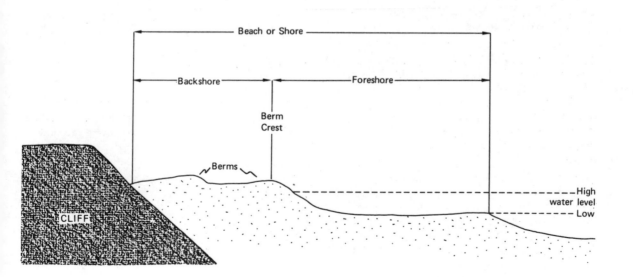

13. 24 Beach and sea-level and some related terms (based on F. P. Shepard).

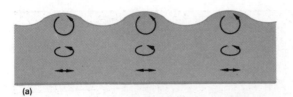

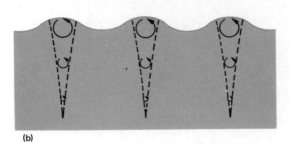

13. 25 (a) Circular movement of particles in wave-form transformed to horizontal oscillation at the sediment-surface in shallow water, compared with (b) movement of particles dying out in depth when unhindered by the floor.

beaches are distinctive in composition because of the local source rocks. There are laminae in many of the beaches around the Tay estuary which contain 80% or more of garnet derived from the garnet-bearing metamorphic rocks of the Highlands. Elsewhere in the world, local supply prescribes beaches of a different composition. Volcanic islands made up entirely of basalt and related rocks cannot supply quartz, and their beaches are black accumulations of basaltic fragments or their constituent minerals, pyroxene, olivine and feldspar. And many tropical shores dominated by corals have nearby beaches made up of debris broken from the reefs. Where quartzose sand and rock debris is in short supply organic material may fill the deficiency. Some very attractive beaches in NW Scotland are silvery-white and sparkling from the calcite fragments derived from the shells of various invertebrates living off shore.

Whatever the composition, the constant working and re-working of the deposits by the waves leads to an excellent separation of grains of different size. Frequency curves of beach sands therefore show a high degree of sorting. There are some exceptions to this; for example, beaches with nearby rocky headlands may well show an admixture of pebbles derived from the headlands and thrown up on to the beach during storms. Not only does size-sorting take place, but there is also a separation of grains according to density. Thus the sands of the Tay estuary have garnets separated from the less dense quartz and concentrated into laminae. Other beaches form important sources of economic elements, for example titanium, when their minerals have been selectively sorted by the sea.

The feature which is perhaps the most obvious on beaches is rippling. Especially noticeable at low tide, ripples often cover the whole or the major part of the beach area. They are commonly of the straight or nearly straight-crested type, run uniformly for long distances parallel to the shore and can be seen forming by movement of the sand in the shallow sea. Water movement is complicated by both tidal and wave action, but the direction of movement is usually towards and away from the land and the ripple crests are transverse to the current direction. Wave action is usually dominant. In shallow water the circular motion of the particles which give rise to the wave form is transformed by the interference of the floor into a to and fro movement (Fig. 13. 25). Under this motion the sand is organised into symmetrical (wave or oscillation) ripples. If the oscillatory movements are not completely symmetrical (and this is the usual case) asymmetrical ripples will be formed with their steep (lee-side) slopes facing up or down the beach. Some modification to the ripple form can be seen during the ebb when the water becomes more and more shallow. The tops of the ripples can be skimmed off, or tiny terracettes can be cut into the sides of the ripples. Tiny runnels may also be formed as the water finally drains out of the mounds of sand. Around any large obstacle the regular flow of water is broken up, and the ripple crests may show a complicated pattern. The flow may be so affected that ripple marking is inhibited altogether. The concentration of flow and therefore of energy in the vicinity of the obstacle gives rise to scouring, and a hollow is left around the boulder. Such scours may be characteristic of flat beaches where the ripples have not formed. If there is a scatter of pebbles, shells or other small objects over the sand, the water as it returns seawards in the backwash of each wave is diverted around each obstacle. Again, the concentration of flow around the obstacle produces a

13. 26 Sedimentary structures on beaches:

(a) Straight-crested ripples running parallel to the shore. Ripple-tops flattened in shallow water during falling tide. Sea-weed 'stalks' rotated into a direction normal to the ripple crests.

(b) Rill marks. Anastamosing pattern of small runnels formed by water draining down slope of beach, from bottom left to top right. Keys indicate scale.

(c) Swash marks. Tiny mound of sand or light debris left in an irregularly crenulated line along the beach at the limit of each wave. The line is clearly marked on this photograph because the water has not yet drained away.

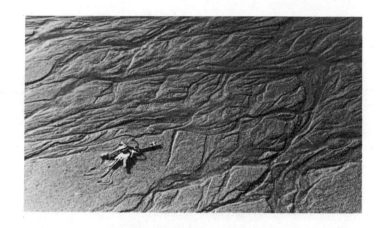

(d) Crescent marks, here formed by scouring by the back-washing waves around two small (3 cm) pebbles.

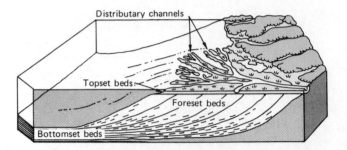

Distributary channels

Topset beds

Foreset beds

Bottomset beds

13. 27 Idealised delta showing topset, foreset and bottomset beds; part of the topset beds form sub-areal portion of the delta cut through by distributary channels.

series of scours which diverge from the obstacles and have a typically crescentic form (Fig. 13. 26d). As water is diverted around each obstacle the lines of flow form a diverging criss-cross pattern. This flow pattern is imposed on the underlying sand as a series of low diamond-shaped or rhomboid ripples.

As the tide recedes and the foreshore emerges, sea-water drains from the beach along lines which are made up of diverging and anastomosing patterns known as rill marks. During this time of emergence, organic activity becomes very obvious. The lug-worm forms string-like mounds of sand, and just alongside each mound there is a small depression caused by the same animal ingesting the sand; a crab scuttles sideways forming a complicated surface marking; a winkle slowly drags itself over the surface leaving a distinct groove behind as a record of its laborious passage; sea-gulls, oyster-catchers and people walk along the edge of the sea leaving footprints which could be fossilised for posterity.

· The structure of the beach can be seen in vertical sections cut into the sand. Sometimes differences in grain-size and composition are so slight that no structure can be seen. Where different grains are present it is apparent that a very fine lamination is characteristic of this environment. This is not surprising because current strength and direction changes so frequently in response to tides and winds. The laminae are also frequently inclined. This cross-lamination is distinctive in that it is made up of long, straight, planar laminae which can dip both inshore and out to sea, sometimes at rather small angles. Curved cross-laminae with

very variable directions of dip are found in the dunes backing up the upper foreshore. These dunes are formed by the wind driving the beach-sand inland. The grains are eroded from the shore as it dries out. Left behind protecting shells or pebbles are elongate mounds of sand ('sand-shadows') which give the beach a distinctive appearance.

Trace fossils, footprints or snail or crab-trails may be the only organic record on the beach, but usually there will also be a collection of shells lying on the surface. Many of these are fragmentary— a death assemblage resulting from the constant re-working by the sea. Some will show a tendency to lie in certain directions, like the long razor shells for example strung out parallel with the length of the beach. Within the sand there may be communities of different shells which live at different depths. These may have been left in positions of life, producing a 'life-assemblage'—a record of a community and the habitat of each organism

The life assemblages will obviously be made up of marine organisms. The death assemblages will be predominantly of marine shells but there may be one or two bits of land plants, some fresh-water animals and even the remains of land-living vertebrates and invertebrates.

In ancient sediments that have formed in an inter-tidal beach environment we may look for a number of features. Some individual features are diagnostic—the rill marks for example—but for certainty it is better to examine the whole assemblage: the well sorted, laminated sands organised into planar cross-bedded units; the frequent straight-crested ripples, some of which are symmetrical; rhomboid ripple marks, crescent marks, perhaps even rain-prints; a death assemblage of scattered and broken shells; some horizons burrowed and perhaps containing shells in a position of growth; frequent trace fossils—footprints and trails.

We have begun our analysis of environments with a study of beach features mainly because most of us are familiar with these areas, and because the sediments can be easily studied. Because of their critical position, forming a narrow zone between land and sea, beach sediments tend not to be well represented in the geological column. They either suffer erosion or they form only a minor part of a marine succession. It will therefore be instructive to turn to another set of conditions.

Deltaic Sediments

Any sediment which escapes the flood plain is dumped at the mouth of a river when the velocity of the water is reduced at its meeting with the sea. As long as marine currents are incapable of moving that sediment the pile will continue to build out into the sea. The growing mound is made up of a flat upper portion (the topset beds) a steeper middle portion (foreset beds) and a lower, distal portion (bottomset beds, Fig. 13. 27). The topset beds tend to build up to sea level, and at any one time a flat sub-aerial portion grades down into a shallow sub-marine portion. The delta takes on a typical fan-like shape in plan as the river breaks up into a number of distributary channels which meander across the topset beds. The upper topset area will be differentiated into channels and inter-distributary regions rather like the flood plain, and sedimentation will be similar to the flood plain. Coarse-grained channel deposits pass laterally through sediments deposited on the inside of river meanders into swampy, fine-grained peaty deposits. In vertical section there is a gradual reduction in the grain-size of the sediments from the mouths of the distributaries through the foreset beds to the bottomset beds. The foreset beds show a rapid alternation of rippled fine-grained sands and clays, and the latter become dominant further into the basin (Fig. 13. 28). Another important change takes place. The fresh or brackish water conditions of the upper delta give way to truly marine conditions in the bottomset beds. This is recorded in the shells preserved in the sediments.

At any one time the sequence of sediments in the growing delta pile will consist of fine-grained

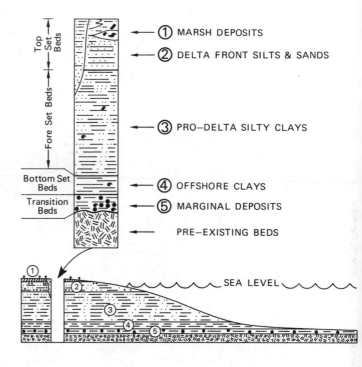

13. 28 **Lithological changes through a single episode of delta out-building.**

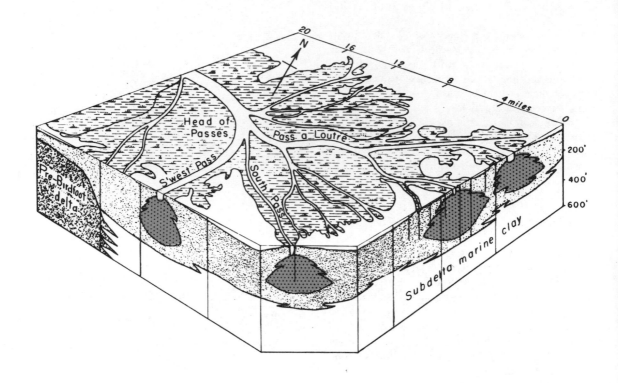

13. 29   Sedimentary 'framework' of the Mississippi (based on Fisk).   The diagram shows lenticular sand bodies (colour) formed by distributaries in the inter-distributary sediment of carbonaceous clays.   The vertical scale is exaggerated 30 times.

clays with marine shells at the base, passing up into sands and silts which grow in importance through the foreset beds. The sequence may then pass into channel sands in the region of the distributary or, more generally, into the fine-grained sediments of the interdistributary swamp.

Such a sequence can be found in the region of the active sub-delta of the Mississippi at the present time (Fig. 13. 29 & 30). Now consider the growing sediment-pile over a period of time under conditions of overall subsidence in the basin, such as obtain in the Mississippi delta. At present almost all of the discharge of the river is concentrated in the present-day Balize sub-delta. Subsidence is going on, but the supply of sediment is great enough to offset this subsidence and the delta is building out into the Gulf of Mexico. Subsidence is indicated by the fact that the rate of growth from the 1 million tons of sediment per day is slower than would be expected if no subsidence was occurring. As the distributaries are extended, the slope of the river becomes lower and lower. This means that during a flood, if the river banks are broken upstream (i.e. crevas-

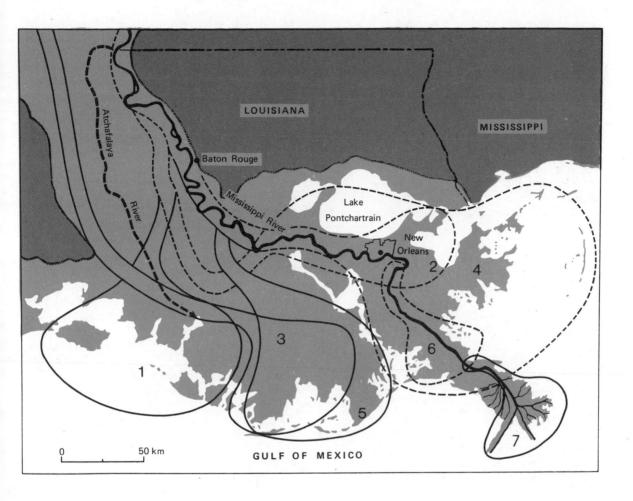

13. 30 Overlapping sub-deltas of the Mississippi (1-7). The oldest, sub-delta 1, is estimated at 3 000 years, sub-delta 3, 1 500 years, sub-delta 5, 1 000 years and the present Balize sub-delta 7 began over 100 years ago.

sing occurs), the river may successfully find a shorter way to the sea on the margin of the present sub-delta.   The crevassing is likely therefore to be successful.   A present sub-delta (say, sub-delta *A*) will be abandoned and a new growth point or sub-delta will be established.   A new sequence of sediments will be built up in sub-delta B.   The area of sub-delta *A* is now starved of sediment and build-up is not able to off-set subsidence.   The area is therefore gradually overrun by the sea.   Over the deltaic sediments a marine sequence is built up, as in the area to the

east of the Balize sub-delta—in the region known as the St. Bernard sub-delta.   In this region the sea is reworking the top part of the delta sequence, and the resulting sands and clays have an assemblage of marine shells.   Over the last 3000 years 'delta switching' of this kind has caused the building of a number of sub-deltas, and there is a natural limit to the life of the present sub-delta. Diversion of the main discharge from the present main channel began about ten years ago.   Complete switching would have occurred into the Atchalfaya River had it not been for the construction of a special barrier. The present sub-delta has had its natural life prolonged, but continuous channel control will be needed if it is to be indefinitely maintained.

Natural sedimentation in deltas therefore comprises periods of active out-building interspersed with periods of stagnation and marine transgression. The succession is characterised by cycles in which marine beds are followed by a coarsening succession from bottomset to foreset beds and, at the top, a return to finer beds with perhaps peats in the inter-distributary areas. The cycle then repeats as the sea sweeps in after 'delta-switching'.

THE PRESENT AS THE KEY

It is now appropriate to examine examples of ancient sedimentary sequences in terms of Lyell's well-known aphorism: 'The present is the key to the past.' We should remember that modern experimental and computer techniques make to-day's 'present' particularly informative.

Carboniferous Successions

Carboniferous rocks form some of the best-known sequences in Britain. They are widespread in the Midland Valley of Scotland, over much of the north of England and the northern Midlands, South Wales and the Bristol Channel area and Devon and Cornwall (Fig. 13. 2). Economically they are of the first importance, and the presence of coal (and to a slightly lesser extent ironstone) in Carboniferous measures has had a decisive effect on the social and economic life of Britain over the last two centuries.

If we consider the rock sequences in general during the early part of Carboniferous times we find considerable variation from the southern to the northern parts of Britain. In Devon and Cornwall the rocks (Culm Measures) are predominantly shales with thin cherts and limestones. Occasional fossils indicate marine conditions, and there are intercalated spilitic (submarine) lavas and pyroclastic rocks. The area was evidently part of the geosynclinal area which developed during Upper Palaeozoic times and stretched through SW England across to Europe. Eventually this would form the Hercynian mountain belt. To the north the rocks in South Wales and the Bristol area are thick limestones of varying sorts but all indicating, from their fossil content, shallow-water conditions.

A common limestone is that which is made of organic remains with many corals and brachiopods. Crinoid debris is also abundant (Fig. 12.2). Other limestones are finer grained, associated with shales containing lamellibranchs, and probably represent sheltered lagoonal conditions as distinct from the open sea environment of the coral-brachiopod limestones (Fig. 13. 2).

The Carboniferous Limestone succession thins to the north, and it is probable that a land mass separated the S. Wales basin from N. Wales at this time. In the latter area, and in the S. Pennines, clear shallow sea produced the same type of deposit. But the limestones give way again northwards to a shaley succession which is thicker and has a goniatite (an extinct group of cephalopods with coiled shells) fauna rather than one of corals and brachiopods. This was a subsiding trough with deeper water conditions. The northern margin of this trough was fringed with reefs, and a thinner succession accumulated over the area of the northern Pennines. The rocks here form part of a facies known as the Yoredales, consisting of a cyclic succession of limestones, sandstones, shales and coals which will be discussed in further detail below. In general however the Yoredales clearly represent an influx of terrestrial debris which is generally lacking further south. In Northumberland and Scotland a Yoredale facies follows a thick succession known as the Calciferous Sandstone Group. This is a very varied group, and sediments are restricted in their lateral distribution, suggesting the presence of a number of separate basins of deposition. This is especially the case in Scotland. Here for example the 'Oil Shales', a fine-grained series of sediments rich in carbonaceous material from which oil may be distilled, occur in a very small area west of Edinburgh. At the same time, in the west of the Midland Valley, volcanic eruptions produced the thick Clyde Valley basalts.

Overall, the clear seas in the southern part of Britain give way to terrestrial conditions in the North. The Midland Valley was made up of a separate series of basins. Some Highland material was deposited in the Midland Valley area; other debris was transported farther south, possibly by a large river in the area of the present North Sea.

Yoredale Cycles

The Yoredale sequence has attracted attention since the early part of last century, when it was

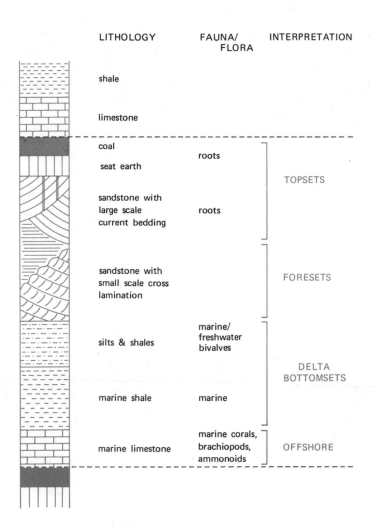

LITHOLOGY	FAUNA/ FLORA	INTERPRETATION

shale

limestone

coal

seat earth — roots

sandstone with large scale current bedding — roots

TOPSETS

sandstone with small scale cross lamination

FORESETS

silts & shales — marine/ freshwater bivalves

marine shale — marine

DELTA BOTTOMSETS

marine limestone — marine corals, brachiopods, ammonoids

OFFSHORE

13. 31 Lithologies and faunas in an idealised Yoredale cycle, with an interpretation in terms of parts of a deltaic environment.

recognised as being made up of a number of cycles of sedimentation. In broad outline each cycle tends to consist of a limestone followed in upward sequence by calcareous shales with marine fossils, shales and siltstone passing up into sandstone which gives way to finer-grained sediment until a fine-grained seat-earth (a fossil soil) is overlain by a coal seam. The cycle then begins again. From the fossils it is clear that the lower portion of the cycle represents marine conditions, and the upper part, fresh or brackish swamps at about sea level (Fig. 13. 31).

If we take the model of the out-building delta as described above from present-day examples, the sequence finds a ready explanation. Suppose we have an area recently inundated by the sea. Terrigenous debris is absent and limestone is built up from the abundant fauna. Now consider the effect of a delta encroaching over the area. Bottomset beds will accumulate first, consisting of silts and muds with perhaps some admixture of marine shells, but the mud will discourage the previously flourishing clear-water forms. The bottomset beds will be followed by silts and clays of the foreset beds which become coarser upwards. If the area is in the neighbourhood of a distributary, sands will be introduced until the sediment builds up to sea level; if the area lies between distributaries, the accretion of sediment will be slower and consist mostly of fine-grained sediment. In both cases sea level will be approached and plants will colonise the area as soon as the water is shallow enough. The growing plants will flourish long enough to produce a peat which will form a coal seam after burial. The action of the plants growing in the underlying mud will modify that sediment into the seat-earth which we now find. A further transgression of the sea will submerge the area and bring in a new limestone to begin a new cycle (Fig. 13. 31).

Most geologists are satisfied that the model of delta sedimentation explains the type and sequence of sediments found in the Yoredales, though considerable effort is being devoted to recognising other marginal environments like lagoons, beaches and other inter-tidal areas which we might expect to find with a delta complex. There is also continuing discussion regarding the mechanism which produced the periodic marine transgressions.

Some workers take the evidence from deltas like the Mississippi, and suggest that the process of 'delta-switching' provides an inherent mechanism in the delta which ensures cyclic development. Suppose after a period of delta out-building, crevassing takes place upstream. The sediment-starved area can no longer overcome the regional subsidence which must be taking place. So the sea rolls in to begin a further cycle. At a later date crevassing brings the sediment back and a new cycle is built up.

Other workers link the cycles with a periodic down-dropping of the area. Instead of supposing uniform subsidence over a long period of time, it is suggested that a sharp drop, perhaps connected with faulting, caused a marine transgression. Supply of debris from the land eventually tended to overcome the transgression and delta-growth occurred, culminating in emergence and the growth of trees. A further Earth movement, causing sudden subsidence, brought back the sea and the limestone.

These movements are thought by some to have taken place on a regional scale perhaps affecting blocks like the northern Pennines, and there is no suggestion that these cycles would be expected to be contemporaneous over areas like continents or even between continents. But there are other geologists who believe that world-wide changes of sea-level might have been responsible for some, if not all, such cycles. It is conjectured, for example, that the glaciations which affected the southern continents in Carboniferous times would cause, like the Pleistocene glaciation of NW Europe and N. America, a periodic raising and lowering of sea level. On this hypothesis marine transgression occurred during an inter-glacial period. The lowering of sea level during a glacial period augmented clastic sedimentation to produce widespread emergence and the development of swamps.

It seems to me that the 'delta-switching' inherent in the delta situation provides the most plausible explanation, but whatever the truth of the situation, the Yoredale cycles underline the fact that, while we may draw maps for different times, the fundamental causes of the different geographies are to be looked for in the Earth's crust and mantle, that is, in the tectonic conditions affecting the region under consideration. The Yoredales and other similar cycles of the Carboniferous System reflect delicate conditions of crustal instability and supply of sediment over very long periods.

Central Coalfields of Scotland

How can other details such as texture or grain-size be used to bring refinements to our task of recreating the past?

Dr. R. O. Muir of the Chelsea College of Science and Technology examined a group of sandstones in the Central Coalfield area of Scotland. The thickness of the succession suggests that the basin of deposition was much the same as the present outlines of the coalfield. Values of grainsize show a general falling off from north to south, and in this direction roundness figures also increase slightly. Add to these observations the elongation of grains and the directions of cross-lamination and ripple marks (pointing to a southerly-directed flow) and it seems evident that the basin was being filled from a land mass in the Highlands. The nature of the sorting in the sandstones suggests predominantly fluvial conditions, although in places the high values of sorting point to the local development of beaches. The composition of the sandstone gives clear evidence of the source rocks during the period. In size and character most of the quartz grains compare closely with those of the Dalradian Quartzites and Schistose Grits of late Precambrian age presently exposed just north of the Highland Boundary Fault; the remaining quartz grains may have come from plutonic rocks. Granitic intrusions are also suggested by the nature of the common heavy minerals.

CONCLUSION

Clearly the building up of a palaeogeographical map is a slow process involving the detailed analysis of rock sequences as we now find them over the surface of the Earth, but for many parts of the world we now have such maps for different geological ages. We can therefore give our time-traveller a briefing about much that he might expect to find, were he to set off for Carboniferous or even more remote times. Time-travelling in this sense may not be possible for any one of us, but we have our imaginations. As the poet Stephen Spender suggests:

Different living is not living

In different places,

But creating in the mind a map.

FURTHER READING:

C. O. DUNBAR AND J. RODGERS 1957. Principles of Stratigraphy. *Wiley.*

G. M. BENNISON AND A. E. WRIGHT 1969. The Geological History of the British Isles. *Arnold* (Paperback).

D. H. RAYNER 1967. The Stratigraphy of the British Isles. *Cambridge University Press.*

F. J. PETTIJOHN 1957. Sedimentary Rocks, 2nd Ed. *Harper.*

L. J. WILLS 1951. Palaeographical Atlas. *Blackie.*

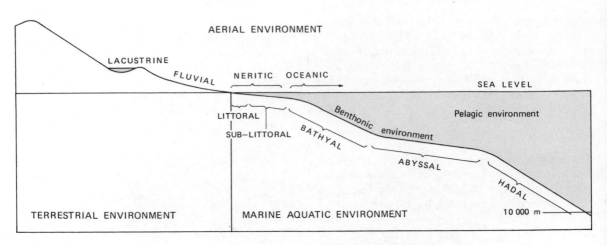

14. 1   Some of the environments occupied by life at present.

Zoologists and botanists can study at first hand the development of in-dividuals from embryo to adult (*ontogeny*) and make inferences concerning the ancestry (*phylogeny*) of a particular species. That an explanation of evolution in the Darwinian sense of 'natural selection' is feasible in principle is now completely accepted, but many details of descent remain to be established. The problem facing the biologist investigating present-day organisms is that he cannot check his inferences concerning ancestry of species 'in the flesh'. The only way to check the family trees constructed by biologists and to assess rates of evolutionary change is to consult the fossil record.

In this chapter, Professor M. R. House of the University of Hull first outlines some of the evolutionary concepts which have arisen out of studies of the fossil record, and then gives a brief summary of life through the Phanerozoic (*literal meaning:* showing life); about the last 570 million years of Earth history. Inevitably, such a summary must use the names of many groups of organisms; as far as possible examples of these are given in the text or illustrations. For fuller descriptions the reader is referred to the texts by Buchsbaum and Romer quoted at the end of the chapter.

The last section of Professor House's contribution examines the possible reasons for the evolutionary 'crises' and 'explosions' previously described. There is as yet no conclusive evidence to favour particular causes for the past fluctuations in evolutionary rates. It is salutary to remember, at a time when man is modifying the environment and causing the extinction of many species, that in the past no group was entirely immune from the effects of changes in evolutionary rates.

14 Evolution and the Fossil Record

by M. R. HOUSE

Our present views on the evolution of life are based on studies of comparative morphology, physiology, biochemistry, embryology, genetics, population dynamics, biogeography and other analyses of living things. But the only way in which the actual order and pattern of evolution can be studied over any length of time is by examining the incredibly rich and varied remains of past life preserved in sedimentary rocks.

SOME 570 MILLION YEARS AGO, at the base of the Cambrian, marine organisms suddenly became varied and abundant, and from that time onwards the fossil record is well documented. Faunas have undergone considerable evolutionary changes, and these have been used to establish a detailed relative time scale of many distinct faunal time zones. Much work is still needed to attach radiometric 'absolute' dates to this zonation of sedimentary rocks, but where firm estimates can be made, faunal zones are frequently found to be of the order of one million years each, or shorter, even in the Palaeozoic. In terms of resolving power this is far more precise than current radiometric techniques allow, and gives a firm framework for the study of evolution in time.

Present land and sea areas offer a great diversity of environments, and similar ones existed even in the Precambrian. Some of these are illustrated in Figure 14. 1. Much of evolutionary diversification is related to the adaptation of organisms to the myriads of ecological niches available in these environments. Our knowledge of the fossil record is most complete for marine areas, where regular sedimentation readily covers shells and skeletons after death. Land areas are typically areas of erosion, and here the record is often incomplete. It is important to stress that the fossil record is almost wholly of plant and animal hard parts, but as such structures are for support and protection, fairly precise reconstruction of soft parts can often be made. Evidence of the habitat of past organisms is usually provided by the nature of the rock in which the remains are found and the palaeoecological setting in which they occur. The distribution of the facies illustrated in Figure 14. 1 has varied considerably in time, but in broad terms palaeogeographic patterns are known, and the influence of palaeogeographic changes can be taken into account in analysis.

Occasionally, unusual preservations give evidence of soft parts, or of soft-bodied organisms, and certain localities have provided full evidence of faunas and floras in environments not typically preserved in the rock record. From certain Tertiary beds at Geiseltal near Halle, East Germany, Voigt described the preserved soft tissues of several animals, including even the patterns of striped muscle fibres and retinal cell structures. Some Devonian silicified plants from Rhynie in Aberdeenshire show not only detailed cell structure, but evidence of fungal filaments and nuclei within the cells. Impressions of jelly-fish and other soft-bodied organisms are not uncommon in the fossil record, and often the burrows, footprints and other structures left behind by organisms enable reconstructions to be made, even when the organism itself has not been preserved; these are called *trace fossils*. Further, the many different types of preservation which may occur preserve different aspects of the original structure. Fossil lake deposits often preserve rich details giving evidence of the flora and fauna as a whole. Thus the Purbeck Beds of the Dorset late Jurassic have yielded evidence not only of fish, turtles and dinosaurs, molluscs and ostracods (microscopic arthropods with bivalve shells), but also of tree trunks and leaves, of a rich insect life and of small tree-climbing primitive mammals, which enable the broad ecology of the time to be reconstructed. Another late Jurassic fauna of unusual excellence and diversity is that from the marine lithographic stone at Solenhofen in southern Germany, in which the first bird, *Archaeopteryx*, was found. A much earlier celebrated fauna, remarkable especially for bizarre arthropods, is that from the Cambrian Burgess Shale of British Columbia, described in the last century by Walcott. Such localities supplement the more typical evidence and enable estimates of the diversity of life at past times to be made. New discoveries are continually adding to the picture.

EVOLUTIONARY CONCEPTS ARISING FROM THE FOSSIL RECORD

The evolution of present day faunas and floras, as extended into the past by the fossil record, is illustrated in the colour chart between pp. 200-201. There is great diversity in the past, and each small group has provided work for many lifetimes. It is useful to discuss here some evolutionary concepts which have grown out of these palaeontological studies to illustrate how the fossil record has contributed to evolutionary theory.

Explosive Evolution

One of the most remarkable features of the fossil record is the evidence that at certain times there was very rapid morphological diversification. This evidence has given rise to such terms as 'explosive evolution', 'evolutionary bursts' and others, although it is realised that this diversification took many millions of years to be accomplished. It is rapid evolution only in contrast to the more typical gradual developments.

At the base of the Cambrian, marine life suddenly became widespread—this is the first 'explosive' event in evolution documented in the fossil record. Diversity is indicated by the range of types first known at this level, even if the lower part of the Cambrian System (covering perhaps 30 million years) still needs precise palaeontological work to establish the order of their appearance. Groups met for the first time in the Lower Cambrian comprise nine phyla of marine invertebrates. These are the Protozoa,

Coelenterata, Archaeocyatha (see Fig. 12. 1), Porifera, Bryozoa (or Ectoprocta), Mollusca, Brachiopoda, Arthropoda and Echinodermata. Some of these may be represented in the Ediacara fauna of southern Australia, which may be latest Precambrian in age. Perhaps the most significant step represented here is the acquisition of some form of hard parts, either supporting spicules, skeletal structures or protecting shells. At first chitino-phosphatic minerals predominated, but later calcium carbonate became by far the commonest form of tissue mineralisation. It should be noted that these early forms were mostly shallow-water suspension or detritus feeders. Predatory carnivorous groups with hard parts are not found, and this suggests that the reasons for the development of hard exoskeletons were other than for physical protection.

In the case of the early Cambrian diversifications many groups are involved, and whilst this may represent the primary evolutionary radiation of the invertebrates leading to the establishment of the modern phyla, the relationships are not documented in detail. This is probably because most organisms had wholly soft-tissued progenitors and intermediates of which little record is known.

The sudden evolutionary radiation of individual groups is often very much better documented. A period of 'explosive evolution' for fishes was the later Silurian and Devonian, and over a similar period for the early vascular plants. For the Reptilia the early Mesozoic radiation was 'explosive', for the flowering plants (Angiosperms) the sudden rise to dominance was a mid-Cretaceous event. The rise of the Mammalia to major importance occurred in the earliest Tertiary, although they have a long earlier record. The majority of modern mammalian groups first appeared in the Palaeocene and Eocene (lowest Tertiary) and this must have been a period of extremely rapid diversification.

Adaptive Radiation

It has long been recognised that periods of sudden diversification of a group are usually related to the exploitation by that group of some new advantageous circumstance. This may be the evolution of an advantageous physical or physiological character, or advantageous changes in the environment which can be brought about either by changes or extinctions in other groups, or by palaeogeographic changes. Often these three factors are inextricably mixed and it is difficult to select which is the most important, although the first is usually crucial. The result of the intimate relation between evolutionary diversification and exploitation of environmental change is that the diversification is often clearly adaptive in nature. This is well illustrated by the placental and marsupial Mammalia, where the classification into Orders is almost wholly one of habit (insectivores, rodents, whales, carnivores, primates etc.) reflecting the initial adaptive radiation established early in the Tertiary. Similar situations occur in the Reptilia, where the evolution of the shelled egg in the Permo-Carboniferous, giving greater freedom from water, made possible the exploitation of land. Eventually the other main environments were exploited, the sea by creatures such as the ichthyosaurs and plesiosaurs, and the air by pterosaurs and pterodactyls.

Similar situations occur in the invertebrates, where specialisations in habit play a clear rôle in the evolution of most groups. For example, the exploitation of the infaunal (burrowing) habit almost defines the irregular echinoids such as *Micraster* (Fig. 14. 4). The development of gas-filled chambers giving buoyancy and hence mobility gave the early cephalopod mollusca (Nautiloids) an immense advantage over their contemporaries who were confined mostly to the sea floor: hence the rapid relative rise of the nautiloid cephalopods in the Ordovician.

At more lowly levels the fossil record provides countless examples of adaptation to environment and particular habits, such as the great variety in appendages of Palaeozoic eurypterid arthropods for habits from walking and crawling to swimming and predation (Fig. 14. 2); or the variety in detailed patterns of filter feeding among the Brachiopoda. For extinct groups it is often difficult to assess the functional rôle of certain characters, but the growing evidence confirms that function and morphology are intimately related and of major importance in establishment of groups.

Evolutionary Rates

The periods of rapid evolution which have been discussed so far contrast with the more typical record of evolution. G. G. Simpson in his classic book *The Major Features of Evolution* (1953) grouped evolutionary rates into fast, normal and

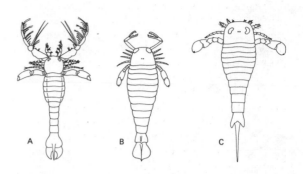

14. 2 Examples of morphological adaptation for swimming, grasping and walking among Palaeozoic eurypterids. A: Megalograptus from the Ordovician of Ohio (×0.05). B: Pterygotus from the Devonian of Germany (×0.05). C: Eurypterus from the Silurian of the Baltic coast (×0.2).

slow as a crude division. Haldane in 1959 introduced a unit to measure the change with time of morphological characters and named it the *Darwin*, but few attempts to express rate of change in quantitative terms have yet been made, although there is ample scope in the fossil record for such analyses.

Some types of calcareous algae seem to have persisted from well back in the Precambrian to the present day. Some genera, notably the brachiopod *Lingula* and some bivalve molluscs, seem to have remained unchanged from the early Palaeozoic to the present—a period of nearly 400 million years. Among plants, the living horse-tail *Equisetum* seems represented in the Carboniferous Coal Measures by *Equisetites*. By contrast, the bulk of present-day mammalian orders appear to have arisen within some 20 million years in the early Tertiary (Fig. 14. 3).

During periods of adaptive radiation, evolution is typically fast: extremes of diversification are produced, but usually the forms are short-lived and extinction rates are high. There is reason to believe that fast evolution occurs when an advantageous character is exploited and hence external selection pressures are low with regard to other groups in the environment. Competition within the group, on the other hand, may well be high, so the characteristic result of rapid evolution is the rapid specialization in particular habits and functional styles of life.

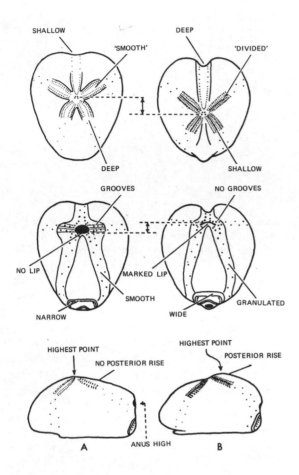

SHALLOW   'SMOOTH'   DEEP   'DIVIDED'

DEEP   SHALLOW

GROOVES   NO GROOVES

NO LIP   MARKED LIP   SMOOTH   GRANULATED

NARROW   WIDE

HIGHEST POINT   NO POSTERIOR RISE   HIGHEST POINT   POSTERIOR RISE

A   ANUS HIGH   B

14. 4 A: Micraster corbovis, an early form. B: Micraster coranguinum, a late form (both ×0.7). It will be seen that changes took place in a number of characters during the evolution from early to late forms. These have been interpreted as a response to more efficient burrowing in the sediments of the sea-floor of the Chalk sea.

After such periods evolution is usually slower, following adaptation to entrenchment in an appropriate ecological niche. The fossil record frequently shows a pattern rather like a toasting-fork: early rapid evolution into many different stocks separated by ecological habit and resultant morphology, and later stabilisation and continuance for varied times of these stocks. An example of this is illustrated for the Mammalia (Fig. 14. 3).

One of the few attempts to analyse changes of this sort quantitatively was made in 1949 by Westoll for the *Dipnoii* or lung fish (in *Genetics, Palaeontology and Evolution*, edited by Jepsen, Mayr and Simpson). This group showed its earliest radiation in the Devonian and is still represented to-day by relict genera in Australia, South Africa and South America. In this case the evolution was analysed over a period of almost 400 million years.

Progressive Evolution

Here some of the features of evolution shown in a genus or small group of genera will be mentioned. One of the first such detailed studies was made in 1899 by Rowe using *Micraster*, a genus of heart urchin (Echinodermata) found commonly in the English Chalk of late Cretaceous age (Fig. 14. 4). Rowe assembled large collections from successive horizons in the Chalk and analysed the successive changes. He showed that these were regular in a number of characters of the test when populations as a whole were studied, but that, at any one level, any individual might be primitive in certain characters whilst being advanced in others. These changes are illustrated in Figure 14. 4. Thus independence of characters in evolution was established. Rowe's work was published in 1899, two years before the rediscovery of Mendel's work on the pea which enabled a genetic explanation to be given to this phenomenon. Subsequently Nichols made comparative studies of *Micraster* and the living *Echinocardium* and demonstrated that some of the evolutionary changes in *Micraster* were related to progressive adaptations to a burrowing habit.

Another such well-known classic in evolutionary palaeontology is that of the horse family, but it is too detailed to be repeated here. It is reviewed in masterly fashion by the vertebrate palaeontologist G. G. Simpson in his book *Horses* (1951), and Figure 14. 5 illustrates the main changes in size, limb structure and teeth. T. H. Huxley, Darwin's famous protagonist, was one of the earliest to collate the European evidence of horse evolution, but this proved to be very incomplete in the light of other (particularly North American) evidence. This serves to emphasise the need for worldwide studies of small groups in order to establish accurately the history of any group.

Another approach to this problem was made by the present writer in studies of the Devonian ammonoid cephalopod *Tornoceras*. Material was

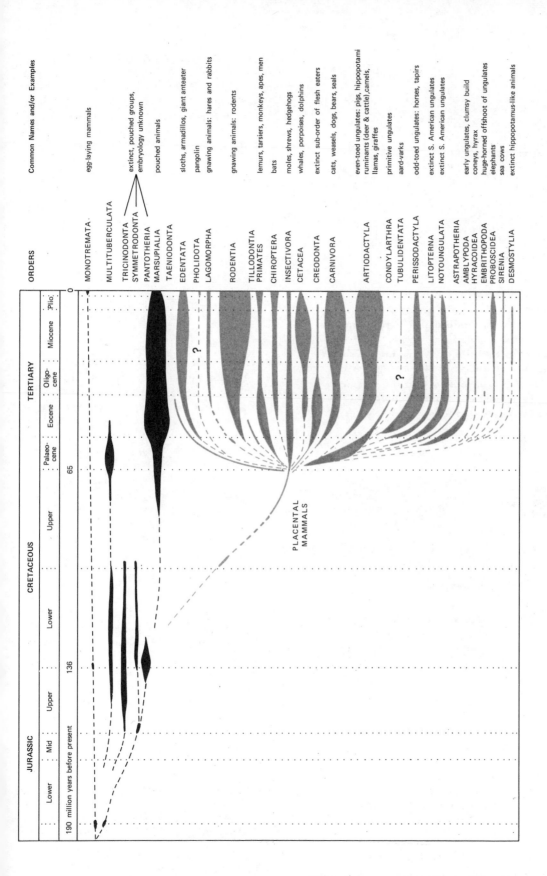

14.3 Evolutionary chart of the Mammalia, showing the long record of primitive groups and the sudden radiation in the early Tertiary (in part after Romer).

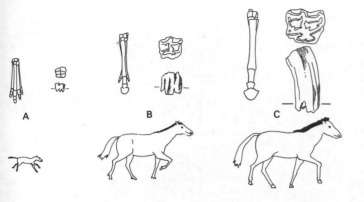

14. 5 The evolution of the horse. A: the lower Tertiary (Eocene) Eohippus, about the size of a fox terrier, with a four-toed pad foot using all digits (probably adapted for forest life), and teeth suitable for leaf biting. B: the intermediate middle Tertiary (Oligocene) Merychippus with a three-toed foot. C: the modern horse, Equus, with single-toed foot, adapted for fast running on open prairies, and teeth adapted for grass grazing.

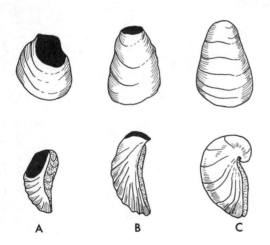

14. 6 The evolution of the oyster Gryphaea in the early Jurassic. A: an ancestral oyster, Ostrea, with a large attachment area on the left valve (marked in black). B: an early Gryphaea showing a reduced attachment area and slightly incurved left valve. C: an advanced Gryphaea, attached in earliest stage only and with an extremely incurved left valve (× 0.6).

available, mostly from New York State, of faunas from a number of successive horizons. The pyritic preservation was such that it was possible to study the growth in each assemblage from the earliest to adult stages, i.e. their ontogeny. Thus an analysis was made of the changes of ontogeny through the phylogeny of this close-knit group of ammonoids. In some characters a progressive pattern of morphological change could be established. On the other hand, overall shell form appeared to vary erratically with time, although the pattern was reasonably consistent at any one horizon. It was inferred that shell form was particularly subject to variation through the interaction of genetic constitution and environment (usually termed *phenotypic variation*). There is need for more studies of this type in the fossil record—and there is no lack of scope.

Evolutionary Convergence

One result of environmental selection pressures operating over long periods is that different groups at different times and places may give rise to very similar forms. Good examples are provided by the gross similarities between diverse groups adapted for active swimming among fish, reptiles such as Mesozoic ichthyosaurs, and mammals such as the Tertiary and Recent whales and sea cows. Sabre-tooth tigers of the Tertiary of South America were marsupial mammals, whilst the present-day tigers are placental mammals. In the Tertiary, especially of North America, there are marsupial rats, but the modern rodents are placental mammals. These examples of convergence are especially interesting where the marsupial forms were geographically isolated from the more advanced placentals.

In these cases different stocks converged towards what must have been the optimum form for their particular modes of life, whether marine active predation, quadrupedal active carnivorous habit, or the typically rodent gnawing mode of eating. Typical gross form associated with ecological habit for a given animal size and mode of life, suggests that there is a 'paradigm' or theoretically appropriate form for given circumstances, and that selection pressures tend to work towards this by eliminating those organisms least approximating to it.

Richer examples of convergence come from invertebrate groups in the fossil record, but here the groups involved are generally less familiar to the non-specialist. *Fusulinids*, giant protozoans from

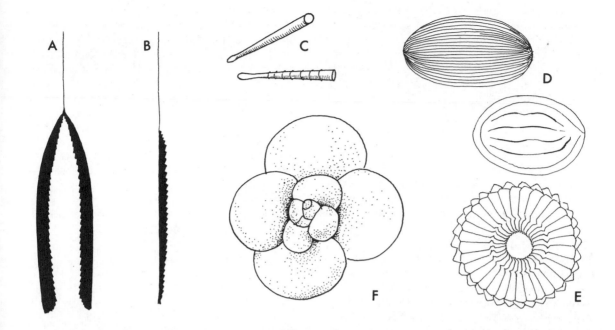

14. 7 Examples of planktonic organisms which have been common in the upper levels of the oceans in the past. A: Didymograptus, a graptolite from the lower Ordovician, probably living attached to floating seaweed (×1). B: Monograptus, a graptolite common in the Silurian (×1). C: Styliolina with small needle-like shell, common in the early and middle Devonian. D: two types of planktonic ostracods (Arthropoda) common in the late Devonian (×20). E: a Cretaceous coccolith (Algae) of which the Chalk is largely composed (×7000). F: Globigerina, a representative of the protozoan group common in the Tertiary and present seas (×100).

the Carboniferous and Permian, often show a spindle-shaped calcareous endoskeleton (skeleton sheathed by the body of the animal) up to 20 mm or more in length (or more often the size of wheat grains). Quite another protozoan group, the *alveolinids* in the Cretaceous and Tertiary, are in gross form convergent with the fusulinids but can be distinguished by their internal structure.

Particularly good examples of convergence have been recorded among fossil corals, brachiopods and molluscs. In one group of the latter, the ammonoids (which form the basis of much of the detailed time zonation of the fossil record), there is such a diverse variety of ornament, such as ribs, nodes, spines and of shell form, from loosely to tightly coiled, that it has often been remarked that plaster casts of quite unrelated forms would be indistinguishable, particularly if they did not show the suture lines (Fig. 14. 11). Yet detailed work on these has elucidated the phylogeny and relations of such forms.

Iterative Evolution

A number of examples are known in which a single stock, or closely related stable parent lineage, has given off a number of short-lived stocks at different times, each stock resulting in closely similar members. One of the classic cases of this is seen in the oysters, a group which first appears in the Triassic. In the true oyster, *Ostrea*, the two valves or shells are somewhat similar in shape and form. From this stock on a number of occasions a form termed *Gryphaea* arose in which the right valve became concave and the left valve markedly convex (Fig. 14. 6). Such stocks possibly arose once in the Triassic, twice in the Jurassic and perhaps twice in the Cretaceous. These concavo-convex oysters are usually placed in the genus *Gryphaea* although everyone realises that it arose from the *Ostrea* stock several times and is therefore a polyphyletic form genus. This particular form appears to be an adaptation to life on a muddy sea floor with fewer numbers of individuals than usual in typical oyster banks; but it would appear that this niche, or the Gryphaea adaptation, did not permit permanent occupation—hence the succession of abortive stocks. Other good examples of this are among the ammonoids, where discus-shaped shells are produced in a large number of unrelated stocks.

Evolutionary Relay

This term, coined by the French scientist Arambourg, draws attention to a natural corollary of evolutionary competition, namely that other groups come to occupy the ecological niches occupied by stocks which have become extinct or severely reduced. The succession of dominance, among vertebrates, of the Amphibia, Reptilia and Mammalia is essentially a succession of this type, although the habitats occupied are complex. Each evolving group shows examples of this, and the relay is often seen to operate as a result of replacement by different stocks of the same group, or by quite different groups.

Such a succession has probably taken place in every ecological niche open for exploitation. In Figure 14.7 one example is shown, that of the upper levels of the oceans inhabited mostly by small planktonic and pelagic organisms. In this case a succession of utterly unrelated stocks has replaced others. These conclusions are based on the remains of organisms with hard parts—there may well have been large numbers of soft-bodied plankton as well, as there are to-day.

Evolution of Communities

We owe especially to the Danish zoologist Petersen the recognition that particular ecological niches on the sea floor are normally occupied by a few species only. These generally live in such a relation to each other that they do not compete directly, but live as a community, often to their mutual benefit. Examples of such communities have long been recognised in the fossil record, usually associated with particular sedimentary settings and occupying discrete palaeogeographical areas. In the preceding section attention was drawn to the successive occupants of the upper levels of the seas in the past, but in that niche, and invariably in others, there are many organisms involved and their inter-relations are complex. That being so, it is surprising that some such communities are relatively long-lived, such as the non-marine bivalve phases of the Coal Measures to seaward of the deltaic swamp areas. Ziegler of Chicago University has in recent years made a notable contribution to the study of Silurian communities, especially in Wales and the Welsh Borderland, and has been able, using such criteria as the displacement effects of contemporaneous submarine lava flows, to demonstrate, at least approximately, the depth zones characterised by given communities. Normally, of course, communities are controlled and limited by many factors. Work in this fascinating field still has a long way to go; but the development of communities is an extremely important factor in evolution, especially for sea-bed organisms and for all terrestrial ones (see Chapter 12).

Continental Drift and Isolating Mechanisms

The importance of various isolating mechanisms in evolution has been amply demonstrated by work on modern ecology and genetics. During the period of major organic evolution with which we are here concerned, isolation of terrestrial organisms was effected on a major scale by the break-up and separation of the continental masses.

In the late Palaeozoic there is considerable evidence to indicate that all existing continents were united in a single mass called *Pangaea* (see Chapter 15). Subsequent break-up took place in two stages, the first of which was the separation of a northern *Laurasia* (comprising Europe, Asia and North America) from the united southern continents (including Antarctica) which formed the land mass known as *Gondwanaland*. The break-up of both Laurasia and Gondwanaland occurred during the Mesozoic, and it is clear from modern work in the oceans, which has led to the formulation of the sea-floor spreading hypothesis (Chapter 16), that this separation took place at different rates in different places, and indeed is still continuing. It should be remembered, however, that shallow seas covered parts of these continental masses.

Effects of continental drift are noticeable in the past distribution of terrestrial and fresh-water organisms. Thus the supposedly fresh-water Devonian fish *Bothriolepis* is almost world-wide in distribution. But in the Permian the flora of the southern continents, or Gondwanaland, is almost entirely endemic. For example, *Glossopteris*, a seed fern, is extremely widespread and abundant and is not known in more northerly areas, where the typical flora of the European Coal Measures

The Earth, photographed on 21 January 1968 by NASA's ATS 3, a satellite in synchronous orbit 38 000 km above the Pacific Ocean. South America and the United States are plainly visible; over the SE Pacific, the cloud pattern indicates the rise and convection to the north-west of air warmed by the sea. The swirling appearance of the depression over the Antarctic results from the Coriolis effect. The dusk 'terminator' (the diffuse boundary between dark and sunlit areas) can be discerned at upper right.

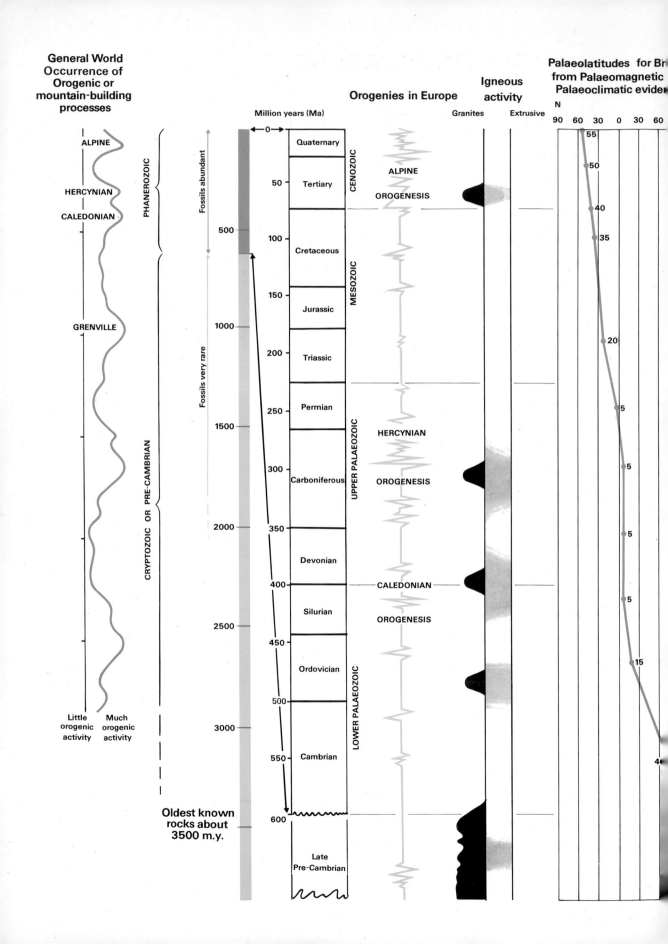

THE FOSSIL RECORD

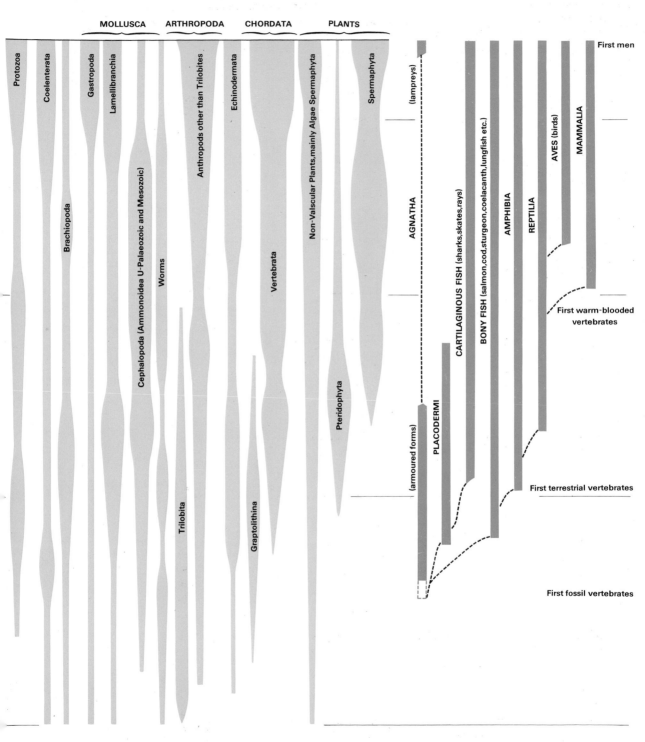

MOLLUSCA · ARTHROPODA · CHORDATA · PLANTS

Protozoa · Coelenterata · Brachiopoda · Gastropoda · Lamellibranchia · Cephalopoda (Ammonoidea U-Palaeozoic and Mesozoic) · Worms · Anthropods other than Trilobites · Trilobita · Echinodermata · Graptolithina · Vertebrata · Non-Valscular Plants, mainly Algae Spermaphyta · Spermaphyta · Pteridophyta

(lampreys) · AGNATHA · (armoured forms) · PLACODERMI · CARTILAGINOUS FISH (sharks, skates, rays) · BONY FISH (salmon, cod, sturgeon, coelacanth, lungfish etc.) · AMPHIBIA · REPTILIA · AVES (birds) · MAMMALIA

First men

First warm-blooded vertebrates

First terrestrial vertebrates

First fossil vertebrates

Invertebrates

Vertebrates

 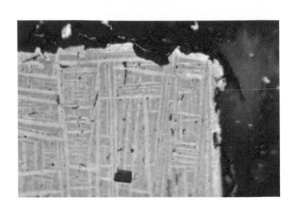

Types of magnetic grains in basalts. Six photomicrographs by Dr. S. E. Haggerty, illustrating the five classes of oxidation states (see p. 255).

Class 1 : a brown titanomagnetite grain (typical diameter 10-30μ) with well-formed edges. This is the lowest state of oxidation.

Class 3 : the process of ilmenite intergrowths formation is now complete, and the titanomagnetite is packed with them.

Class 5 : grains in their highest states of oxidation. One type comprises brown titanomagnetite containing dark rods, long regions of grey pseudobrookite and brighter areas containing mainly titanohaematite.

Class 2 : the first stage of high-temperature oxidation. The titanomagnetite now contains a few lighter-coloured intergrowths of ilmenite.

Class 4 : the intergrowths of Class 3 have been altered to a more reflecting form. The silicate material around the grain is now reddened.

Class 6 : a second type of high-oxidation grain which is actually in a somewhat higher state than the grain shown for Class 5. Grey pseudobrookite is now intermixed with more brightly coloured titanohaematite.

is usual. Wallace, the contemporary and associate of Darwin, drew attention to the line (named after him) which separates distinctive faunas and floras of South East Asia from those of Australia and New Guinea. This appears to mark the junction of two areas joined in pre-Tertiary times. After that separation, primitive mammalian monotremes and marsupials evolved in Australia unaffected by the later evolution of placental mammals in other continents: only man has changed the balance by introducing higher mammals in historic time.

Migration across the Panama Straits appears to have been closed after the late Cretaceous, and a somewhat similar situation to that of Australia occurred so far as marsupial evolution is concerned until the end of the Tertiary, when migration was again possible. Then the more advanced mammalian competitors from the north were almost completely successful in what must be regarded as one of the most spectacular evolutionary battles documented in any detail. Migration across the Bering Straits appears to have been more common in the Tertiary, and this was the route which enabled Old and New World horses to have a common history, but they became extinct in North America after the separation of the straits. This explains the astonishment of Montezuma's forces on seeing the horses and horsemen of Cortez (made the more astounding by their first inference that both were one animal).

sider the morphological diversity represented in the past, but here we have to accept either the subjective views of specialists, or the taxonomic categories which have been erected by many specialists over the last two hundred years. In the latter case there is no clear-cut and accepted definition for the discrimination of species of the past, in the way common interbreeding provides such a method (in theory) for most living groups. For genera and higher taxonomic grades neither the palaeontologist nor the botanist or zoologist has any clear criterion. We thus fall back upon usage, and very uneven that certainly is. These basic problems and uncertainties should be borne in mind, particularly when drawing inferences from the diagrams included here. As in all science, future discoveries will modify our current views.

What is easiest to assess in the fossil record is diversity, even if we rely for assessment of this on taxonomic categories or specialist opinion. Diversity is not the same as success or abundance, but gives us an indication of adaptation to environment as that is expressed in morphological terms.

In the discussion of the succession of Phanerozoic life which follows, attention will be drawn particularly to periods of significant extinction and appearance of new groups, and in a final section some of the possible explanations of these fluctuations will be reviewed.

ANALYSIS OF THE FOSSIL RECORD

It would be misleading to suggest other than that the fossil record is extremely incomplete. There is also the problem of reconciling our knowledge of the fossil record with that of present-day organisms. Workers in these two fields use different criteria for identification, yet there is close agreement between classifications based on present-day hard parts of organisms and those of fossils (as is revealed in Figure 14. 4). Therefore, although the bold features of the course of organic evolution are clear, there are many gaps. This is not the only factor which makes rigorous analysis of the record difficult. Only the marine record is well documented and accordingly receives more consideration here. Moreover, we can assess abundance of organisms in the past only in a subjective way. It is easier to con-

Precambrian Life

The record of life in the Precambrian is considerably longer than the period from the time life first appears abundantly at the base of the Cambrian to the present. Supposed algal structures have been recorded from the Fig Tree Series of Swaziland, which may be 3500 million years old, and algal type stromatolites from the Bulawayo Limestone of Rhodesia may be over 3000 million years old. A varied microfossil assemblage is known from the Gun Flint Formation of southern Ontario which is about 1900 million years old. Algae certainly are not uncommon fossils in the late Precambrian, but the evidence for animal remains is weak until the very late Precambrian. In the later part of the Precambrian, trace fossils, resulting from the burrows or tracks of organisms, are found in many parts of the world.

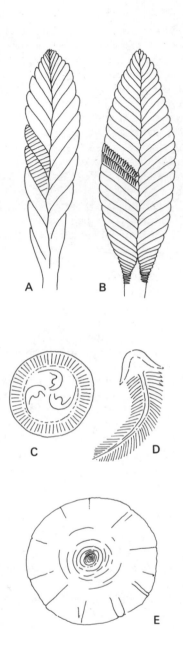

A rich fauna from the Pound Quartzite of Ediacara in South Australia is the best-documented Precambrian metazoan assemblage, but it appears to be very late Precambrian indeed, and Lower Cambrian fossils occur in the overlying Ajax limestone. A considerable diversity of form is represented in the Ediacara fauna, and whilst there is some uncertainty as to the precise relationships of some of these fossils, they have been interpreted as including medusoid (jelly-fish) and pennatulid coelenterates (sea pens), annelids (worms) and others of uncertain phylum. Twentysix species occur, not including trace fossils, and some of these are shown in Figure 14. 8.

One of the commonest forms in the Ediacara fauna is *Rangea*, an elongate subdivided leaf-like structure some 20 cm in length. This genus is also known in South West Africa, and a closely similar form, *Charnia*, occurs in the Precambrian of Charnwood Forest, Leicestershire. Thus elements of the Ediacara fauna seem to have a very wide distribution, even if there is some question of their exact relation to the base of the Lower Cambrian.

Cambrian

The early Cambrian faunas are impressive in their diversity, relative abundance and wide distribution, and contrast strongly with the poverty of the Precambrian record. Throughout the succeeding Phanerozoic there is a rich faunal link with the present. Much has still to be accomplished in the detailed correlation of Lower Cambrian deposits throughout the world. It is customary to emphasise the abrupt nature of the Cambrian appearances, but it should be recalled that the Lower Cambrian lasted perhaps 30 million years, and precise documentation of the succession has not been made.

Even so the evidence is impressive. Nine invertebrate phyla certainly are first recognizable in the Lower Cambrian: the Protozoa, Coelenterata (perhaps in the Ediacara fauna), Archaeocyatha, Porifera, Bryozoa, Mollusca, Brachiopoda and Echinodermata. In contrast to the Ediacara fauna, all these had hard parts. From an evolutionary point of view this radiation laid the foundations of the Invertebrata, and all phyla appearing here, apart from the Archaeocyatha, are still abundant in present seas (Figure 12. 1). Over 900 species of Lower Cambrian fossils have been described, and even within phyla there is

14. 8 Sketches illustrating late Precambrian fossils. A: Charnia, from Charnwood Forest, Leicestershire (×0.35). B-E: fossils from Ediacara, Australia. B: Rangea (×0.35). C: Tribrachidium (×1). D: Spriggina (×1). E: Cyclomedusa.

evidence of considerable diversity. Lower Cambrian metazoans were filter, suspension, or mud feeders mostly living on the sea bottom, and some may have been partially burrowing in habit.

Conditions in Cambrian times were apparently quite equable: limestones and reefs of calcareous algae and especially archaeocyathids are widespread, the latter in regions as far apart as Greenland, Siberia and Antarctica. This evidence stands in contrast to the late Precambrian evidence of an almost world-wide glaciation. The climatic amelioration and the spread of shelf seas may in part account for the sudden diversification of marine life at that time.

Some features of invertebrate evolution in the Phanerozoic are illustrated by Figure 14. 9, which plots the number of genera of various common fossil invertebrates (all with hard parts). For the Cambrian the relative importance and, particularly, the rapid rise of the trilobites (an extinct group of Arthropods) is particularly remarkable, and in detail shows evidence of competition and replacement of one trilobite group by another. In the Lower Cambrian there is evidence of a provincial distribution pattern for the genera *Olenellus*, *Callavia* and *Redlichia* (Fig. 14. 10).

Ordovician

Between the Ordovician and Cambrian, there was gradually brought about a major change among the occupants of the shallow marine environment. This is illustrated in Figure 14. 9 by the major diversification in a large number of common groups, most of which had more meagre histories in the Cambrian. The groups most affected are those with calcareous hard parts. The rugose and tabulate corals and stromatoporoids, all coelenterates, achieve world-wide distribution by the close of the Ordovician: somewhat similar in diversification and distribution are the Bryozoa. These forms appear commonest in regions of calcareous sedimentation, suggesting widespread shallow, clear and warm shelf seas. Exploitation of the environments presented in shelf seas seems also to account for diversification in other groups (notably the Brachiopoda).

Two groups which show great evolutionary diversification do so by the exploitation of environments previously not exhaustively colonised. The planktonic graptolites evolved from bottom-dwelling forms early in the Ordovician, and the evolution of this group is as well documented as

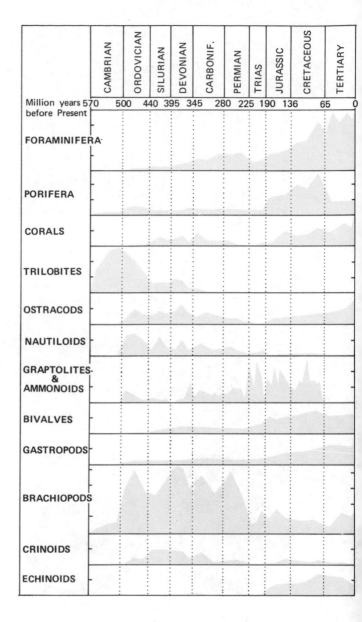

14. 9 **The number of genera recorded in the past for certain common marine invertebrates with hard parts. The scale divisions represent fifty genera.**

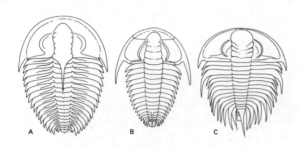

14. 10 Lower Cambrian trilobites (Arthropoda) which characterise different provinces. A: Olenellus found in N. America, Greenland and Scotland. B: Callavia, found in eastern N. America and Europe. C: Redlichia, found in Asia and Australia.

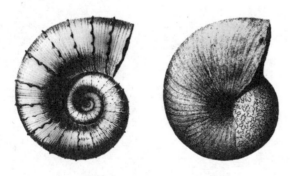

14. 11 Two types of ammonite common in the Mediterranean Mesozoic. Left: Lytoceras. Right: Phylloceras (part of the shell is missing, showing the elaborate and frilled sutures). (Both about × 0.2).

that of any extinct group; the various morphological stages enabling one to divide the Ordovician and subsequent Silurian into about 40 successive faunal zones and to recognise, in many cases, world-wide correlation.

Another remarkable evolutionary development by the nautiloid cephalopods was the development of gas-filled chambers, enabling them to rise from the sea floor. The stocks derived from the initial radiation appear to represent various means of overcoming buoyancy problems.

Silurian

The stocks of typically shallow-water marine invertebrates established during the Ordovician for the most part continue unchanged into the Silurian, and it is doubtful whether the data allow any analysis of the slight fluctuations in generic diversity which can be recorded (Figure 14. 9). There is, for example, a net fall in trilobite genera from the early Ordovician to the Silurian, but this may well mean that the remaining stocks, although showing less morphological diversity, were nonetheless in command of significant environmental niches. It would appear generally that Silurian marine conditions were not dissimilar to those of the Ordovician, but the better development of reefs is one significant change.

Egyed and others have estimated that between 50 to 75 × 10^6 km^2 of present-day continental areas were covered by sea during the Silurian. This is the maximum extent of the sea during the Palaeozoic, and at that time areas were open for shallow water invertebrate exploitation. Since then, the seas have retreated from the continents to their present positions, but not without significant fluctuations, as will be remarked later.

In two important respects an undercurrent of organic change is present in the Silurian: firstly in the clear origins, before the end of the period, of early vascular land plants; secondly in the appearance of true fish, although agnathous fish are known in the Ordovician. Already these two groups appear to have begun colonisation of the fresh-water environment.

Devonian

Traditionally the line between the Lower and Upper Palaeozoic is drawn at the Silurian-Devonian boundary. The major floral and faunal changes do not take place in the marine environment, but the rapid establishment of vascular plants on land, giving by the Middle Devonian the first forests, and the rise to great importance of the fish, present two of the most spectacular evolutionary diversifications known in the fossil record.

In the Lower Devonian there is some evidence for a provincial distribution of faunas. An Old World Province includes Europe and the Urals, western North America and eastern Australia and New Zealand. An Austral province includes South Africa and South America, Antarctica and the Falkland Islands. Eastern North America and northernmost South America are referred to as the Appalachian Province. These provinces are mostly distinguished by Brachiopoda. The Austral province was probably a high latitude one at that time, as suggested by the presence of glacial deposits, and by the rarity or absence of corals and cephalopods.

In the early Devonian there is a decline and apparently an extinction of graptolites, which until this time were so important among the marine plankton. As the graptolites decline, the ammonoid cephalopods appear, apparently derived in the middle Lower Devonian from certain nautiloids. There is no evidence that the ammonoids were successful in competition with graptolites, but they certainly subsequently exploited deeper marine waters. From their appearance in the Lower Devonian the ammonoids were one of the most successful of marine groups, evolving rapidly; they are found in a number of different rock types, probably as a result of their shells drifting after death. It is the extraordinary evolution of this group which provides the most detailed means of correlating marine rocks from the Devonian to the close of the Cretaceous, when they became extinct. But even during the Devonian the ammonoids suffered from substantial extinctions. At the close of the period they were almost completely extinguished, and only very few genera survived to form the root stocks for subsequent evolution.

As a result of a marked burst of evolutionary activity in the early and middle Devonian, corals diversified rapidly and became world-wide in distribution. But quite suddenly, at the close of the early Upper Devonian, extinctions almost eliminated the stock (Fig. 14. 9). Thus the extremely widespread reefs and associated coral and stromatoporoid limestones disappear in the Upper Devonian, and although later Devonian rocks are widespread in many parts of the world,

reef corals have not been recorded in them. Even the deeper- and colder-water solitary corals are not common. The same break is marked in the brachiopods, which had become extremely diversified in the Lower and Middle Devonian.

Throughout the Devonian major trilobite groups, most of which were established in the early Ordovician radiation, became extinct. It was a progressive decline, and the extinction at the close of the Devonian of the phacopid trilobites left only a meagre remnant surviving in the Carboniferous.

The Old Red Sandstone deposits of the Devonian represent the first extensive terrestrial deposits known in the Phanerozoic, and the extensive evolution of vascular plants and fish exploited the environments presented. By the later Devonian we find the first amphibians, and vertebrate colonisation of the dry land had begun. The Middle Devonian contains the earliest insects, and in 1961 Rhodendorf reported the first winged insect from the Russian Upper Devonian, indicating the first known colonisers in the aerial environment. Thus the Devonian was a period of great evolutionary changes in many groups.

Carboniferous

The early Carboniferous saw a return to widespread coral distribution over extensive areas of the world. Since the colonial corals found at this time are very different from those of the Middle and early Upper Devonian, a period of sudden evolutionary diversification from the surviving stocks is indicated. This proves to be the last major burst of the Palaeozoic rugose corals, after which they decline and become extinct at the close of the Permian period.

Other groups characteristic of shallow waters also show substantial diversification, particularly the brachiopods and especially the crinoids (Echinodermata), which are responsible for forming enormous thicknesses of crinoidal limestones in many areas (Fig. 12. 2).

After an initial high rate of evolution at the beginning of the Carboniferous, little diversification occurred during the rest of the period. Already in the early Carboniferous there is further evidence of diversification of the land plants. Groups well represented in the Devonian continue, and yet larger trees were produced among the Pteridophytes, but particularly important is the continued diversification of the seed ferns

(Gymnospermae). These culminate in the floras of the Upper Carboniferous Coal Measures, widespread in a belt from eastern America, Britain and Europe to the Donetz basin in Russia. This famous Carboniferous flora is comparable in many ways to present-day mangrove swamps, and so suggests an equatorial forest environment.

Associated with the Coal Measure floras are rich amphibian and fish remains. In the uppermost Carboniferous the primitive cotylosaur reptiles appear, but they are as yet insignificant. The insect faunas, especially of such celebrated localities as Commentry in northern France, are extremely rich, and include the remarkable dragon-fly, *Meganeura*, which had a wing span of 0.73 m—one of the largest insects known. Other air-breathing arthropods are present, such as spiders, millepedes and centipedes.

The typical later Carboniferous Coal Measure environments of Northern America, Europe and the Ukraine are widespread, but extensive calcareous marine deposits occur in other areas (central North America and Russia). In this environment the fusulinids (large protozoans) already referred to abounded, and there was a rich and varied shallow-water fauna of invertebrates, wholly absent from Coal Measure areas, which provides the continuum for the Lower Carboniferous calcareous marine faunas into the Permian. Notwithstanding this, the late Carboniferous appears to be a period of relative evolutionary stagnation.

Permian

Many marine invertebrates show marked resurgence and diversification during the Permian, notably the fusulinid Foraminifera (largely responsible for the generic Palaeozoic peak on Figure 14. 9). Echinoderms diversified greatly, and some remarkable new forms arose among the Brachiopoda and Ammonoidea. There was, in general, little to indicate the bleak times ahead.

For terrestrial colonisers the picture is perhaps enhanced by the widespread New Red Sandstone conditions (desert and fluvial) with an accompanying improvement in the record of this type of environment. Certain amphibians were well developed in size and variety, although reduced at the close of the period. Reptile evolution was substantial, including several orders of the mammal-like reptiles (Therapsids) commonest in the mid- and late Permian. Habits were both carnivorous and herbivorous. Therapsids were widely distributed, and good successive records are known, especially from the Beaufort Beds of South Africa, and from Russia. Insect faunas now include bugs, cicadas and the first true beetles.

Earlier, attention was drawn to the contrast between the northern hemisphere Coal Measure floras and those of Gondwanaland, where the *Glossopteris/Gangamopteris* flora occurs which, whilst often abundant, is restricted in variety. Association of Gondwanaland plants with glacial deposits shows that this contemporary flora is the high latitude equivalent of the Coal Measure flora.

Changes in the floras are apparent with a progressive reduction in pteridophytes and seed ferns, and a rise in conifers which, by the end of the period, were the chief woody trees.

Permo-Triassic Boundary

This is the division between the vastly different faunas of the Palaeozoic and Mesozoic. It corresponds approximately to a major period of extinction, especially for marine animals, and so this boundary has been the subject of great interest. Unfortunately, continuous marine sections through the boundary are few, but the best have been studied in detail, revealing once again the contrast between typical Permian elements and those of the Triassic, although the exact nature of the transition is still imprecisely known. The broad effects of these extinctions are shown in Figure 14. 13.

The closing stages of the Permian have been called a 'crisis in the history of life'. The fusulinid foraminifera, trilobites, eurypterids, rugose corals and three out of four bryozoan groups disappeared, with groups of echinoderms and brachiopods. Among the molluscs, the ammonoid cephalopods only just survived into the Trias, although others remained almost unscathed.

Even though the period over which the extinctions occurred is considerable, they far outweigh appearances, making this the most spectacular extinction period in the fossil record (Fig. 14. 13). Against this must be set the limited floral changes which took place, but between the Permian and Triassic some 75% of amphibian and more than 80% of reptile families disappeared.

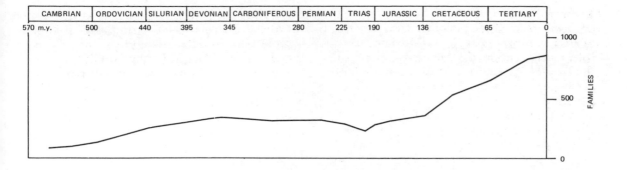

14. 12 The numbers of families of all groups at successive times in the Phanerozoic.

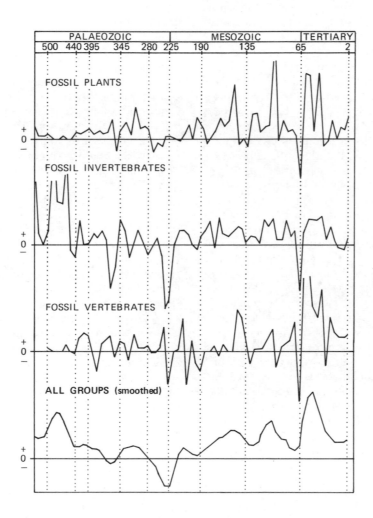

14. 13 The net results of extinction and appearances of new groups at stratigraphical boundaries (mostly stage) during the Phanerozoic. Based on 2 526 mixed taxa from all environments.

The net result of these extinctions, whatever the cause, was to leave many marine environments and ecological niches open to exploitation and re-colonisation during the Mesozoic. Herein lies the crux of the contrast, for these environments could only be colonised by survivors.

Triassic

The characteristic features of Mesozoic invertebrate and vertebrate faunas (if not of floras) are established within the Trias or earliest Jurassic. The history of marine organisms through the Mesozoic is one of increasing diversification, and this is indicated, at generic level, by the graphs in Figure 14. 13, but it is also true at the family level (Fig. 14. 12).

What then are the distinctive elements of the Mesozoic invertebrate faunas already shown in the Triassic? The records bearing on this are mostly from the Middle and Upper Trias, for the Lower Triassic record is very sparse. The bivalves come to rival or surpass the brachiopods in the fully marine environment, and of these the entry and success of the oysters is noteworthy. Among Mollusca the belemnoids, although known probably back to the Devonian, begin to be important. Scleractinian corals with aragonitic skeletons, in contrast to the calcitic hard parts of Palaeozoic rugose corals, form a new distinctive element. The more complex-sutured ammonoids (Fig. 14. 11), descended from the survivors from the Permian, initiate spectacular evolutionary radiations. Among the larger neritic animals, crustaceans become the dominant arthropods (although ostracods developed further) and simpler echinoderms replace their elaborate Palaeozoic ancestors. When related to the late Permian extinctions, the faunal transformation is considerable. In many areas, transgressing seas opened wider environmental opportunities, but many groups took long to re-establish themselves as important elements.

Changes in vertebrate faunas were also considerable. Fishes with more modern aspect became common. The reptiles had produced the earliest dinosaurs of quadrupedal and bipedal gait. The first ichthyosaurs and plesiosaurs found in the Triassic establish that diverse marine habits were being explored. Gliding pterosaurs occur near the Triassic/Jurassic boundary at this time, together with the first undoubted mammals.

During the Triassic the land floras, too, take on a Mesozoic aspect with the entry of the palm-like cycadeoids (the Mesozoic is often called the 'Age of the Cycads') and the extension of the ginkgos.

Thus, by substantial evolutionary radiation in some groups and by expansion in others, a new and distinctive pattern of life had been established by the end of the Triassic which was to continue to the close of the Cretaceous.

Jurassic

If the Triassic saw the establishment of faunas of Mesozoic aspect, the Jurassic saw its extension. As illustrated in Figure 14. 9, almost all common marine invertebrates show evidence of progressive diversification, as the wealth of fossil remains in the English Jurassic rocks affirms. Again the ammonoids vigorously evolved, and apparently several groups replaced others in the period, suggesting the operation of Arambourg's relay. Coral and sponge reefs are in places well developed, but an important palaeogeographical distinction, so far as Europe is concerned, is the continuance of a belt of deeper water from Spain eastward along a Mediterranean to Himalayan line. (The Tethys Ocean, Fig. 15. 19.) Shallower waters to the north of this were periodically colonised, largely in relation to changes in sea level. On land dinosaurs flourished and pterodactyls developed flight. By the end of the Jurassic the first bird, *Archaeopteryx*, is known with feathers modified from reptilean scales. Rare mammals occur, but they were as yet inconspicuous. Another group, the coccoliths (Fig. 14. 7e), appear in the lowest Jurassic and by the close of this period coccolith limestones occur, heralding the major development of such rocks in the Cretaceous.

Cretaceous

No major faunal or floral break separates the Jurassic from the Cretaceous, but there are substantial palaeogeographic changes. In the seas the coccoliths reach their acme, and another planktonic group, the globigerinid foraminifera (Fig. 14. 7f) appear and progressively increase in importance. In clear, shallow water, reefs are composed of the strange giant rudist bivalves, which mimic the morphology of corals. The teleost fishes begin their spectacular radiation.

On land the bizarre dinosaurs include the largest known reptiles, and marine reptiles also attain enormous proportions.

Animal life therefore shows further developments of the pattern established in the Trias and continued in the Jurassic.

The evidence of plants is very different, and the mid-Cretaceous saw the entry and most spectacular rise to dominance of the flowering plants (Angiospermae). A recent survey gives the following documentation of families :-

Uppermost Upper Cretaceous (*Maestrichtian*)	67
Lowest Upper Cretaceous (*Cenomanian*)	50
Uppermost Lower Cretaceous (*Albian*)	4
Lower Upper Cretaceous (*Barremian*)	1

The terms in brackets are stages of the geological column, based on successive assemblages of fossils.

All of these families exist to-day, although many others were established during the Tertiary. It is the sudden nature of their appearance which is remarkable. Bees and butterflies radiate in parallel with the flowering plants.

Cretaceous/Tertiary Boundary

Certain faunal characteristics of the Mesozoic are brought to an abrupt halt approximately at this boundary. Here the ammonites disappear, as do some bivalve molluscs (including the rudists), the belemnite cephalopoda, and some families of bryozoa, echinoids and planktonic foraminifera.

Perhaps the most spectacular change is in the reptiles, which are so substantially reduced that the term 'ruling reptiles' can never again be applied to them. Exact timing of individual extinctions is difficult, but nonetheless it is clear that a decimation of the reptiles occurred approximately at the Mesozoic-Tertiary boundary.

This boundary therefore clearly marks another major crisis in the history of life. As was the case with the Palaeozoic-Mesozoic boundary, the break is emphasised by the new groups which rise to importance in the succeeding period—the Cenozoic.

Tertiary

The enormous diversifications and dominance of the Mammalia is by far the most striking feature of this period. They now enter the sea and the air and on land become the ruling vertebrates. In the seas the major rise of teleost fish continues, and sepioid cephalopoda (e.g. squids) are another distinctive element. Floral changes on the other hand are not at all marked, although the record shows diversification of the angiosperms in the early Tertiary. Again the faunas progressively colonise habitats and environments largely vacated as a result of late Cretaceous extinctions. The time gap gives no evidence to suggest that the primary cause of the changes is successful competition against the vanquished stocks.

The divisions of the Tertiary—Eocene, Oligocene etc.—(Fig. 14. 3) were named by Lyell to indicate the successively more modern aspect of Cenozoic faunas. Some of the major changes during this period include the spectacular radiation of the planktonic foraminifera (globigerinids), which oust the coccoliths from their relative importance in the upper levels of the oceans. Little radiation occurs in the sponges, brachiopods, gastropods and bivalves, but there is a substantial increase in insects and a steady one in ostracods.

For vertebrates, however, the early Tertiary shows a most remarkable diversification of the mammals (Fig. 14. 3) comparable to that of fish in the Devonian and reptiles in the Permo-Trias. What is remarkable is the suddenness of the take-over. Good late Cretaceous mammal faunas are known in a number of areas, but they are of primitive types. Yet by the Eocene mammals have radiated, and specialised stocks have developed, exploiting as wide a range of environmental niches as they do to-day. By the Eocene, primates, rodents, carnivores, a great variety of herbivores, whales, sea cows and bats are known (in fact a number of these existed in the Palaeocene). The time scale available for this radiation is considerable (about 10 million years), but this does not make it less spectacular, and gives evidence of the high rate of morphological adaptation when selection pressures from other groups are low during the exploitation of new environments. Equally spectacular is the rise of the birds in the early Tertiary, although the record available is certainly not very complete.

In this chapter there is not room to document the known, detailed changes in the content and distribution of Tertiary faunas and floras. There is evidence that with the onset of colder conditions, and the accompanying contraction of climatic belts towards the Pleistocene ice age, extinction and redistribution of faunal and floral elements occurred. Indeed it is likely that equilibrium has still not been reached—and the impact of man is a relatively recent additional factor in this respect.

INTERPRETATION OF FLUCTUATIONS

We now turn to speculations. In science it is quite usual for the most interesting problems to be those we have not yet solved. The major problems in the interpretation of the fossil record of evolution lie in the causes of evolutionary diversification and extinction. A good deal may be learnt by studying the impact of certain factors on present-day organisms; interpretation of past events, however, is another matter.

The relative fluctuations of individual groups during the Phanerozoic has already been outlined. Another wide assessment, based on the net result of appearances and disappearances of some 2500 taxa, is illustrated in Figure 14. 13 from the analyses of Cutbill and Funnell published in *The Fossil Record* (1967). This emphasises in a different way the fluctuations at higher classification levels, and for more groups, than those illustrated in Figure 14. 9.

Factors affecting living organisms may be classified as biotic and physico-chemical. Biotic factors include those inherent in an individual, or an individual group, as well as the limitations imposed on them by other organisms, either in terms of food supply, direct or indirect competition, or the broad environment provided by other organisms. Physico-chemical factors include temperature, chemistry, availability of materials, topography, radiation and extra-terrestrial controls, and other physical changes which may take place such as earthquake shock waves and sea level changes.

From time to time authors have championed one factor above all others as a cause of evolutionary change. For example, increased cosmic radiation (resultant upon collapse of the Earth's magnetic field during polarity reversals) has been put forward as a cause of extinctions. But, as will be seen from the preceding discussion and as has become clearer with recent work, there are no documented examples of specific universal decimation in all environments which would support such a view (see Chapter 18). Rather it seems that a large number of factors probably operate. However, as Bramlette has emphasised, there is no weaker link for marine life than the plankton supply, which would be particularly vulnerable to such radiation.

Let us consider first the evolutionary radiation near the base of the Cambrian (see colour plate). Rudwick has drawn attention to the possible relation of the evolutionary radiation to climatic amelioration after the late Precambrian glaciation, when invertebrates may have diversified in restricted (and hitherto undiscovered) areas. Others would lay stress on chemical reasons, notably the rise in biogenic oxygen which reached a threshold only at this time (see Chapters 10 and 12). Arguments suggesting unusual chemical composition of the seas have less to commend them in view of the Precambrian stromatolite record. Palaeogeographic changes, especially if providing widespread shelf seas, would create an immense new area of environments for colonisation. There are other plausible explanations as well.

The extinctions at the close of the Palaeozoic do not show a precise relation to the widespread glaciation of the late Palaeozoic, but it was probably a factor, for example, in lowering ocean temperatures and compacting climatic belts. Fischer has argued convincingly that the widespread Permian salt deposits, aided by pockets of dense brine on the ocean floors, could so have reduced the salt content of the seas as to cause extinctions. Others have invoked climatic changes. Newell of Columbia University has emphasised the possible effects of changes in sea level (positively for the Cambrian and Ordovician radiations, negatively for the late Palaeozoic). The detailed record does not suggest that the late Palaeozoic extinctions were sudden in time or that several factors contributed.

A somewhat similar picture is presented by the changes at the Cretaceous-Tertiary boundary, where the extinctions may cover a considerable period of time. Here palaeogeographic changes seem important. The products of long Mesozoic stability (the low-lying tracts of shelf seas and alluvial plains) were particularly vulnerable to change. For instance, there is evidence for a withdrawal of the sea in continental areas in the late

Cretaceous, and so, for example, dinosaurs living in inland swamps would have found their position a particularly compromising one.

MacAlester of Yale University has recently drawn attention to the remarkable correlation between the extinction rates of organisms and their free oxygen requirements for metabolism. He takes this as an indication that fluctuations in atmospheric oxygen in the past could have important repercussions on life. And so another, and attractive, hypothesis is added to our speculations (this is further discussed in Chapters 10, 11 and 12).

The dangers of dogmatism in this field of interpretation will be apparent, but many recent writers have emphasised the major effects which sea level changes would have. This is particularly important in the light of modern views on the development of the oceans. The sea-floor spreading hypothesis (Chapter 16) demands movement of the ocean floors with associated volcanism. This is unlikely to have taken place without topographic changes of the ocean floor and hence changes in sea level.

Darwin, in *The Origin of Species*, was obliged to state that the fossil record added little evidence to support his contentions, but the century of palaeontological work since then has documented in extraordinary detail the course of the evolution of life. Unfortunately the palaeontologist is often so concerned with the practical applications of his work for economic purposes, that there remains much work in analysing and interpreting the fossil record.

The study of fossils will always remain a major branch of the Earth Sciences. Fossils give us the most accurate means we have for time discrimination over the last 600 million years; fossils give the key to palaeo-ecological interpretation of past sedimentary rocks, enabling us to determine palaeo-environments and palaeo-geographies. Finally, fossils give the prime evidence of the extraordinary evolution of life from small beginnings to the wealth of living things to-day. And within this evolution lie the beginnings of Man himself.

FURTHER READING:

F. H. T. RHODES 1962. The Evolution of Life. *Pelican*.

J. F. KIRKALDY 1963. The Study of Fossils. *Hutchinson*.

K. P. OAKLEY AND H. M. MUIR-WOOD 1967. The Succession of Life through Geological Time. *British Museum (Natural History)*.

A. S. ROMER 1968. Man and the Vertebrates. *Pelican*.

The concept of continental drift is now over 40 years old, but it is only recently that it has come to be widely accepted among Earth scientists. In the 17th and 18th centuries, when the outlines of the continents had become sufficiently well known, many scientists, such as Francis Bacon (in 1620) commented on the similarity of the shapes of the coastlines on either side of the Atlantic. But the Comte de Buffon (in 1749) flatly denied the fit of Africa and South America, and suggested that the Atlantic had been formed by the sinking of the mythical 'Atlantis' and later erosion of the continents by sea currents. Indeed a 'catastrophic' origin for many geological phenomena was in vogue during this period due to the continuing influence of religious dogma.

By the end of the 19th century the geology of the southern continents was sufficiently well known for an Austrian geologist, Eduard Suess, to suggest that Africa, South America, Australia and India were once part of a super-continent, for which he coined the name *Gondwanaland* (after a region of India called Gondwana). In 1910, a German meteorologist, Alfred Wegener, proposed that at the beginning of the Mesozoic era (which we now know to have started about 200 million years ago) Gondwanaland was united with Eurasia and North America to form a single super-continent, which he called *Pangaea*. Wegener proposed that this vast continent began to break up at the beginning of the Mesozoic era by a process of lateral crustal movement. To support his theory, Wegener marshalled considerable geological evidence, including similarities in geological structure, distribution of rock types and fossils—and even the similar shapes of the coastlines on either side of the Atlantic. At the time the geological evidence was not unequivocal, and even more damaging for Wegener's drift theory was the pronouncement by contemporary geophysicists that the properties of the Earth's crust were not such that large lateral continental displacements could occur.

The controversy continued unresolved for over forty years and, ironically, was rekindled in the 1950's and 60's by the geophysicists, who this time produced evidence in favour of drift. This new evidence came from geomagnetic studies, both of the continents and the oceans, and in turn stimulated a more rigorous objective and quantitative re-study of the geological 'match' between continents. In this chapter Dr. A. G. Smith of the Sedgwick Museum, Cambridge, summarizes some of the evidence favouring drift including geomagnetic data (palaeopole investigations), geological similarities between the southern continents, and computer estimates of the topographic fit of continents, including those bordering the Atlantic. The evidence to be described in this chapter, together with that from the oceans described in Chapter 16, comes as near to 'proving' Continental Drift as is possible for any geological hypothesis.

15 Continental Drift

by ALAN GILBERT SMITH

Whether the present continents have moved large distances relative to one another was, until a few years ago, an extremely controversial issue in geology. The reader may, like most Earth scientists, find the evidence given here virtually conclusive, but contrary views still prevail.

TO-DAY THE GREAT MAJORITY of geologists accept as a fact that the present distribution of the continents results from the break-up and joining together of previous distributions of the continents. This process, known as *continental drift*, has probably acted for a very long time, perhaps ever since the Earth first had a crust.

Why there should have been such a prolonged and heated discussion, followed by such a rapid change of opinion, forms an absorbing episode in the history of science, whose outlines only will be sketched here. In part this change has been brought about by new evidence, particularly by sea-floor spreading data (Chapter 16) and polar wandering curves from different continents (this chapter); but much of the evidence discussed here was available thirty years ago, when the controversy was at its peak. It consists of geological similarities between the joins of two continents of similar shape.

SOUTH AMERICA AND AFRICA: A CASE STUDY

Fitting the Continents together

A glance at any atlas or globe shows the Atlantic coasts of South America and Africa to have a roughly similar shape. However, the coastline is not a very significant geological feature because its shape changes when the sea level rises or falls relative to the level of the land. Over periods of the order of a few million years the coastline is a very unstable feature that constantly fluctuates in shape and position. Thus the similarities in shape of any two coastlines may generally be regarded as of no particular significance. (Imagine with the aid of an atlas the changes in the coastline of South America and Africa that would take place were the sea level to rise by 200 metres.) The real edge of a continent lies somewhere on the continental slope, where the sea bottom falls comparatively rapidly from somewhere near sea level down to the deep ocean floor, about 4000 metres below the surface (Figure 15. 1a).

The similarity in shape of a contour drawn on the continental slope of eastern South America and the same contour drawn on the west African slope is quite striking. Is this a chance similarity, or is it significant? Geologists have approached this problem by estimating the best fit between the two continental edges and comparing the geology across the join. But what is the edge of a continent? How does one estimate the best fit? And how does one describe how to bring one continent to the other so that a map may be made of the best fit?

Carey, a Tasmanian geologist, constructed a large globe, traced off the submarine contours of one continent on to a large transparent cap, moved it to the other continent, and obtained the best fit by eye. A map was drawn from a photograph of the best result. This may seem an inexact way to investigate the problem, but the eye is a particularly good estimator of shapes, and the result obtained by Carey is not significantly different from that obtained by computer (Fig. 15. 2 and 15. 3a). The principal drawback of his method is that it does not describe how to make the fit in numerical terms, nor does it indicate which of the submarine contours is the best fit.

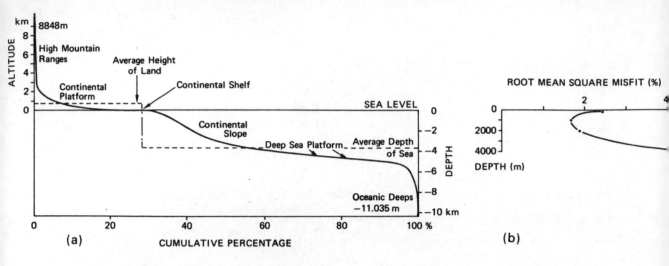

(a)

ALTITUDE

km
8848m

High Mountain
Ranges

Average Height
of Land

Continental
Platform

Continental Shelf

SEA LEVEL

Continental
Slope

Deep Sea Platform

Average Depth
of Sea

DEPTH

Oceanic Deeps
−11.035 m

CUMULATIVE PERCENTAGE

(b)

ROOT MEAN SQUARE MISFIT (%)

DEPTH (m)

15. 1a Profile showing percentage of the Earth's surface lying above a certain height. Two levels predominate: one on the continents, the other on the deep ocean floor. The edge of the continent, defined as half-way down the continental slope, lies at a depth of approximately 2 000 metres (about 1 000 fathoms).

15. 1b Misfit variation with depth for the submarine contours between South America and Africa. The 1 000-metre (approximately 500-fathom) contour gives the best fit. It is only slightly better than the 2 000-metre contour.

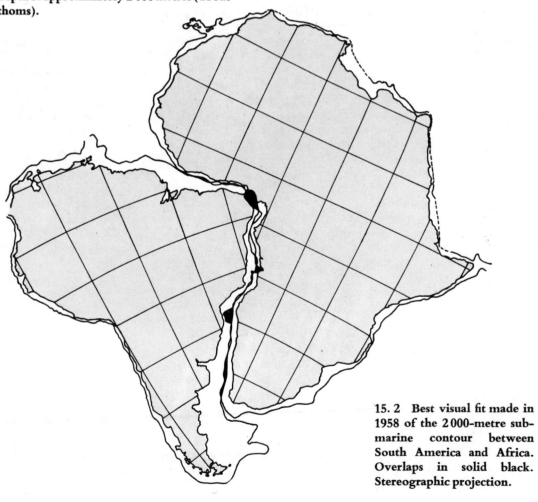

15. 2 Best visual fit made in 1958 of the 2 000-metre submarine contour between South America and Africa. Overlaps in solid black. Stereographic projection.

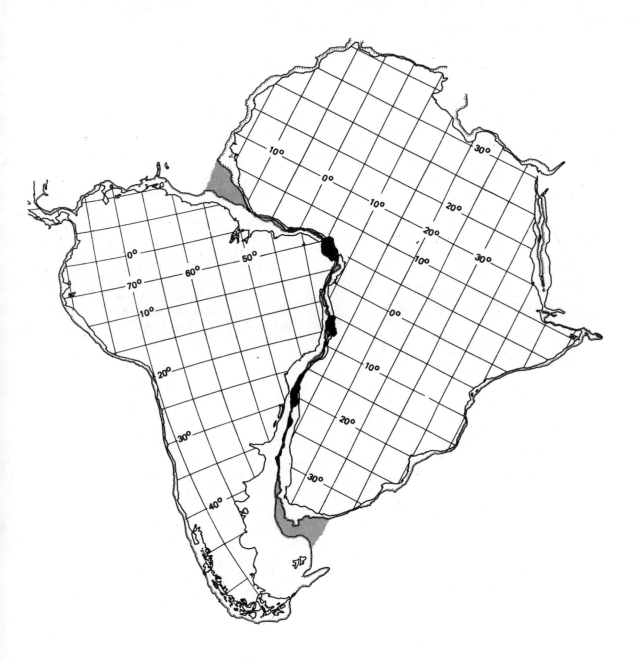

15. 3a   Computer fit of the 500-fathom (approximately 1 000-metre) contour between
South America and Africa, made in 1965.   The map was drawn by hand from numerical
data, using Mercator's projection, with Africa in its present position.   Overlaps in solid
black, gaps in colour.

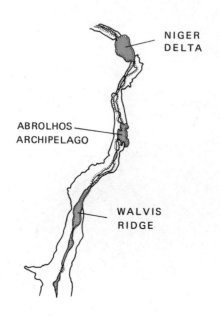

NIGER
DELTA

ABROLHOS
ARCHIPELAGO

WALVIS
RIDGE

15. 3b The principal overlapping areas in Figure 15. 3a, here shown in colour.

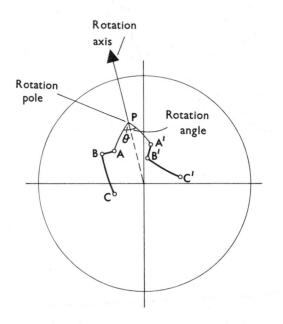

Rotation axis

Rotation pole

P

Rotation angle

θ

B   A

A'
B'

C'

C

15. 4 Rotation about a pole on a spherical surface. Rotation about the pole P through the angle θ moves ABC to A'B'C'.

The computer method depends on applying a theorem in spherical geometry to the problem. One way of stating the theorem is that if two identical shapes are drawn on the surface of a sphere, they can always be brought into coincidence by turning one or other of them through a certain angle about a certain axis through the centre of the sphere. Where this axis intersects the surface is known as the *rotation pole*. The angle is known as the *rotation angle* (Fig. 15. 4).

Sir Edward Bullard, Professor of Geophysics at the University of Cambridge and author of Chapter 4, suggested applying this theorem to the problem of fitting continents together, and he and Everett, one of his research students, wrote a computer programme that would give the rotation pole and rotation angle needed to make the best fit. The best of all possible fits may be thought of as the one giving the smallest overlaps and gaps between the two contours. Using the best data available in 1963, Everett examined in detail the fit between the submarine contours of South America and Africa. He showed that the 500-fathom (approximately the 1000-metre) contour was a slightly better fit than the 1000-fathom contour. Both fits were much better than those of the 100-fathom or 2000-fathom contours (Fig. 15. 1b). One could use these results to *define* the continental edge as the best-fitting contour, which would make the 500-fathom line the edge of the continent. The theoretical edge is half-way down the continental slope, or at a depth of about 2 kilometres (approximately 1000 fathoms), and is generally not far from the best-fitting contour.

The average misfit was about 90 kilometres, which means that on average there is a gap or an overlap of this size between the two 500-fathom contours in their best-fit position. The rotation pole was 40°N and 30.6°W, close to the western Azores, and the rotation angle was 57°. Thus if one traces the outline of South America on to a spherical cap, and turns the cap through an angle of 57° about a point near the Azores, then one will obtain the same result as Everett—try it on a globe with a strip of tracing paper.

The latitudes and longitudes of points on South America in their new position alongside Africa may readily be found from spherical geometry, and it is then a simple but very tedious job to make a map (Fig. 15. 3). The map seems to distort South America, but the distortion is caused only by the projection used, and in no way reflects a real distortion of South America relative to Africa.

The excellent match between the submarine contours strongly suggests that South America and Africa were once joined together. If they were, one expects to be able to trace some geological features across the join. Such features will be older than the separation of the two continents. Any features that overlap the join presumably formed after some separation had occurred. These inferences were made long ago, and the similarity in geology between South America and Africa provided one of the many arguments used in 1912 by Wegener, a German meteorologist, to develop the ideas of continental drift in a convincing manner. These ideas were later elaborated, particularly by du Toit, a South African geologist, and some of the evidence used by both of them will be examined below.

Overlapping Areas

The three main areas of overlap on the fit are: the Niger delta and northeast Brazil; an area off Brazil on which lies the Abrolhos Archipelago and part of west Africa; and an area off the coast of southwest Africa and part of South America (Fig. 15. 3b). Geophysical and geological evidence suggest that the Niger delta is a thick wedge of sediments younger than 50 million years; the age of the bedrock of the Abrolhos Archipelago is not known; the overlapping area of southwest Africa occurs where an oceanic structure known as the Walvis Ridge intersects the African continent. This ridge is no older than the oldest sea-floor in the South Atlantic, and is therefore almost certainly much younger than 200 million years. Where it joined the continent it could have caused the uplift of the submarine part, thereby extending the submarine contours seaward beyond their original positions and causing the overlap. In other words, none of the evidence shows these features to be older than the Mesozoic or Tertiary periods. If the continents were once united, they must have separated by at least 200 kilometres about 50 million years ago to allow the Niger delta to form.

Matching the Geology across the Join

When matching the geology across the join of two continents, one is faced with the problem of which geological features are significant evidence for original continuity and which are not. The kinds of evidence one looks for are such things as

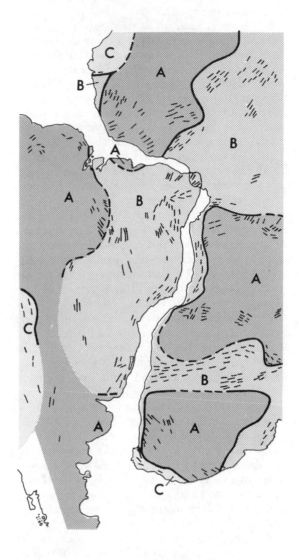

15. 5 Age provinces and structural trends of South America and Africa plotted on Figure 15. 3b. A = ages older than B; B = ages in range 550 ± 100 million years; C = ages younger than B. The structural trends are the trends of folds and structures in the metamorphic basement and are shown by short lines. The age provinces and structures in northeast Brazil and West Africa match quite well. Those in other areas match less well, probably because they have not been studied in such detail.

similarities in the stratigraphical succession, in the fauna and flora preserved in it, and in the orogenic belts that cross the join.

In a general sense all of these features may be thought of as the end result of a complex environment that has acted on various original materials to produce the rocks, fossils, and structures one sees to-day. So, when one uses the similarities between two geological features as evidence that the continents were joined together, one is really saying that there were similar environments and similar starting materials on either side of the join.

As a generalization, the environments may be divided into two kinds: surface environments, and subsurface environments. The first are reflected in fossils and sedimentary rocks, the second in orogenic belts (see Chapter 20). Similar surface environments today consist mostly of belts parallel to latitude lines. In some cases—for example, the tropical climatic zone—they stretch around the widest part of the globe. If these act on similar starting materials they will give rise to similar deposits.

For example, tropical rain forests occur in parts of West Africa and eastern South America to-day, and the deposits that are forming in these areas, in the Amazon and Niger deltas, when looked at by geologists millions of years hence will probably closely resemble one another. Yet it is clear that the deposits are not being laid down on a single continent. In this case their apparent similarity would be a misleading indicator of the proximity of the two continents, but would show that the two deltas had each been positioned in the tropical zones of to-day. A detailed study of the fauna and flora buried in the sediment would show some differences, which might suggest the deposits had not been laid down near each other. In other words, the similarity of sedimentary rocks can be used as an argument in favour of original contiguity only if the similarities extend to the detailed properties of the sediments, and only if the similarities are not of the kind one expects to find over large areas. Because many of the arguments used to support continental drift have relied on comparing the *superficial* similarities among sedimentary rocks, they have been controversial and inconclusive.

By contrast, the subsurface environments reflected in orogenic belts occur as rather narrow, roughly linear zones on the surface. They are commonly only 1000 kilometres in width and have sharply defined boundaries. If one finds that, after making a geometrical fit, the orogenic

belts of the same age are aligned across the join, then this is a persuasive argument for the reality of the join. When such arguments were used by du Toit, the age of the orogenic belts was very poorly known. The advent of isotopic dating (Chapter 2) has enabled these belts to be dated relatively precisely. Several 'provinces' showing the same range of isotopic ages and similar structural trends have been recognized on the eastern border of South America and the western border of Africa, particularly from the work of Patrick Hurley at Massachusetts Institute of Technology. When these belts are plotted on the best fit between the two continents, there is a good match across the join. These data provide some of the best evidence supporting their original continuity (Fig. 15. 5).

Figure 15. 5 suggests the two continents were joined together about 500 million years ago. If one rejects this evidence, one is forced to postulate that extraordinarily similar deep-seated processes may occur within continents even when the ocean between them shows no sign of structures with the same trend or of the same age. If one accepts this evidence and the evidence from the overlapping areas, it suggests that South America and Africa separated from each other some time in the interval from 500 to 50 million years ago. Can we be more precise than this?

The Permo-Carboniferous Glaciation

On top of the eroded remnants of these orogenic belts lies a series of sedimentary rocks. The best known of them consist of glacial deposits formed mostly in the time interval of 350 to 250 million years ago. In Africa these extraordinary deposits are known as the *Dwyka Series*, and extend as far north as the present equator. They resemble in all respects the blanket of glacial deposits left by the Pleistocene ice age in the northern hemisphere. Erratic blocks carried by the ice lie strewn over a large area of South Africa, mimicking the boulders of Shap granite and Norwegian igneous rocks that are scattered over parts of England to-day. Ice erosion has created deeply grooved and scratched surfaces in the bedrock of valleys, exactly like those one finds on the glaciated valleys of Britain. No geologist disputes the glacial origin of the bulk of the Dwyka Series, and similar undisputed deposits are found in South America.

For several reasons glacial deposits generally provide much better evidence for the original arrangement of continents than do any other kinds of sediment. Firstly, they form in polar regions, where the latitude belts are smallest, and consequently where the errors in repositioning the continents are least. Secondly, at any one time the boundary between glacial deposits and other kinds of deposits is much sharper than boundaries between sediments formed in other environments, for example between tropical and sub-tropical environments. This is because the boundary between a stable ice-sheet and adjacent areas is a phase change (from ice to water) dominated by only one variable, the melting temperature of ice, rather than depending on several variables, such as rainfall and wind pattern as well as temperatures. Lastly, ice transports blocks over great distances, and the blocks are commonly large enough to enable the source areas to be unambiguously identified.

Nevertheless, many geologists have argued that the Permo-Carboniferous glacial deposits were caused merely by similar conditions in South America and Africa during Permo-Carboniferous time, rather than by the result of glaciation on a single continent. What do detailed investigations reveal? Martin, a South African geologist now working in Germany, has studied the glacial deposits of South America and Africa. In the Parana basin in eastern Brazil (Fig. 15. 6), the fossil boulder clays (known as *tillites*) are very uniform and widespread. Therefore they are unlikely to have been deposited by localized valley glaciers and were probably deposited by a large ice-sheet. The direction of the ice flow was inferred from wrinkles, folds, and thrusts in the underlying sediments which were soft enough at the time of glaciation to have been deformed by the shearing action of the overriding ice. The ice flowed from southeast to northwest, that is, from a source area southeast of the present Brazilian coast. The average thickness of the deposits is in many places as much as 600 metres, or about ten times the average thickness of the Pleistocene glacial deposits in northern Europe, formed during the last ice age. Where is the source area for this enormous blanket of glacial material? Those geologists who do not accept continental drift are forced into the following kind of argument:

Suppose the source area lay between the present coast and the continental edge. Then the volume of glacial deposits implies that about 500–

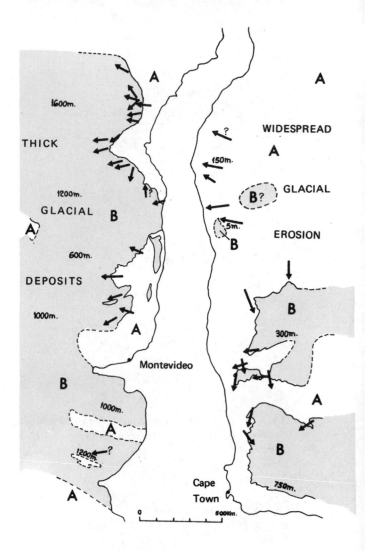

15. 6 **Distribution and thickness in metres of Permo-Carboniferous glacial deposits in South America and Africa plotted on Figure 15. 3b. Arrows indicate directions of ice flow. A = no glacial deposits; B = glacial deposits.**

800 metres of solid rock has been scraped off the land that once occupied this area. But ice-sheets generally do not grind down solid rock in this manner. They may modify pre-existing topography and reduce its general height, but only after a far greater time than that available are they likely to strip off such a thickness of bedrock. Thus the source area must have lain beyond the

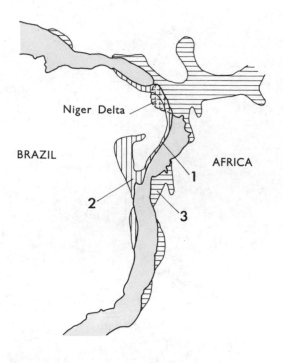

15. 7 Cretaceous and Jurassic sedimentary basins in West Africa (horizontal shading) and Brazil (vertical shading) plotted on Figure 15. 3b. Basins: 1 Sergipe; 2 Bahia; 3 Gabon.

present continental edge and have extended into the deeper parts of the Atlantic Ocean. But there is no trace of such a vanished source area to-day, and one must postulate that a large piece of continent has disappeared in some unspecified manner off eastern Brazil.

How such a vanishing act could be performed is an even more difficult problem than envisaging how continental drift occurs. From geophysical evidence, the Atlantic off eastern Brazil is floored to-day by normal oceanic material. This means that some 40 km of continental crust, postulated as the source area for the Permo-Carboniferous glacial deposits, must have been converted in some way to 5 km of oceanic crust and 35 km of upper mantle. An alternative explanation would be complete removal! Arguments of this kind in science are known as special hypotheses and are commonly suspect.

The alternative view is that the source area for the glacial deposits lay in Africa. The glacial deposits of southwest Africa are sparsely distri-

buted and consist mostly of glacial deposits preserved in valleys that were in existence prior to the glaciation. Though the deposits are meagre, there is abundant evidence for ice erosion. Where they are not covered by younger deposits or modified by later erosion, the valley walls and floors are scratched and locally polished. Rocks have been plucked into characteristic glacial shapes. The direction of ice flow is from east to west. The scarcity of glacial deposits in southwest Africa, coupled with abundant evidence for ice erosion, can be interpreted as meaning that southwest Africa was covered by an actively eroding ice-sheet which dumped its load farther west. This is precisely what the hypothesis of continental drift requires. Furthermore, in Brazil the load contains erratic blocks of distinctively coloured quartzite, dolomite and chert (a siliceous sediment) that cannot be matched with any known source rocks in Brazil. Yet very similar rocks occur in some of the tillites of southwest Africa, and in some cases the source rock can be located there. This evidence does not prove original continuity between Africa and South America, but it does show that it is more reasonable to postulate a much smaller separation between the two continents about 300 million years ago than exists today.

Exactly the same conclusions follow from a detailed analysis of other sediments in eastern Brazil and western South Africa. The available data suggest that the continents were still close to each other, and probably joined together until at least Upper Triassic time (about 200 million years ago). Furthermore, the Triassic rocks, consisting of wind-blown desert sands covering vast areas, are completely different in character from the older glacial deposits. The change in climate recorded by these rocks suggests either that the climatic belts of the world have changed their positions relative to one another, or that South America and Africa moved from high palaeolatitudes near the south pole into lower and warmer latitudes nearer the equator, some time in the interval between the Permo-Carboniferous glaciation and the formation of the Upper Triassic deserts.

Dating the Break-up

The geological evidence, taken as a whole, indicates that the two continents broke up some time in the interval between 200 and 50 million years. Can one tell from the land geology when

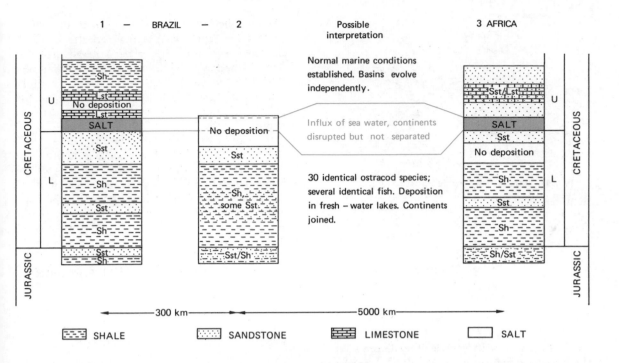

15. 8   Generalised Jurassic and Cretaceous successions in the basins shown on Figure 15. 7.

the break-up began? Obviously, this information, if it exists, is mostly likely to be found along the continental margin.

In Lower Cretaceous time (about 135 to 100 million years ago) several sedimentary basins formed along the continental margin of Brazil and West Africa (Fig. 15. 7). In general the succession in these basins starts with sandstones and shales deposited in fresh water about 135 to 110 million years ago (Fig. 15. 8). The fish and ostracods (small arthropods with a bivalve shell) in these rocks are extraordinarily similar. For example, at least thirty of the ostracod species are common to the basins in eastern Brazil and the Gabon basin in Africa. Eggs of present-day ostracods resembling some of the fossil species become infertile after immersion in sea water. It is highly probable that this was also true of the fossil species themselves. Thus there was probably a fresh-water connection between the two basins in part of Lower Cretaceous time.

On top of these fresh-water deposits are beds known as evaporites, formed by the precipitation of minerals from evaporating sea water. The evaporites are roughly between 110–100 million years old in both countries. The rocks following on the salt deposition are still marine, but the similarities between the successions in each basin diminish with time.

These marine deposits are the first such deposits known to have been laid down in eastern South America and western Africa since Upper Carboniferous time (about 300 million years ago). Unlike the earlier marine deposits they are confined to the present continental margin, rather than being widely distributed or confined to troughs at a high angle to the margin. The fresh-water/ evaporite/marine sequence is accompanied by a progressive decrease in the similarity between the two successions. All of these facts are consistent with the break-up of the two continents in Lower Cretaceous time (about 100 million years ago). It is not possible to prove from the land geology alone that this is what actually occurred; only deep drilling of the Atlantic or sea-floor spreading data can verify this interpretation (Chapter 16).

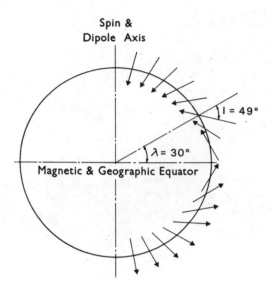

Spin &
Dipole Axis

I = 49°

λ = 30°

Magnetic & Geographic Equator

made to the raw data, but a sufficient number of rocks have been analysed magnetically to show that the main assumptions are generally correct. The method is capable of giving palaeolatitudes reliable to within 10 degrees, or better. Unfortunately the method tells one nothing about the palaeolongitude of the rock at the time of its magnetization.

In spite of this drawback one may use palaeomagnetism as a tool for verifying continental drift. In fact, the difficulties of accounting for the palaeomagnetic results by assuming that the continents in past times had been fixed in their present-day relative positions provided the first piece of geophysical evidence for continental drift. What does one expect to see in the palaeomagnetic data for South America and Africa? Because of the glacial deposits in the Permo-Carboniferous

15. 9 Variation with latitude of inclination due to the magnetic field of a centred dipole parallel to the Earth's axis of spin.

Palaeomagnetic Evidence

The Earth's magnetic field to-day is very similar to the field that would be produced by a bar magnet at the centre of the Earth, aligned parallel to the Earth's spin axis (Chapter 4). Of course, there is no such thing as a bar magnet in the Earth's core, but the fact that the field behaves as if there were means that the lines of force emerge at the Earth's surface at angles that are unique for a particular latitude (Fig. 15. 9).

This angle is known as the inclination, I. If L = latitude, then the relationship between I and L is:

$$\tan I = 2 \tan L .$$

When sediments containing iron minerals are laid down, or when igneous magmas crystallize, they will in general take on a weak magnetization whose direction parallels that of the Earth's magnetic field at the time. By assuming that this field has always resembled that of a bar magnet parallel to the Earth's spin axis, it is possible to determine the palaeolatitude and pole position from the direction of magnetization in a rock. There are of course important corrections that need to be

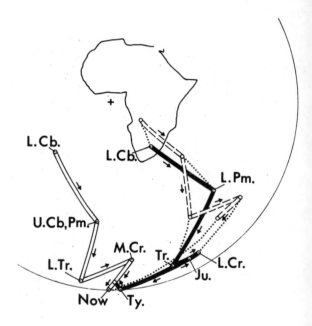

L.Cb.

L.Cb.

L.Pm.

U.Cb,Pm.

M.Cr. Tr.

L.Tr.

L.Cr.

Ju.

Now Ty.

15. 10 Polar wandering curves for South America and Africa plotted on a stereographic projection. Three curves are shown: those for South America (unbroken double lines) and Africa (black line) in their present-day positions; the third for South America in its best-fit position next to Africa (broken double lines). Cb. = Carboniferous; Pm. = Permian; Tr. = Triassic; Ju. = Jurassic; Cr. = Cretaceous; Ty. = Tertiary; L. = Lower; U. = Upper. Dotted line joins poles of same age.

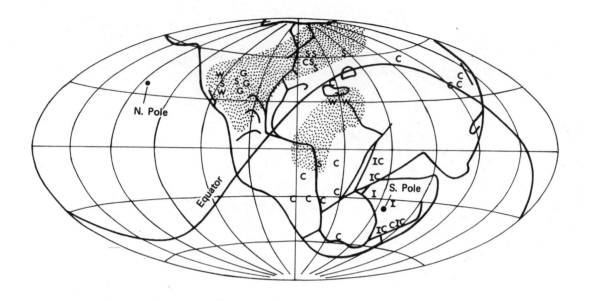

15. 11 A map of the world in Permian time made in 1924. The Pole and equator were estimated from the distribution of wind-blown sands (W and dotted pattern); rock-salt (S); gypsum (G); coal (C); and glacial deposits (I). Compare with Figure 15. 19, noting that this figure shows the world and not just one hemisphere.

period one expects South America and Africa to have been near the pole and also to have had latitudes that correspond (within the limits of error) to those of the best geometrical fit. During Triassic time one expects the continents to have had the same relative palaeolatitudes, but the desert deposits suggest they were closer to the equator. The Jurassic geology has not been mentioned in this chapter, but since break-up probably did not occur until mid-Cretaceous time it is only in Upper Cretaceous or later time that one expects the palaeolatitudes to differ.

Though as yet sparse, the available data confirm these predictions in a remarkable way. When plotted on the best-fit reconstruction, the Carboniferous, Permian and Triassic poles lie within overlapping circles of error. None of them have overlapping circles of error in their present-day positions. The palaeolatitudes of Permian and Carboniferous time place the southern parts of both continents within the Antarctic circle of that time, that is, within 23.5 degrees of the pole (Fig. 15. 10). The Triassic poles from Africa and South America place both continents forty degrees closer to the equator. There are as yet no reliable

data for the Jurassic pole of South America, but the Cretaceous poles are the first poles on the best-fit reconstruction that do not have overlapping circles of error. Such a separation suggests the continents moved apart in the Triassic to Cretaceous time interval.

Thus the conclusions drawn from the geological evidence are borne out by a completely independent geophysical technique. This concordance between two completely different disciplines supplies the best kind of 'proof' of continental drift. When the sea-floor spreading and plate tectonic evidence (Chapters 16 & 19) are also added to the argument, the original fitting together of the two largest southern continents in Upper Palaeozoic time, and their break-up in Mesozoic time may be regarded as established facts.

GONDWANALAND

Having considered the evidence for the fit of South America and Africa, one may go on to consider evidence for the fitting together of other continents. Long before isotopic dating had been

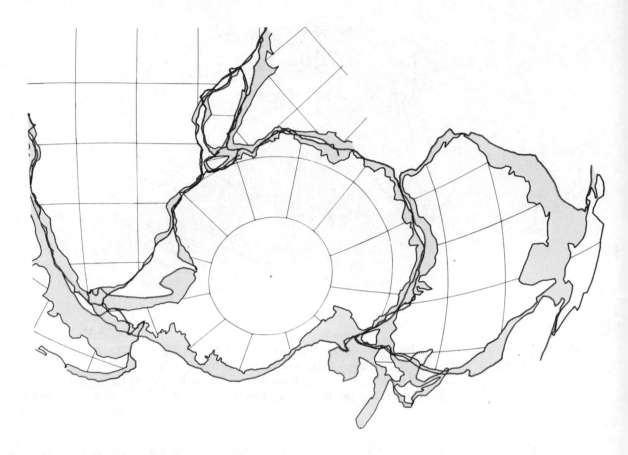

15. 12   A possible geometrical fit of the southern continents at the 500-fathom (approximately 1 000-metre) contour.   Antarctica was fitted at the 1 000-metre contour.   Lambert equal area projection.

developed and prior to the discovery of palaeo-magnetism, of sea-floor spreading and of tectonic plates, Wegener, following Suess, proposed that all of the southern continents had once been joined together into a super-continent known as *Gondwanaland*. He used the distribution of Permo-Carboniferous tillites in other parts of the world to support his view (Fig. 15. 11). Du Toit enlarged on Wegener's work and marshalled an impressive array of facts in support of the former existence of a super-continent. However, the evidence was always controversial and never conclusive. Even to-day the shape and extent of Gondwanaland are uncertain.

If one starts from the assumptions that Gond-wanaland was a single continental area without any oceanic areas within it and that all the main pieces are recognizable, then one may apply the best-fit techniques to the problem of putting it

together again. One can think of this as a 'Humpty Dumpty' problem, though one hopes to be in a somewhat better position than were all the King's horses and all the King's men. Even so, there is no unique geometrical solution, as one may readily verify for oneself with a globe and cut-outs of the continents.

What are the pieces in this puzzle? They certainly include South America, Africa, Arabia, Madagascar, India, Ceylon, Antarctica, Australia and New Zealand. South America and Africa have already been reassembled. The Gulf of Aden and the Red Sea between Arabia and Africa need to be closed. The match of the Red Sea coasts is extraordinarily good. But one cannot use the coastline as the best-fitting contour unless there is some independent evidence for regarding it as the edge of the continent rather than the 500-fathom line used for the South America/Africa fit.

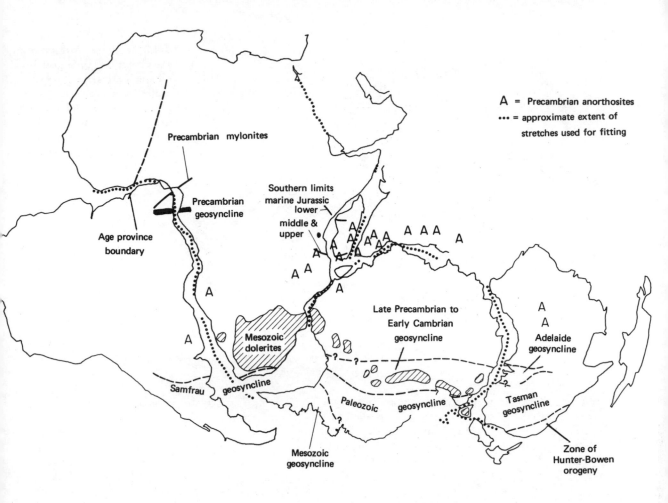

15. 13 Some geological features plotted on Figure 15. 12. Except for the features between West Africa and Brazil, the data plotted have been taken from maps of individual fragments. Though interesting, the alignment of Precambrian anorthosites may not be significant evidence for the fit.

However, the crust seaward of the present Red Sea coasts seems to consist mostly of basaltic rocks added to both sides of the Red Sea rift since it started to form 25 or more million years ago. Why the original edges of the rift have been added to in this way, and why they should stand so high is not understood.

By fitting South America-Africa-Arabia together one obtains one large piece of the jigsaw puzzle. A second large piece is obtained by fitting together Australia, Antarctica and India. Here the geology must be used as a guide to the fit. In very general terms, the geology of Australia consists of old Precambrian rocks in the west that pass eastward into progressively younger north-trending belts (Fig. 15. 13). Exactly the same general

pattern exists in Antarctica from northeast to southwest. Thus one looks for a geometrical fit that matches up these well-defined changes. The contour giving the best fit is assumed to be the 500-fathom (or 1000-metre) line because there is no evidence that the edges of Antarctica and Australia differ from the edges of South America and Africa. The next problem is to decide where to place India. The present geological evidence is indecisive. Adding it to northwest Australia creates a second large piece in the jigsaw that cannot be fitted against the first large piece without giving rise to gaps or large overlaps (Fig. 15. 12). Since one is seeking a fit with no gaps or overlaps this solution is rejected. Instead, India has been added to Antarctica. The second large piece and the

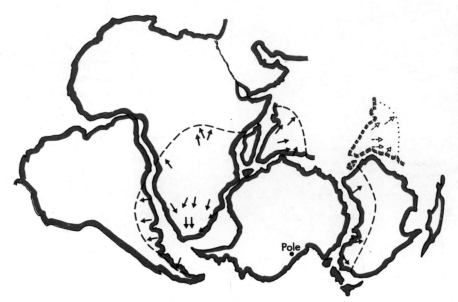

Pole

first large piece can now be combined, leaving a gap of the correct shape and size to take Madagascar and Ceylon. Finally, New Zealand can be attached to Australia and Antarctica, though the resulting fit is not a very good one (Fig. 15. 12).

Gaps and Overlaps in the Gondwanaland Fit

It is an unproven assumption that no gaps existed in Gondwanaland. Such gaps would have consisted of small deep-sea areas rather like the present Black Sea or even deep lakes rather like Lake Baikal located within the present Eurasia supercontinent. The gaps in the fit are small, and in fact all of them can be attributed to missing pieces or to uncertainties in the positions of the known pieces. For example, the Gulf of Aden has not been fully closed. To close it one must break Africa or Arabia somewhere. One obvious place to do this is at the African rift valley, an area known to have been the site of active faulting for the past 25 million years. When more is known about the geology of this region it will probably be possible to eliminate this gap. Similarly the small triangular wedge between Madagascar, Africa and India will neatly accommodate the Seychelles. These islands, consisting in part of Palaeozoic granite, lie in isolation in the northwest Indian Ocean, and are almost certainly part of a former continental area.

The gaps and overlaps of the Antarctic peninsula and South America and Africa occur in regions that have been highly deformed in the past 200 million years. Thus the present shape of Antarctica is not the shape it had when joined to the other two continents. It is not known precisely what its original shape was, nor is it clear how to find out what it was, but there is no reason to suppose that the gaps and overlaps existed in the past.

The embayment between western Australia and eastern India is bordered by Precambrian rocks. In Australia these structures are truncated by the present continental margin. There was undoubtedly a piece or pieces joined on here at one time. Some of them may lie in the present Indian Ocean, forming some of the plateau-like areas in it that are shallower than 500 fathoms but lie below sea level. Other pieces may lie in the highly deformed zones of southeast Asia. None of these possibilities can yet be tested with the available data. The gaps and overlaps between Australia and New Zealand are also attributable to the deformation of New Zealand since the break-up of Gondwanaland.

In other words, it is possible to reassemble all the pieces into a single supercontinent, and to account for the gaps and overlaps in the fit by referring to plausible geological explanations.

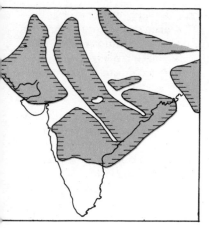
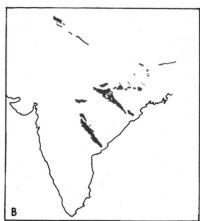
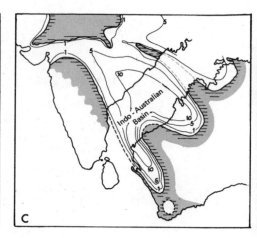

15. 15   (A). A possible interpretation of areas of erosion or non-deposition (coloured) and areas of deposition (blank) during the Permian period in India.   Compare with C.

(B). Outcrop areas of Permian rocks in India to-day.

(C). An alternative interpretation of the areas of erosion or non-deposition (coloured) and deposition (blank, thicknesses in thousands of feet) during the Permian period in India.   The apparent match between the Permian deposits of India and Australia suggests the two continents were joined in Permian time.

Geological Evidence for the Reassembly

The present geological evidence for the reassembly is not as good as it was for South America and Africa, and some of it has already been used to make the reassembly.   For example, the match of the Palaeozoic mountain belts in Antarctica and Australia has been used to suggest how they were joined together.   Nevertheless, the same pattern of belts continues into Africa and South America (Fig. 15. 13).   North of these belts lie the relatively undisturbed sedimentary rocks that include the Permo-Carboniferous glacial deposits discussed earlier.   When those glacial deposits of Upper Carboniferous age are plotted on the fit (Fig. 15. 14) they form exactly the grouping that one might expect from a large polar ice-cap.

In Jurassic time there is a good match between the outcrop pattern of dolerites (mostly intrusive basaltic rocks) from South Africa across Antarctica to Tasmania.   The southern limits of marine Jurassic rocks in East Africa and Madagascar also line up on the reassembly.   Finally, the distribution of Precambrian anorthosites—rocks consisting almost entirely of calcic plagioclase feldspar—forms a roughly arcuate pattern extending from South America, across Africa, through Madagascar and India to Australia, though such a match may be fortuitous.   Thus, except for the link between India and Antarctica and between Australia and New Zealand, there is positive evidence that Gondwanaland existed as an entity until Jurassic time (about 190 million years ago).

The Problem of India's Position

The reader may have been convinced of the correctness of the Gondwanaland reassembly, which is essentially that of du Toit, made over 30 years ago.   Nevertheless, there is doubt about the proper position of India.   Ahmad, an Indian geologist, has argued that the distribution and thickness of Permian rocks in the sedimentary basins of southeast India and northwest Australia are very similar (Fig. 15. 15).

Although his map suggests that the Permian sediments of India cover a large area, their present outcrop is in fact limited to a few trough-like zones.   As he notes, some previous geologists have seen in these narrow zones the original form of the Permian rocks of India.   If they had been deposited in narrow troughs, then the evidence for fitting India against Australia loses much of its persuasiveness.   Perhaps the main general objection against fitting these two continents together is that the Upper Carboniferous tillite pattern (Fig. 15. 14) does not match across the join.

15. 16 Reconstruction of Gondwanaland based entirely on palaeomagnetic data. The method assumes no relative movement has occurred among the continents during Cambrian to Permian time. Any disagreements between Figures 15. 12 and 16 can be attributed to uncertainties in the palaeomagnetically determined positions of the fragments, rather than to real differences between the two.   Equal area meridional projection.

Other Evidence for the Reassembly

The sea-floor spreading pattern will eventually enable the reassembly to be verified (Chapter 16). A recently published fit based entirely on palaeomagnetic data also supports the main features of the reassembly (Fig. 15. 16) and does not support the placing of India against northwest Australia. Once again, it is the agreement between two entirely different kinds of evidence that gives the best available proof of the former existence of Gondwanaland.

LAURASIA

The geometrical fit of the northern continents is quite good, and the reassembly forms a supercontinent known to geologists as *Laurasia*. It includes North America, Greenland and Eurasia. To make a fit at the 500-fathom contour it is necessary to remove Iceland, to omit oceanic ridges connecting Greenland with Europe, and to leave out part of an area shallower than 500 fathoms between southern Greenland and Labrador.   To fill up all the spaces north of France and Newfoundland it is necessary to retain the Rockall Bank (Fig. 15. 17).

Except for the area between Greenland and Canada, all of these omissions and also the retention of the Rockall Bank may be justified by presently available data.   Iceland and the oceanic ridges are believed to be features mostly younger than 70 million years, and Rockall Bank has recently been shown to have the seismic characteristics of continental crust.   In other words, the northern continents may be fitted together without gaps and probably formed a single unit some time before the Tertiary period (70 million years ago).

The pre-Mesozoic geology includes many orogenic belts.   The two youngest of these appear to be continuous on the geometrical fit.   The older Caledonian belt extends from Spitsbergen, through Greenland and Scandinavia, and south to Britain.   It was deformed roughly in the interval 600–370 million years ago, and lines up with the Appalachian system of Newfoundland and the northeastern United States (Fig. 15. 18).   The younger Hercynian (or Armorican or Variscan) belt, deformed approximately 370–260 million years ago, extends west from central Europe, across southern Britain, and lines up with a belt of similar age in Newfoundland and northeastern North America.   This match suggests the northern continents were a single unit some time in the interval 260 to 70 million years ago.

The sedimentary history of the northern continents contains nothing so distinctive as the glacial deposits found in Gondwanaland, nor is it known to include the freshwater/evaporite/marine succession suggested to indicate the break-up of Africa and South America.   Until the mid-1960's most geologists and geophysicists regarded the geometrical fit and the matching of the orogenic belts as inadequate evidence for continental drift. Even to-day it is still possible to argue that it is unnecessary to appeal to continental drift to account for the present world-wide continental distribution of most sediments and fossils.   Knowing that drift has occurred, we see from these data that, except in unusual circumstances, it is not possible to argue convincingly for the former joining together of two continents by using evidence from surface processes alone.

Gondwanaland and Laurasia may be joined by fitting northwest Africa and North America together. Wegener referred to the single continent so formed as *Pangaea*. This fit is the worst of all the geometrical fits, particularly in the Caribbean and Mediterranean regions. There is excellent palaeomagnetic evidence for the rotation of small continental fragments in these regions. Figure 15.19 shows an attempt made to reassemble the continental fragments around the Mediterranean, but no attempt has been made to improve the fitting together of the Caribbean region. The result is somewhat different from Wegener's reconstruction. It shows a large wedge-shaped area, presumably representing a former ocean, between the eastern ends of Gondwanaland and Laurasia. Such an ocean, known to geologists as the *Tethys*, has also been postulated from the distribution of fossils in the adjacent continental areas.

The rotation of Spain to fill up the Bay of Biscay is supported by palaeomagnetic and geological as well as sea-floor spreading evidence. The positions of the other fragments in the Mediterranean should be regarded as speculative and not yet proven.

If the Hercynian and Appalachian orogenic belts represent the disappearance of a former ocean between two colliding continents (Chapter 20), then presumably North America and Africa were joined together at the end of this orogeny about 260 million years ago. That is, the northern and southern continents joined together along the Appalachian belt, which has its counterpart in northwest Africa (Fig. 15.18). In eastern North America the next episode after the orogeny was the deposition of thick lacustrine and terrestrial sediments in fault-bounded troughs, together with the eruption of lava flows about 200 million years old. Some geologists see this episode as a forerunner of the break-up of Africa and North America, in much the same way as the rift valley system of east Africa today could be interpreted as beginning the break-up of the African continent. Thus Africa and North America may have been joined together for perhaps as short a time as 50 million years, and the existence of Pangaea may have been restricted only to part of Permo-Triassic time (about 280–190 million years ago). Gondwanaland itself probably formed about 500 million years ago, and Laurasia, west of the Urals, about 370 million years ago.

15.17 The fit of North America, Greenland and Europe at the 500-fathom contour. Iceland and the oceanic ridges in the North Atlantic have been omitted, but Rockall Bank has been retained. Conical projection.

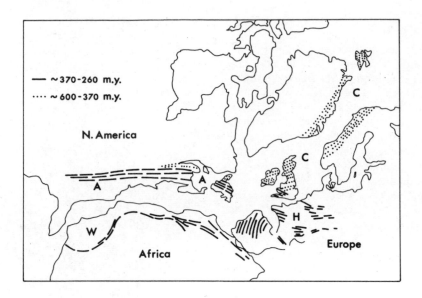

15. 18   The youngest pre-Mesozoic orogenic belts plotted on Figure 15. 17, with the Bay of Biscay closed and Africa added to North America at the 500-fathom contour. The older Caledonian-Appalachian belt (C) is shown as dotted lines parallel to the general trend of the structures in it; the younger Hercynian-Appalachian belt (A, H) and a belt of similar age in West Africa (W) are shown as dashed lines. Greenland and northwest Scotland, with North America already joined to them, had been added to Europe about 370 million years ago, whereas Africa and the remainder of the southern continents were probably not joined until about 260 million years ago.

CONCLUSIONS

Because the continents fit together so well, it is possible to draw important conclusions about the nature of the Earth, about continents and their margins, and about orogenic belts.   The first conclusion is that the Earth probably has not undergone any large-scale expansion or contraction in the past 200 million years.   Large-scale expansion was, until quite recently, thought to provide an alternative explanation for the origin of ocean basins.   It was argued that the continents had once formed a continuous shell covering a much smaller Earth, that expansion occurred and that oceanic crust filled the gaps between the continents as they separated from one another on the larger globe.   However, had so much expansion taken place it is unlikely that one could fit them back together on a globe of the present size, because their shapes would have been considerably distorted during expansion and adjustment to the larger globe.

The second conclusion is that during drift the continents behave as rigid pieces except where cut across by orogenic belts.

The geometry also shows that if orogenic belts within continents result from collisions of the continents, then the collision takes place in a way that eliminates all traces of the original ocean between them, otherwise, for example, there would be unfilled holes along the length of the Caledonian chain.   This is a very remarkable fact, for it is highly unlikely that the edges of the continents before collision have just the shape needed to close all the space between them.   Why colliding continents behave in this way is not known, but it probably has much to do with the process causing continents to drift in the first place.

OLDER CONTINENTAL REASSEMBLIES

Before the Pangaea of Permo-Triassic time was formed, one must envisage a different mosaic of continental fragments reassembled in a different way.   It is fairly clear where to snip Pangaea and so outline the continental fragments needed for, say, the reassembly of 370 million years ago: one merely cuts it along the lines of the youngest orogenic belts within Pangaea.   The youngest of these is the Hercynian belt between Africa and the

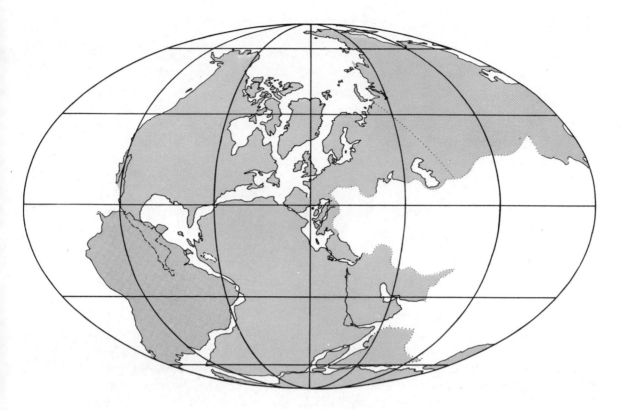

15. 19 A possible arrangement of the continents in Permian time. The reassembly combines Figures
15. 3 a, 12 and 17, and orientates them so that the palaeomagnetic and geographic poles coincide. It needs
to be modified in the Caribbean, where overlapping areas exist; in the Arctic; and probably in Asia east of the
Urals—the dotted line across Eurasia. The fit of the pieces around the Mediterranean is speculative and not
yet verified. The dotted line marking the northern boundary of the southern continents and the southern
boundary of the northern continents marks the approximate limits to the Tethyan ocean floor. Australia
lies in the extreme bottom right of the map. Mollweide projection, grid lines at thirty degree intervals.

northern continents, and also the Urals. Palaeomagnetism can be used to orientate the fragments
into the correct palaeolatitudes but there is at the
moment considerable uncertainty in the relative
palaeolongitudes.

The uncertainty exists for at least two reasons.
Not only has all sea-floor older than 200 million
years probably been destroyed, but the older
continental margins have been changed by the
younger orogenic belts that have formed on them
and one cannot fit them together precisely with a
computer. Thus one will never know the pre-
Mesozoic positions of the continents as exactly as
one will, in principle at least, know their Mesozoic
and Tertiary positions. In this sense the pre-
Mesozoic Humpty Dumpty problem is perhaps

worse than was Humpty Dumpty's. Nevertheless, no one foresaw the incredible advances that
have taken place in the Earth sciences in the past
decade, and it may be that one only needs to look
at the Earth in a fresh way for the present difficulties to vanish, or at least to diminish greatly. If
this ever happens, new problems just as daunting
as the present ones will doubtless arise, but then
that is how science progresses.

FURTHER READING:

PATRICK M. HURLEY 1968 (April).   The Confirmation of Continental Drift.   *Scientific American.*

ALFRED WEGENER 1929.   The Origin of Continents
and Oceans, 4th Ed.   *Methuen.*

It is fair to say that almost all Earth scientists are now convinced that continental drift has occurred.   Indeed, some would even go so far as to say that it is no longer a hypothesis, or even a theory, but a fact. Be that as it may, the combined evidence in favour of continental drift is extremely convincing, and no particular piece of evidence is more convincing, or more crucial, than that from palaeomagnetism.

But while palaeomagnetic directions have been instrumental in 'proving' continental drift, they also form the critical evidence in favour of another hypothesis concerned with the mobile Earth—sea-floor spreading.   Just over ten years ago, the ocean floors were still regarded as the oldest and most dormant parts of the Earth.   By 1960, however, improved techniques— many of them developed during, and because of, World War II—had enabled scientists to examine, for the first time, the detailed topography of large areas of ocean floors. The results were as surprising as they were impressive. For the ocean floors were revealed not entirely as dull, flat regions but as possessing 'mountain ranges' comparable with the most impressive mountainous regions on the continents.   It later emerged that these ranges, or 'ridges', were not isolated features but formed part of a world-wide and almost continuous network, and that, furthermore, they were seismically active.   From this quickly developed the idea that the ocean floors, far from being dormant and old, were, in parts, the most active and youngest regions of the Earth.

We now know that the ocean floor ridges are regions at which material from the Earth's mantle rises and becomes new crust which slowly spreads away from the ridge axes.   How this idea developed, the evidence which supports it, and some of the implications of sea-floor spreading are described in the following article by Dr. F. J. Vine who, by his own work, was largely responsible for convincing Earth scientists that the ocean floors really are young and mobile.

16 Sea-Floor Spreading

by F. J. VINE

The formulation, extension and final confirmation of the concept of sea-floor spreading occurred within a single decade, the 1960's. In one giant leap forward, this single, revolutionary and elegant hypothesis provided us with a much greater understanding of the Earth. It simultaneously revived the older, but largely rejected, concept of continental drift and paved the way for the more modern and highly successful concept of plate tectonics.

THE CONCEPT

The hypothesis which we now know as 'sea-floor spreading' was first formulated by the late Professor Harry Hess of Princeton University in 1960. Hess's idea derived from attempts to synthesise the results obtained by marine geologists and geophysicists such as Maurice Ewing and his co-workers at Columbia University. In the 1950's these scientists had made two very fundamental discoveries about the deep ocean floor. The first was made with a seismic refraction technique, adapted for use at sea. By setting off explosives, comparable to depth charges, and deploying near-surface receivers, or 'hydrophones', up to 40 kilometres away, it became apparent that seismic waves which penetrate the oceanic crust are refracted, and then transmitted with a velocity characteristic of the Earth's upper mantle, at a mere 12 km below sea-level. This implies a crustal thickness of 6 or 7 km, which contrasts with a typical, although highly variable, thickness of 30 to 40 km of crust beneath the continents. The second discovery was that the mid-ocean ridge or rise, first documented in the North Atlantic, is in fact a feature of all the major ocean basins, and its crest may be traced out on the basis of *bathymetry* (sea-floor topography) and earthquake epicentres (see Fig. 16. 1). The world-encircling mid-ocean ridge system is a topographic rise, or swell, thousands of kilometres in width and tens of thousands of kilometres in length. Its crest stands 2 to 3 km higher than the flanking abyssal plains, and in the Atlantic it is characterised by a median rift valley and active volcanism.

Hess postulated that the mid-ocean ridges are situated over the rising limbs of convection currents in the Earth's mantle, and that the thin oceanic crust is nothing more than a surface expression of the mantle, derived from it by simple chemical modification, and continuously created by a process of lateral accretion or 'spreading' away from ridge crests. Carried to its logical conclusion this concept implies the formation of the whole of the Atlantic and Indian Ocean basins by a process of 'spreading', and the rifting and drifting apart of the surrounding continents, as suggested previously by the proponents of continental drift. From consideration of the earlier ideas regarding the age of the initiation of drift in the Atlantic area (based on the geologic record of the surrounding continental margins) Hess suggested that the sea-floor might be spreading at a rate of approximately 1 cm per year per ridge flank. Generation of new oceanic crust at ridge crests at a lateral rate of a few centimetres a year is sufficient to account for the formation of all the present deep ocean floor within the past 200 million years or so—less than five per cent of geologic time. Thus, having postulated the formation of approximately two-thirds of the present surface area of the Earth within so short and recent a period of Earth history, Hess was faced with essentially two alternatives: either the Earth expanded, or old crust is destroyed at the same rate at which new crust is generated at ridge crests. Hess felt that the necessary increase in volume of the Earth—by a factor of five within

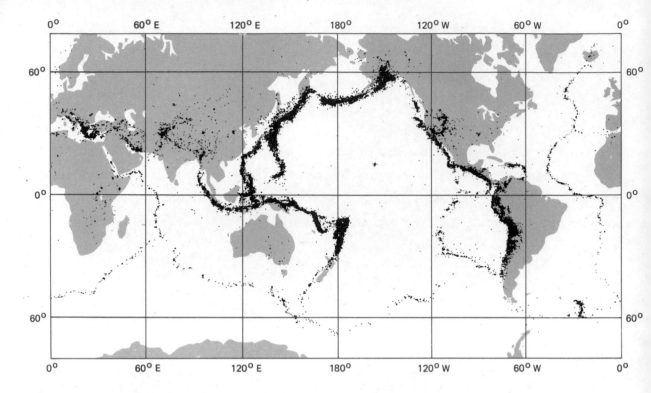

16. 1 The seismicity of the Earth is revealed by this epicentre plot of approximately 30 000 earthquakes which occurred during the period 1961–1967 and were located by the Coast and Geodetic Survey of the U.S. Environmental Sciences Services Administration.

the past 200 million years—was highly implausible, and instead suggested that the trench systems of the Pacific hemisphere (Fig. 16. 2) are the sites of the descending limbs of the mantle-wide convection currents (Fig. 16. 3). He postulated that in these areas oceanic crust is thrust down into, and largely resorbed by, the mantle.

An integral part of Hess's original hypothesis was that oceanic crust is little more than hydrated mantle, the water required for hydration being derived from the mantle itself, and the reaction being limited to ridge crests and shallow depths by the fact that volatiles would only be brought up in the vicinity of mantle upwelling and that the reaction is only possible at temperatures below about 500°C. Beneath the trench systems the crust reverts to mantle as a result of heating and dehydration; and the water released would find its way into the hydrosphere. Deep-sea sediments which accumulated on the oceanic crust

during its passage from ridge crest to trench system would presumably be accreted to the continents or island arcs adjacent to the trenches. Thus if spreading and drift have taken place throughout geologic time, part or all of the water of the oceans may have been derived from the mantle. Hess postulated that the resulting steady increase in depth of the oceans has been matched by concomitant thickening of the continents by vertical accretion. In this way the continents will always have maintained essentially the same 'freeboard' with respect to sea-level.

In summary, Hess suggested that, despite their great age and apparent permanency, continents have been, and are being, passively drifted apart and together on the backs of mantle-wide convection cells (Fig. 16. 3). In contrast the ocean floors are young and ephemeral features of the Earth's surface, constantly being regenerated at ridge crests and destroyed in the trench systems.

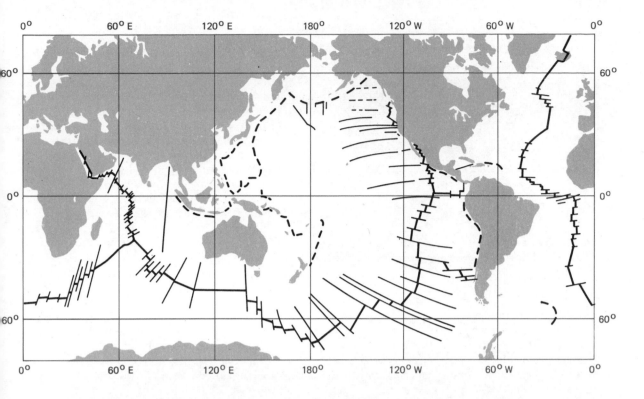

16. 2 Summary of the major topographic and structural elements in the ocean basins, for comparison with Figure 16. 1. Thick dashed lines indicate deep-sea trenches; thick solid lines, the crest of the mid-ocean ridge system; and thin solid lines, major fracture zones or faults.

SUPPORTING EVIDENCE

The occurrence of earthquakes along the crest of the mid-ocean ridge system, the dearth of sediments at ridge crests and the active volcanic islands associated with the crest of the Mid-Atlantic Ridge are all readily explained by Hess's model. Moreover, the ocean basins as a whole contain a remarkably thin veneer of sediments and small number of seamounts if recent rates of accumulation and formation are extrapolated over the whole of geologic time. In 1960 Hess was also able to state that no material greater than about 100 million years in age had ever been recovered from the deep ocean floor or truly oceanic islands. Thus many basic geological facts were in accord with Hess's idea; in addition he was able to interpret the then recent results of geophysical measurements at sea in terms of his model.

The flow of heat through the ocean floor from the Earth's interior was first determined in the 1950's by means of a thermistor probe, which measured the temperature gradient in the sediments, and nearby coring, which facilitated measurement of the thermal conductivity of the sediments. Results obtained in this way revealed that the heat flow through the ocean floor is in general comparable to that determined previously for the continents, i.e. on average 6×10^{-2} joule m^{-2} s^{-1}. Over the mid-ocean ridges, however, it is, in places, several times this value. Hess argued that these anomalously high values reflect the emplacement of hot mantle-derived material in the vicinity of ridge crests.

In addition to high values of heat flux, marine geophysicists had also discovered that ridge crests are sometimes characterised by anomalously low seismic wave velocities in the upper mantle. This was ascribed to thermal expansion and micro-

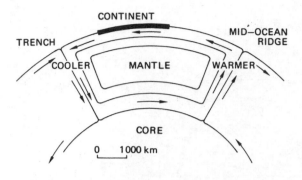

16. 3 Mantle-wide convection with resulting sea-floor spreading and continental drift as envisaged by Hess.

fracturing associated with the upwelling mantle, both effects producing a reduction in the seismic wave velocities and density of the mantle material. The idea that enhanced geothermal gradients due to upwelling beneath mid-ocean ridges give rise to less dense material in the upper mantle is a very reasonable one, since it provides an explanation for the fact that no significant disturbance in the Earth's gravitational field is observed on crossing a ridge. The excess mass of the ridge itself must therefore be compensated by a corresponding mass deficiency at depth; that is, thermal expansion, and probably a small degree of partial melting, within the uppermost mantle give rise to the topographic expression of the ridge.

A MAGNETIC TAPE RECORDER

Although there is no disturbance or 'anomaly' in the Earth's gravitational field over ridges, there is, typically, a disturbance in the Earth's magnetic field. During and after World War II, fluxgate and proton-precession magnetometers were developed to measure magnetic fields to one part in 10^5. By 1960 both types of detector had been towed at sea-level by oceanographic research vessels, and two surprising and enigmatic results had been obtained. In the North Atlantic an increase in the Earth's magnetic field amounting to 1 or 2 per cent of its regional total intensity had been recorded over the median valley at the ridge crest. This is the opposite to what one would predict on the basis of the deficiency of magnetic

material represented by the valley itself, if the oceanic crust is magnetised in the present direction of the Earth's magnetic field—that is, by induction. Moreover the magnetic anomaly was found to persist where the median valley is absent or poorly developed.

In the northeast Pacific, and not obviously related to the mid-ocean ridge system, linear anomalies had been mapped orientated approximately north-south and bounded by steep gradients between highs and lows (Fig. 16. 4). The discovery of this remarkable pattern, quite unlike any mapped previously over the continents, aroused considerable interest at the time because it was clearly offset by the east-west submarine fracture zones in the area, sometimes by hundreds of kilometres as on the Murray fracture zone. The offsets were originally interpreted to indicate large-scale horizontal or strike-slip movements of the adjacent crustal blocks along the fractures. Although we now know that this is not the only or most probable interpretation of these offsets, it was significant at the time in that most structural geologists had convinced themselves, on the basis of continental geology, that movements of the Earth's crust are predominantly, perhaps entirely, vertical rather than horizontal. Thus suggesting large-scale horizontal displacements of the ocean floor seriously questioned this fundamental assumption and paved the way for the even more heretical concept of sea-floor spreading. The anomalies themselves posed a serious problem as regards interpretation in that they seemed to imply improbable structures, or lateral changes in composition in the oceanic crust, which retained their width and spacing for thousands of kilometres along their length. In addition they do not correlate with variations in the topography or structure of the sea floor as revealed by echo-sounding and seismic refraction experiments. Hess chose not to discuss these magnetic anomalies in his original paper on sea-floor spreading but in 1963 F. J. Vine and D. H. Matthews rectified this omission and in so doing added a rider and modification to the hypothesis.

As we have seen, Hess considered oceanic crust to be hydrated mantle (serpentinite); an alternative hypothesis maintains that it is at least in part, perhaps entirely, derived from the mantle by partial fusion. The low melting point fraction of the Earth's mantle is thought to be basalt, but, despite the fact that both basalt and serpentinite are dredged from the ocean floor, Hess considered that partial fusion is only important in the construction of seamounts, atolls and guyots, that is, submarine volcanoes. Vine and Matthews accepted the concept of sea-floor spreading but maintained that at least the uppermost part of the oceanic crust is derived by partial fusion of the mantle beneath ridge crests and hence made up of extrusives and intrusives of basaltic composition. Of all rock types occurring at the Earth's surface, basalt is particularly proficient at acquiring and retaining an appreciable intensity of permanent (fossil) magnetisation. On cooling below its Curie temperature, after solidifying, it acquires a magnetisation which reflects the direction and intensity of the ambient magnetic field at that time. The intensity of this remanent or permanent component of magnetisation in basalts is invariably greater than that induced by the present Earth's field; hence the fossil magnetisation rather than the induced magnetisation might be all important in interpreting magnetic anomalies due to magnetisation contrasts within the oceanic crust.

To complete their idea, Vine and Matthews made one further assumption: that as new sea floor forms at, and spreads laterally away from, ridge crests, the Earth's magnetic field reverses its polarity intermittently, thus producing avenues of alternately normally and reversely magnetised material parallel to the ridge axis (Fig. 16. 5). The contrasts produced in this way, that is, between basaltic material with opposing directions of magnetisation, could account for both the central magnetic anomaly observed over the ridge crest in the North Atlantic and the linear anomalies of the northeast Pacific. The median rift valley in the North Atlantic, for example, is presumably floored by basalt extruded and intruded during the current (normal) polarity of the Earth's field and is flanked by material emplaced during the preceding, reversed epoch. The great strength of this model is that it accounts for the magnetisation contrasts implied by the magnetic anomalies without invoking improbable lateral variations in the composition or structure of the oceanic crust. Thus Vine and Matthews converted Hess's conveyor belt of oceanic crust, which provided a plausible mechanism for continental drift, into a tape recorder which potentially might provide compelling and detailed evidence for sea-floor spreading.

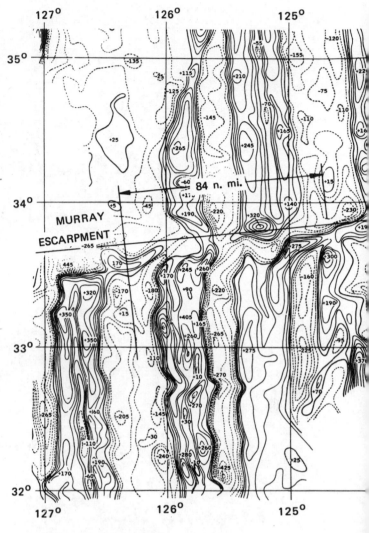

16. 4 Map of disturbances or 'anomalies' in the Earth's magnetic field over part of the area west of California surveyed by the U.S. Coast and Geodetic Survey ship 'Pioneer' in the 1950's. The survey lines ran east-west and were approximately 5 nautical miles apart (1 n. mi. = 1 minute of latitude = 1.854 km). Contour interval = 50 gamma (1 γ = 10^{-5} gauss or 10^{-9} tesla). Note the offset of 150 km along the Murray Fracture Zone.

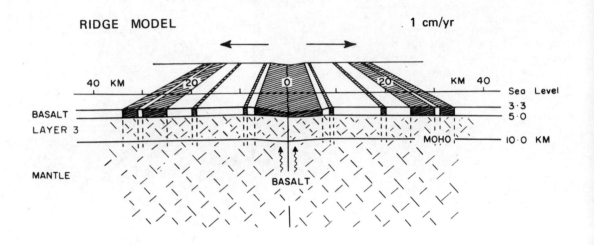

RIDGE MODEL   1 cm/yr

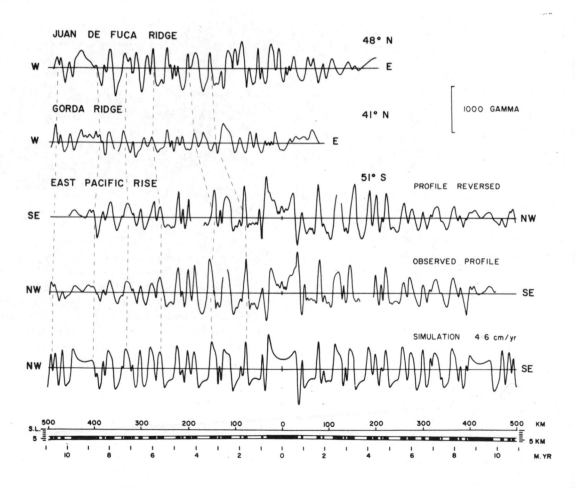

CONFIRMATION

However, in 1963 it was still debated whether the Earth's magnetic field had reversed at all, and those who believed it had could not suggest the times at which it had reversed, even for the most recent past. Thus until more oceanic magnetic surveys became available and the question of geomagnetic reversals could be settled, it was impossible to pursue the Vine-Matthews hypothesis further. At about this time, refinements in the potassium-argon dating technique made it possible to date basaltic lava flows extruded subaerially within the past few million years with considerable accuracy. Orientated samples from many such flows were dated and their direction of magnetisation measured. It was found that all flows of a particular age, irrespective of their geographic locations, indicate the same polarity for the Earth's magnetic field and that both polarities are indicated for different times. By studying a great number of flows from many parts of the world it was possible by 1966 to suggest a geomagnetic reversal time-scale for the past three

and a half million years. This reversal time-scale has been incorporated into Figure 16. 5. In the same year other researchers were studying the fossil magnetism of deep-sea sediments. Where the top few metres or tens of metres of sediment recovered by corers consist of a complete record of pelagic sedimentation for the past few million years, precisely the same sequence of reversals is preserved in the remanent magnetisation of the sediments. This independent work provided striking confirmation of the reversal time-scale derived from the study of subaerial lava flows.

Thus it became possible to re-examine the Vine-Matthews hypothesis. It was found that, if one assumes a constant rate of spreading of a few centimetres per year and the reversal time-scale defined by the above techniques, the observed magnetic anomalies can be reproduced for almost all ridge crests for which magnetic data are available. Moreover, by obtaining a best fit between the observed and predicted anomalies a precise rate of spreading can be deduced. An example of such a simulation for a South Pacific profile is shown in Figure 16. 6. Spreading rates deduced in this way vary from 1 centimetre per year per ridge flank in the vicinity of Iceland to approximately 9 centimetres per year per ridge flank on the East Pacific Rise in the equatorial Pacific.

A literal interpretation of the Vine-Matthews hypothesis implies that for most latitudes and orientations of ridge crests the magnetic anomalies should be roughly symmetrical about the ridge axis, although in most instances some degree of asymmetry is introduced because of the dipolar rather than monopolar nature of the Earth's magnetic field (as, for example, in Fig. 16. 6). As additional profiles of the magnetic field over the oceans became available it was increasingly apparent that the anomalies about ridge crests do show a symmetry and correlation with the reversal time-scale, but that this is more evident in the Pacific, where spreading rates are high (typically greater than 3 cm per year per ridge flank), than in the North Atlantic and north-west Indian Oceans where most of the early work was done and spreading rates are low, i.e. between 1 and 2 cm per year per ridge flank. However, in 1966 a detailed aeromagnetic survey of the Reykjanes Ridge in the North Atlantic became available and amply confirmed these predictions as shown in Figures 16. 7a and 7b and 16. 8.

Opposite:

16. 5 Diagrammatic representation of the oceanic crust at a mid-ocean ridge crest, assuming active spreading at a rate of 1 cm per year per ridge flank, and the geomagnetic reversal time scale as defined in 1966. Shading in the basalt layer indicates normal magnetisation. 10 km horizontally correspond to one million years.

16. 6 An observed magnetic anomaly profile across the East Pacific Rise, together with the profile reversed about its mid-point to demonstrate its symmetry, and a computed profile assuming the applicability of the Vine-Matthews hypothesis for the past 11 m. yr. The profile is also compared with composite profiles across the Juan de Fuca and Gorda Ridges of the north-east Pacific. S.L. = sea level.

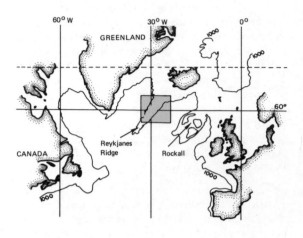

16.7a The location of the Reykjanes Ridge, south-west of Iceland, and the area (coloured) of Figure 16.7b. The 1 000-fathom contour is shown, together with the 500-fathom contours for Rockall Bank.

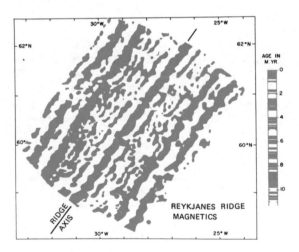

16.7b Summary diagram of magnetic anomalies recorded over the Reykjanes Ridge. Areas of positive anomaly are shown in colour. The central positive anomaly of Figure 16.8 correlates with the ridge axis.

Unfortunately the technicalities of both methods for determining the age of geomagnetic reversals are such that neither can extend the time-scale back beyond 3.5 million years very readily or precisely. One can provide convincing evidence that spreading has taken place at ridge crests during the past few million years, but to what extent is one justified in assuming that spreading has gone on for tens or hundreds of millions of years and formed the whole of the ridges and ocean basins, as Hess postulated?

Fortunately several implications and therefore tests of the model are readily apparent. If the Earth's magnetic field has reversed its polarity throughout the past 200 million years, as it has recently, then spreading should be recorded in the form of linear magnetic anomalies paralleling the ridge crests and symmetrically disposed about them. Moreover, if spreading has been continuous about all ridges, the same sequence of anomalies should be reproduced across all ridge flanks, and indeed the same sequence of anomalies is observed across ridge flanks in all the ocean basins. Profiles across the western flanks of the ridges in the north Pacific, south Pacific and south Atlantic are shown in Figure 16.9. It transpires that the linear anomalies of the northeast Pacific, which were mapped in the late 1950's, are the part of this sequence associated with the west flank of the East Pacific Rise, the crest of which has interacted with the west coast of North America since the time of their formation. The whole or parts of the same sequence of anomalies can be recognised on either side of the East Pacific Rise and Pacific-Antarctic Ridge in the south Pacific, to the south of Australia, on both sides of the ridges in the south Atlantic and southeast Indian Oceans, and to some extent in the north Atlantic and northwest Indian Oceans; but slow spreading rates in the latter two areas make correlations very difficult unless a detailed survey is available. The areas in which the sequence of anomalies has been documented are summarised in Figure 16.10.

The world-wide correlation and symmetry of magnetic anomalies about ridge crests provide compelling evidence for sea-floor spreading and reversals of the Earth's magnetic field throughout an appreciable proportion of geologic time. It became of great interest therefore to try to determine the interval of time represented by these anomalies. With the possible exception of the northwest Indian Ocean and the extreme

North Atlantic the anomaly sequence is complete in all areas in which it has been observed. Thus, unless there have been world-wide stoppages, spreading has been continuous in most areas, but not necessarily at a constant rate. In fact if one assumes a constant rate in any one area one can deduce the implied variations in all others; this is apparent, for example, from the correlation lines on Figure 16.9. In order to assign provisional ages to the magnetic anomalies, oceanic crust and geomagnetic reversals, it was decided to assume a constant rate of spreading in the South Atlantic, since this implied least variation in spreading rates elsewhere. The rate assumed was approximately 2 centimetres per year per ridge flank, i.e. that deduced from the central anomalies; and it resulted in the oldest correlatable anomaly being assigned an age of 76 million years.

Thus very specific predictions were made about the age and history of the mid-ocean ridges across which the magnetic anomaly sequence could be recognised. Clearly the next acid test was to see whether the age and distribution of sediments and the age of the oceanic crust across such ridges conform to this picture. Up to 1968, the year in which the above correlations and time-scale were proposed, the only means available to test these predictions were somewhat inadequate ones—the deep-sea sediment corer and the rock dredge. However, within a year a much more definitive test could be made because of the advent of the U.S. Deep-Sea Drilling Project. The first phase of this so-called 'JOIDES' (*Joint Oceanographic Institutions Deep Earth Sampling*) project occupied 31 sites in the Atlantic Ocean and 53 sites in the Pacific Ocean. Leg III in the southern Atlantic was specifically designed to test the spreading hypothesis and proposed geomagnetic reversal time-scale: eight sites were drilled at points across the Mid-Atlantic Ridge at approximately 30°S and beneath the previously dated magnetic anomalies. At all but one site the whole sedimentary column was recovered and the basalt 'basement' penetrated. The age of the oldest sediment recovered at each site, that is, immediately overlying or incorporated within the basalt, was directly proportional to the distance of the site from the ridge axis (Fig. 16.11). This confirms that spreading has been continuous throughout the Cenozoic and at an essentially constant rate of 2 centimetres per year per ridge flank, precisely the assumptions made in assigning provisional ages to the magnetic anomalies as described above.

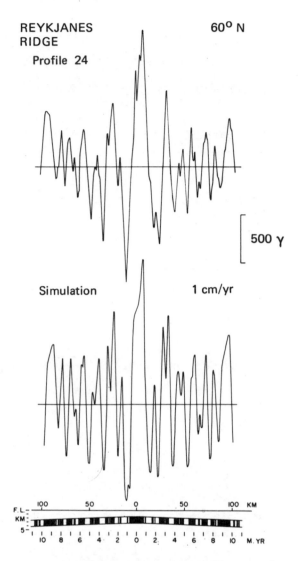

16.8 An observed aeromagnetic profile across the Reykjanes Ridge, compared with a computed profile assuming the reversal time scale derived from the South Pacific profile shown in Figure 16.6. F.L. = flight level.

This remarkable confirmation of the provisional ages assigned to the oceanic crust and geomagnetic reversals enables us to estimate the total area of deep ocean floor formed by spreading during Cenozoic time, that is, the past 65 million years. Such an estimate is shown in Figure 16.10. This leads us to the startling conclusion that perhaps 50% of the present deep sea floor, i.e. one third of the surface area of the Earth, has been created during the most recent 1.5% of geologic time.

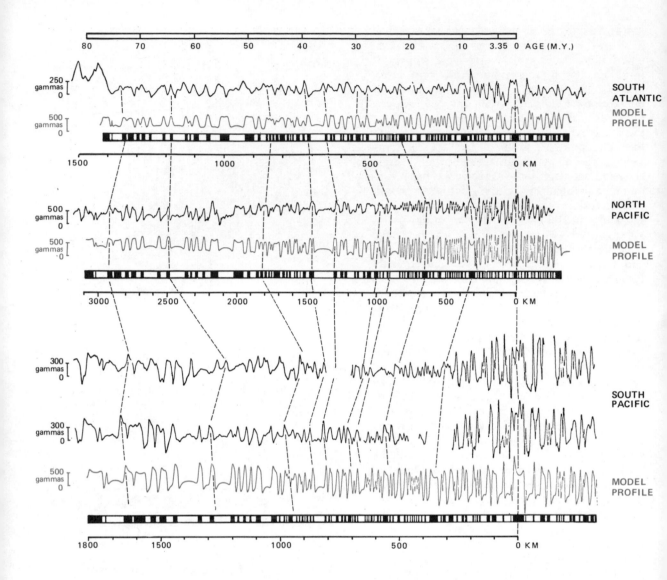

16. 9 Magnetic anomaly profiles across the western flanks of the oceanic ridges in the South Atlantic and North and South Pacific. The observed profiles are compared with computed profiles which assume the same sequence of reversals of the Earth's magnetic field but variations in spreading rate with time in the North and South Pacific. The anomalies have been calibrated with respect to age by assuming a constant rate of spreading in the South Atlantic.

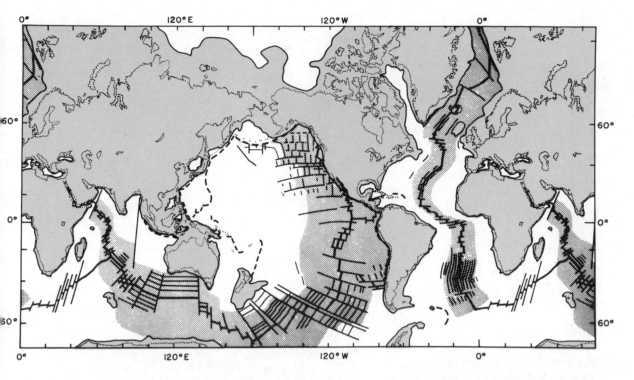

16. 10 Provisional attempt to delineate areas of continental (coloured) and oceanic crust. Within the ocean basins correlatable linear magnetic anomalies, paralleling the ridge crests, are indicated by thin solid lines. Oceanic crust thought to have been formed within the Cenozoic (i.e. the past 65 m. yr.) is shaded. Ridge crests, fractures and trenches are summarised as in Figure 16. 2.

MESOZOIC OCEAN FLOOR

It now seems extremely probable that all remaining oceanic areas are Mesozoic in age—that is, formed between 65 and 225 million years ago. At the time of writing the JOIDES drill ship *Glomar Challenger* had occupied over 100 sites in the Atlantic and Pacific Oceans. Several sites in the western Pacific and western Atlantic were specifically chosen in the hope of sampling some of the oldest sediment in the ocean basins. The oldest sediment recovered so far is middle Jurassic in age (that is about 160 million years old); this was obtained in several holes on JOIDES Leg XI just north of the Bahama Bank in the western North Atlantic.

The results of deep-sea drilling and coring can be extrapolated over large areas by means of seismic reflection profiling. Seismic profilers were developed in the early 1960's and are essentially low-frequency echo sounders. The lower frequency range at which they operate

necessitates more elaborate sound sources and receivers but has the great advantage that it enables one to see details of structure and layering within the sediments and the topography of the basalt basement beneath them. Thus reflecting horizons which have been dated by drilling, or by coring where they outcrop in erosional channels or on fault scarps, can be mapped over vast areas by continuous reflection profiling. The age and lateral extent of all horizons mapped in this way are always consistent with the predictions of sea-floor spreading and continental drift. The older horizons are found to 'pinch out' towards ridge crests precisely as predicted by Hess in 1960. In the western North Atlantic and western Pacific, for example, there is a very pronounced reflecting horizon, now thought to be due to chert layers within the sediments, which correlates roughly with the Mesozoic-Cenozoic boundary. In neither area does this layer extend beneath the Cenozoic sequence of magnetic anomalies to the east.

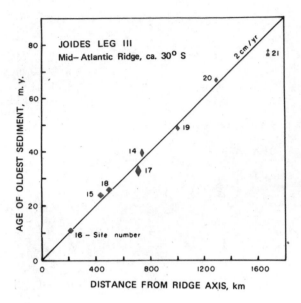

16. 11 The age of the oldest sediment recovered at each site plotted against the distance of the site from the ridge axis. With the exception of sites 17 and 18 all cores were drilled to the west of the ridge axis. The diamonds indicate the estimated error in the age and distance determinations. Basalt basement was not reached at site 21.

Radiometric dating of basalt, serpentinite and related rocks from the deep-sea floor is fraught with difficulties, but such reliable dates as have been obtained, notably from seamounts, are consistent with the sea-floor spreading picture. The few results obtained with a relatively new and more promising technique, fission-track dating, have also confirmed the ages assigned to the magnetic anomalies.

If the unshaded area of the ocean basins shown in Figure 16. 10 was formed by spreading during the Mesozoic era, as suggested above, then one would expect reversals of the Earth's magnetic field during Mesozoic time to be recorded in the form of linear magnetic anomalies over these areas as for Cenozoic crust. However, beyond the sequence of correlatable linear anomalies towards the margins of the South Atlantic and in the central north Pacific the magnetic field is relatively 'quiet' and apparently devoid of such anomalies. For some time this presented something of an enigma, but recently it has been suggested, on the basis of palaeomagnetic studies on land, that there

were very few reversals of the Earth's magnetic field during late Mesozoic (that is, Cretaceous) time and that most of the reversals that did occur are confined to the latest Cretaceous and therefore correlate with the oldest anomalies of the sequence shown in Figure 16. 9. This fits well with the magnetically 'quiet' zones and presumed age of the oceanic crust in the vicinity of Hawaii and along the margins of the South Atlantic. The initiation of drift and spreading in the South Atlantic area is thought to have occurred in Middle Cretaceous time, little more than 100 million years ago, as discussed below. In the western Pacific, southeast of Japan, and in the North Atlantic, off the coasts of the United States and northwest Africa, linear magnetic anomalies have been mapped over even older ocean floor and are thought to reflect geomagnetic reversals during Jurassic, that is, mid-Mesozoic, time. (See Fig. 16. 10 and Table 1.)

Thus the most recent phase of sea-floor spreading and continental drift is recorded in remarkable detail by the magnetic anomalies and the age of oceanic areas. However, for all their clarity and compatibility with drift, the magnetic anomalies only permit us to assign dates for the initiation of drift in relatively few areas; specifically the extreme North Atlantic, south of Australia, and southeast of New Zealand. In order to assign an age to the initiation of drift in any other area one must turn to other criteria, notably from the geological record on the trailing margins of the continents. In drawing up the suggested time-table for Continental Drift given in Table 1, igneous activity, marine transgressions and the formation of evaporite basins on the continental margins have been assumed to be precursors of the initiation of drift in any particular area (see Chapter 15). The extrusion and intrusion of tholeiitic basalts and dolerites, for example, would appear to be characteristic of the initial stages of rifting, and since these rocks can be dated by radiometric techniques they might therefore provide an important indication of the age of this event. Thus the Mesozoic extrusives and dyke swarms along the east coast of the United States give radiogenic dates around 200 million years; the Stormberg lavas and associated intrusives of Rhodesia, South Africa and Swaziland yield Lower Jurassic ages, with the earliest activity occurring in the uppermost Triassic (i.e. around 200 million years ago); extrusives in Brazil, Uruguay and Southwest Africa, on the other hand, give Lower Cretaceous ages (120–135

TABLE 1: A TIME-TABLE FOR CONTINENTAL DRIFT		
	Opening of the Atlantic	*Fragmentation of Gondwana*
0 —		
TERTIARY	(Opening of the Red Sea, Gulf of Aden, Gulf of California, and initiation of Galapagos Rise)	
	Opening of the extreme North Atlantic and the Arctic Ocean. —	Separation of Australia from Antarctica and India from the Seychelle Bank. —
UPPER CRETACEOUS	North Atlantic extends northwards to form the Labrador sea and Bay of Biscay.	Initiation of Pacific-Antarctic ridge between New Zealand and West Antarctica.
100 m.y. —		
LOWER CRETACEOUS	Opening of the South Atlantic	
LOWER JURASSIC	Partial opening of North Atlantic between North America and Africa. —	Separation of Australia + Antarctica from Africa + South America. —
200 m.y. —		

million years); the Deccan Traps of northwest India give potassium-argon dates implying an age of approximately 65 million years, and the oldest activity in the Brito-Arctic or Thulean igneous province, extending from northwest Scotland to east Greenland, is also thought to be Upper Cretaceous to Lower Tertiary in age. It seems probable that the onset of this igneous activity may precede the formation of new oceanic crust and the separation of the continents by as much as 20 million years.

FURTHER MODIFICATIONS

The concept of giant conveyor belts and equivalent tape recorders driven by mantle-wide convection gives a graphic portrayal of the ideas of Hess and Vine and Matthews, but it is now thought to be somewhat misleading in terms of the mechanics and causes of spreading. Our current ideas on the nature and cause of the spreading process have been derived from a better understanding and more detailed consideration of the geometry of spreading and the seismicity of the Earth, these two phenomena being intimately related.

In 1965 J. Tuzo Wilson noted that within the framework of sea-floor spreading an entirely new class of fault or fracture in the Earth's crust is possible. *Transform faults*, as Wilson named them, are strike-slip faults whose seismically active lengths terminate at ridge crests or trench systems. Since ridge crests are considered to be the loci of the creation of oceanic crust, and trench systems the sites of its destruction, it is possible to predict the sense of movement on such faults. Exposures of postulated transform faults on land— for example, the San Andreas fault in California and the Alpine fault in New Zealand—were found to vindicate the implied sense of movement, but many transform faults are only exposed on the ocean floor and had to await sophisticated analysis of the earthquake activity along them before the sense of movement could be determined.

Perhaps the simplest type of transform fault is that which traverses mid-ocean ridges at right angles to the ridge crest and offsetting it by tens, or occasionally hundreds, of kilometres (Fig. 16. 12). The traditional explanation of such a geometry, assuming conservation of crust, was that the ridge crests on either side of the fracture were originally

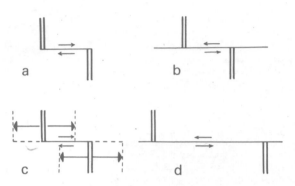

16.12 A transform fault of the right-lateral, ridge-ridge type (a and c) contrasted with a left-lateral transcurrent fault (b and d). The ridge crest is indicated by the parallel bars, and the active trace of the faults by the thin solid line.

collinear and progressively offset by shearing, that is, lateral translation, along the intervening fracture (Fig. 16.12 b, d). Wilson pointed out that the non-conservative process of spreading implies that the offset might be original and unchanging with time, with renewed spreading about the ridge crests (Fig. 16.12 a, c). If this is the case the fault will only be seismically active between the offset points on the ridge crest, and the sense of movement along this active length of the fracture will be the opposite to that predicted by the classic interpretation.

The setting up of the World-Wide Standardised Seismograph Network (WWSSN) in the 1960's enabled seismologists to determine the depth of origin (focus) and geographic location (epicentre) of earthquakes with much greater precision than had been possible previously. Thus, in 1966, L. R. Sykes was able to show that the distribution of earthquake epicentres along mid-ocean ridge crests and the focal mechanisms deduced for some of the larger earthquakes occurring on the transverse fractures confirm Wilson's predictions. Focal mechanism solutions give an indication of the sense of movement on faults and are obtained from the analysis of the first ground motion at many seismograph stations throughout the world.

The prediction and confirmation of transform faulting provided further support for the concept of sea-floor spreading but simultaneously sowed

the seed for its modification. If spreading and drift are driven by mantle convection to the surface, as Hess had suggested, how is it that both the rising and descending limbs of convection cells can be abruptly terminated and offset along transform faults? Mechanically this seems quite unreasonable. For example, the geometry of the ridge-ridge transform faults in the equatorial Atlantic (Fig. 16.2) implies convection cells that would be implausibly narrow compared to their length and height.

When the epicentres obtained from the first seven years' results from the WWSSN were plotted on a world map it was seen that the seismicity of the Earth defines the three structural elements of the sea-floor spreading model, (ridge crests, trench systems, and transform faults) with much greater precision than had been thought previously (Fig. 16.1 and 2). Furthermore the maximum depths at which earthquakes occur beneath trench and mountain systems, the 'sinks' in Hess's model, were seen to be very much greater than those beneath ridge crests, the sources. Whereas earthquakes on ridge crests appear to be restricted to depths of less than 10 or 20 km, earthquakes beneath trench and mountain systems occur at all depths up to a maximum of several hundred kilometres, exceptionally 700 km (Fig. 16.13). Thus the seismicity of the Earth is largely restricted to active ridge crests, transform faults, trenches, and young fold mountain systems. These narrow zones outline essentially aseismic areas of the Earth's crust which appear to behave as quasi-rigid plates which do not exhibit major internal deformation (Fig. 16.13).

In 1967 W. Jason Morgan investigated the possibility of analysing the geometry of spreading in terms of relative movements, on the surface of a sphere, between rigid plates bounded by ridge crests, transform faults and trench systems. The remarkable success of this extension of Wilson's transform fault hypothesis pre-dated the publication of the new epicentre plots and heralded a new formulation of the sea-floor spreading hypothesis. Much of the Earth's seismicity is clearly related to relative movements between quasi-rigid plates, and consequently this new concept is termed *plate tectonics* (see Chapter 19). The earlier terms 'continental drift' and 'sea-floor spreading' are now seen to be inappropriate in that individual plates typically include both continental and oceanic crust, although a few are entirely oceanic (Fig. 16.13). In this vein, and again as a result of

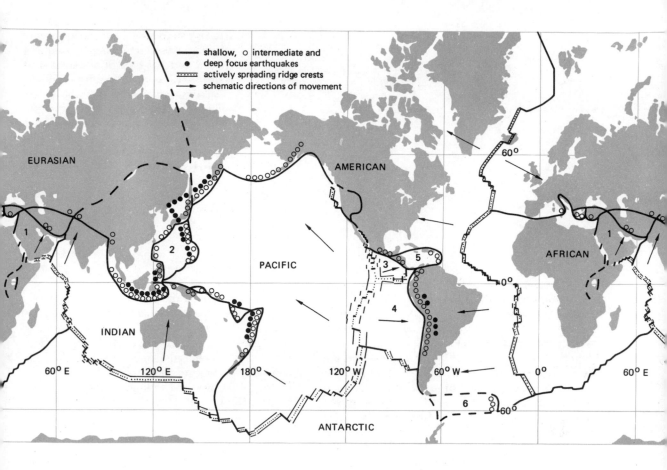

16. 13 Summary of the seismicity of the Earth, and the consequent extent of lithospheric plates bounded by active ridge crests, transform faults, trench systems and zones of compression. The six major plates are named; the following minor plates are numbered: (1) Arabian; (2) Philippine; (3) Cocos; (4) Nasca; (5) Caribbean; (6) Scotia. Spreading rates at ridge crests are indicated schematically and vary from 1 cm per year per ridge flank in the vicinity of Iceland to 9 cm per year in the equatorial Pacific Ocean.

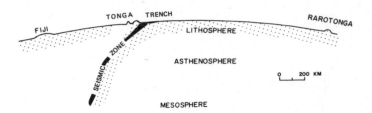

FIJI · TONGA TRENCH · RAROTONGA · LITHOSPHERE · SEISMIC ZONE · ASTHENOSPHERE · MESOSPHERE · 0 200 KM

16. 14 Postulated east-west section through the Tonga Trench, assuming that a relative lack of attenuation of seismic waves corresponds to rigidity. The lithosphere and mesosphere are thought to have appreciable strength, whereas the asthenosphere is believed to be nearer its melting point and might flow more easily over geologic periods of time.

seismological studies, it has been realised that the Earth's outermost skin, or *lithosphere,* which exhibits appreciable strength and rigidity, extends well beneath both continental and oceanic crusts to depths of 50 or even 100 km. At ridge crests the lithospheric plates are thinned by the elevation of the geotherms as a result of mantle upwelling and emplacement; beneath the trench systems the oceanic plate plunges downward, typically at an angle of about 45° to the other plate as reflected in the deep focus seismicity (Fig. 16. 14). Thus the processes of creation and destruction of oceanic plates display an intriguing symmetry and asymmetry respectively.

It is clear that the most detailed and quantitative evidence for the sea-floor spreading hypothesis relates to the mid-ocean ridges. The nature of the trench systems is an equally important aspect of the hypothesis but is apparently more equivocal. Many surface structures in island arcs and trench systems suggest extension of the Earth's crust, as though the cold, descending lithosphere beneath is behaving as a sinker and pulling on the plates above, and indeed this may be the case. Flat-lying sediments in some trenches, and the absence of buckled sediments in virtually all, have led some to doubt this aspect of the spreading hypothesis. However, the degree of buckling in sediment beneath the landward or inner wall of a trench and the extent of slumping of sediments off this wall are still unknown. The most compelling evidence for the underthrusting of island arcs and Andean-type continental margins by oceanic lithosphere comes from studies of the earthquakes occurring at depth.

Global analysis of the geometry of plate movements enables one to predict the current direction and rate of underthrusting in the trench systems, since the direction and rate of spreading at ridge crests can be deduced from the transform faults and magnetic anomalies. The directions predicted in this way correspond to the slip vectors derived from focal mechanism solutions for some of the larger shallow focus earthquakes landward of trenches, and the rates deduced are directly proportional to the maximum depths to which earthquake foci occur. Presumably the latter reflect a thermal relaxation effect as the cold lithospheric slab is thrust down into the hotter upper mantle; in this case one would expect these maximum depths to depend on the rate of underthrusting. Studies of the variation in the velocity and attenuation of seismic shear waves for various ray paths beneath trench systems also confirm the picture of an underthrusting slab of cold lithosphere (Fig. 16. 14).

Hess's concept of convection to the surface was seen to be improbable because of transform faults; his postulate of mantle-wide convection also seems improbable now for an equally simple geometrical reason. A fixed geometry of mantle-wide convection cells in which the rising limbs are correlated with ridge crests and descending limbs with trench systems (Fig. 16. 3) is incompatible with the new and highly successful concept of plate tectonics. The geometry of plate movements is such that only one plate or boundary between two plates can be fixed with respect to some frame of reference such as the Earth's geographical co-ordinates. All other plates and boundaries then move with respect to each other and the frame of reference. Consideration of the plates which incorporate Africa and Antarctica provides a simple example of this point. Both Africa and Antarctica are surrounded on at least three sides by ridge crests and by no trench systems (Fig. 16. 2). Therefore, if either continent is regarded as being fixed, the surrounding ridge crests must clearly migrate away from it as spreading occurs.

Thus despite the vast accumulation of near-surface evidence favouring the hypothesis of sea-floor spreading we are little nearer to understanding the process or processes at depth by which it is initiated and maintained. Many consider that thermal convection in some form is the only known process capable of supplying the energy requirements, but it now seems probable that there is no simple relationship between mantle convection currents and surface features. Convection in this context should be interpreted as the generation of horizontal temperature gradients which result in gravitational instabilities and hence vertical transport of hot and cold material. In this regard the cold descending slab of oceanic lithosphere landward of the trench systems is clearly a source of gravitational energy. That such a sinker exists and is an important driving force is reflected in the tensile nature of intermediate-depth earthquakes beneath some trenches, and in the observation that the only systematic correlation between surface features and the departures of the Earth's gravity field (as measured by orbiting satellites) from that predicted from the figure of the Earth is that between regional gravity highs and the trench systems.

Our present knowledge of the physical properties of the Earth's mantle is far from perfect, particularly with regard to temperature and viscosity, two all-important parameters. Probably convection, if it occurs at all, is only possible in that part of the upper mantle which is nearest to its melting point. This region, sometimes termed the *asthenosphere* (Fig. 16. 14), is probably at depths between 50–100 km and 300–400 km, the latter depth being particularly uncertain. Convection in this shell may produce a net viscous drag on the lithospheric plates above. Alternatively convection 'plumes' originating from greater depth beneath approximately twenty 'hot-spots', such as Hawaii and Iceland, may provide a similar driving force. Finally the elevation of mid-ocean ridges,

presumably resulting from the upwelling of hot, low density, mantle material beneath their crests, produces horizontal gravity gradients within the flanking plates which tend to pull them downhill, i.e. away from the ridge axis.

There is thus no lack of possible driving mechanisms: plates may be pushed apart by mantle upwelling, pulled apart by sinkers beneath the trench systems, carried by convection currents in the asthenosphere or slide under the influence of gravity. Moreover, stresses set up between and within plates may in part determine their motions; this may be particularly relevant in the case of small plates. It seems probable that plate movements are not solely attributable to any one of these effects, but result from a combination of all acting in varying degrees on individual plates.

Thus the new geological and geophysical data acquired during the 1960's amply confirmed the original hypothesis of sea-floor spreading and led to its extension and gradual modification, particularly with regard to mechanism. At the present time there is no serious obstacle to the acceptance of sea-floor spreading and continental drift as facts rather than theories, especially in the light of the results of the deep-sea drilling programme. A measure of the revolution in thinking which took place within Earth science during the late 1960's is the fact that in 1966 only a very small minority of geologists and geophysicists were convinced that such fantasies had ever reflected reality.

FURTHER READING:

SIR EDWARD BULLARD, H. W. MENARD *et al*. 1969. The Ocean. Special issue of *Scientific American*.

H. TAKEUCHI, S. UYEDA AND H. KANAMORI 1970. Debate About the Earth (2nd Ed.). *Freeman, Cooper & Co., San Francisco*.

R. A. PHINNEY (ed.) 1968. The History of the Earth's Crust. *Princeton University Press*.

H. H. HESS 1962. History of Ocean Basins. In 'Petrologic Studies', Buddington volume. *Geol. Soc. Amer., New York*.

F. J. VINE 1966. Spreading of the Ocean Floor: New Evidence. *Science 154*.

J. T. WILSON 1965. A new class of faults and their bearing upon Continental Drift. *Nature 207*.

There is a common misconception about science, which is shared by many laymen and even some scientists. It is that science is primarily, or even solely, concerned with facts. Nothing could be further from the truth. In practice, the scientist struggling to obtain factual data almost invariably has in the back of his mind a hypothesis which he is trying to test, or perhaps alternative hypotheses whose merits he wants to evaluate critically. The accumulation of facts is not an end in itself, but only a means to an end—and were it not so, science would be a dull occupation indeed.

But what happens when established facts contradict each other, or when different facts support mutually exclusive hypotheses? Suppose, for example, that in testing the validity of alternative hypotheses every piece of evidence but one supports Hypothesis A, and the exception supports Hypothesis B. If Hypothesis A is the valid one, the contradictory evidence poses a problem.

In recent years, a situation of this type has arisen in palaeomagnetism—the study of the directions and intensities of magnetizations in rocks. About one half of the world's rocks are found to be magnetized in directions precisely opposite to those which would have been produced by the Earth's magnetic field in its present sense; and almost every line of evidence suggests that this is because the Earth's field has reversed itself many times in the past (Hypothesis A). But one piece of evidence favours the idea that it is not the Earth's field which has reversed but the rock magnetizations themselves (Hypothesis B). It turns out that both hypotheses cannot be correct; and so if Hypothesis A is valid, how do we explain the evidence for B?

So far, this particular conflict has not been resolved. When it is, the answer may turn out to be trivial; but equally, it may turn out to be a growth point in geophysics and the source of a fundamental advance in our understanding of the relationships within the Earth. In the meantime, all we can do is to describe the nature of the dilemma; and this is done here by Dr. Peter J. Smith, Senior Lecturer in Earth Sciences at the Open University. It is an interesting case history of how facts and theories are not always quite as black and white as they are often supposed to be; and it illustrates well the nature of a problem which scientists frequently face.

17 The Oxidation : Polarity Paradox

by PETER J. SMITH

One clear correspondence between two sets of observations stands in the way of a theory that could account for reversed magnetization in rocks. The scientist ignores such conflicts of evidence at his peril: a good paradox marks a hidden truth.

ONE OF THE EARLIEST, and most exciting, discoveries in palaeomagnetism was that some rocks are magnetized in the 'wrong' sense. Their directions of magnetization are not in the direction to be expected from the Earth's present magnetic field, even taking continental movements into account, but precisely opposite. As more palaeomagnetic data accumulated, it became apparent that these reversely magnetized rocks were not freak occurrences but were extremely common; and by the early 1950's it was clear that they accounted for about fifty per cent of all rocks on the Earth's surface.

But how could a rock possess a reversed magnetization? All laboratory experiments—with one exception, to which we shall return later—have shown that whenever a rock is magnetized, the direction of the resulting magnetization is in the direction of the magnetic field which produced it. Yet there must be some explanation for reversely magnetized rocks. There seem to be, in fact, only two possibilities:

1. That the laboratory evidence is correct—and reversely magnetized rocks acquired their magnetizations in the directions of the Earth's field at their time of formation. This would mean that at the time these rocks were produced the Earth's magnetic field must itself have been reversed. This phenomenon is known as *field reversal*. It gives rise to the concept of two geomagnetic field senses—a field in the present sense, known as the 'normal' field, in which rocks become 'normally' magnetized; and a field in the opposite sense, the 'reversed' field, in which rocks become 'reversely' magnetized.

2. That the laboratory evidence is incomplete—and some rocks either acquire a magnetization in a direction opposite to that of the magnetic field, or they acquire a magnetization which, though initially in the same direction as the magnetic field, subsequently reverses spontaneously. These phenomena are known as *self-reversals*.

A few of the early workers in palaeomagnetism did, in fact, suggest that reversely magnetized rocks had been produced in a reversed Earth's field; but it was not a view which received common support. Accordingly, many people tried to test the alternative explanation, self-reversal, and in so doing examined large numbers of rocks in an attempt to detect physical or chemical differences between normally magnetized and reversely magnetized samples. At the same time they began to invent theoretical mechanisms whereby rock magnetizations could possibly self-reverse spontaneously.

SELF-REVERSAL

When, in the early 1950's, systematic studies of reversely magnetized rocks began, Néel proposed several theoretical mechanisms for self-reversal. They were simple and highly idealized, for at that time the idea was not primarily to discover which self-reversal mechanisms, if any, had actually operated but rather to determine, first, whether self-reversal was even possible in theory. To obtain some idea of how it could possibly occur, let us examine in a little detail just one of Néel's mechanisms.

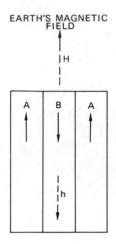

EARTH'S MAGNETIC FIELD

|H

A · B · A

|h

17. 1 A highly idealised mechanism for self-reversal in a magnetic grain. The grain comprises two different magnetic materials, A and B; and the Curie point of A is higher than that of B. During cooling, A first acquires a magnetization in the direction of the Earth's field, H. B then acquires a magnetization in a combined field, the resultant of H together with the reverse field, h, due to A. If h is greater than H, each component will end up magnetized in the direction of the corresponding solid arrow.

Consider, for example, a magnetic grain which contains two different types of magnetic material, A and B, arranged in the form of a 'sandwich' (Figure 17. 1). Then suppose that the Curie point of material A is higher than that of material B. When the magnetic grain cools from a high temperature, during the formation of the rock of which the magnetic grain is a part, the first material to acquire a magnetization will be A, because, during cooling, the Curie point of A is reached first. The magnetization of A will then be in the same direction as the Earth's magnetic field, H.

As the cooling continues, the temperature will reach the Curie point of material B; and B will then begin to acquire magnetism. But notice that B is now cooling not only in the Earth's

magnetic field, H, but also in the field, h, produced by material A which is already magnetized. The lines of force from the two pieces of material A, which act effectively as two bar magnets, indicate that h is opposite in direction to H. If it happens that h is greater than H, material B will then acquire a magnetization opposite to A and thus opposite to the Earth's magnetic field, H. When the rock finally reaches the temperature of its surroundings, the grain will have one component (A) magnetized in the field direction (H) and one component (B) magnetized in the direction opposite to that of the field. If A is subsequently removed—it may, for example, be weathered away, or simply altered chemically to a material which is non-magnetic—the magnetic grain will comprise only material B which is magnetized in a direction opposite to that of the field in which the grain was produced.

This is, of course, a highly idealised mechanism; but it does show that self-reversal is a viable theoretical concept. Since Néel suggested this mechanism, several other possibilities have been proposed, some of which are physically more realistic. However, they all have one thing in common—with one exception, no self-reversal mechanism has been proved to account for a completely reversed magnetization. The exception is a lava flow from Mount Haruna in Japan, which can be induced to self-reverse spontaneously in the laboratory. On the other hand, it should be noted that self-reversal is likely to be very difficult, if not impossible, to detect after the event. The mechanism described above, for example, is irreversible and leaves no trace of its action. Inability to detect self-reversal does not, therefore, necessarily mean that it has not occurred.

FIELD REVERSAL

Although there is no evidence for the common occurrence of self-reversal, there is a great deal in support of reversals of the Earth's magnetic field. We have, in fact, already mentioned one small piece of evidence—the fact that about half of the world's rocks are magnetized normally and the other half reversely. The Earth's magnetic field is produced by motions within the liquid iron core, which are governed by the Earth's rotation. The Earth's magnetic dipole is thus aligned more or less along the rotational axis. But why should it lie along the axis in one direction rather than the other? Clearly, the dipole can only lie in one direction at once but, statistically, each sense is equally likely. We would thus expect the dipole to have been in the present 'normal' direction for about half the time and in the opposite 're-versed' sense for the other half. Accordingly, if the Earth's field has reversed itself in the past we would expect about fifty per cent of the world's rocks to be reversely magnetized—and this is the case. If self-reversal had been dominant, on the other hand, there would be no particular reason why just half of all rocks should possess this property; and the fifty-fifty split in polarities would be an unlikely coincidence.

But this is indirect evidence for field reversal; and there is much stronger and more direct evidence available. Let us suppose that reversely magnetized rocks are due to field reversal—that is to say, a rock is reversely magnetized because it formed in a reversed Earth's field. Then if the Earth's field is reversed at a given time, *all* rocks produced at that time will be reversely magnet-ized; and if at any given time the Earth's field is normal, all rocks produced at that time will be normally magnetized. If, on the other hand, reversely magnetized rocks are due to self-reversal, there should be no time correlation. A rock with the self-reversing property is likely to appear at any time; and if all rocks of a given age are col-lected together, some should be found to be re-versely magnetized and some normally mag-netized.

So what happens in practice? When rocks younger than about 4.5 million years are accurate-ly dated and their magnetic polarities measured, they are found to fall into a well-defined time sequence (Fig. 17.2). It is indeed true that all rocks of a given age have the same polarity—for the past 4.5 million years, that is—and this is the strongest evidence in favour of field reversal. Unfortunately, the well-dated polarity-time scale cannot be extended back beyond about 4.5 million years because the dating becomes too inaccurate to resolve the distinct periods of normal and reversed polarity. However, the pattern of magnetic anomalies over the floors of the oceans suggests that field reversal is responsible for most, if not all, reversely magnetized rocks up to many tens of millions of years in age; and there is no reason to

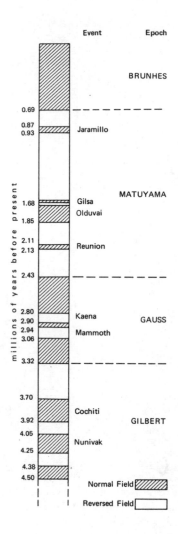

17.2 The polarity:time scale for the past 4.5 million years. Events are short periods within the longer epochs; and events and epochs are given names.

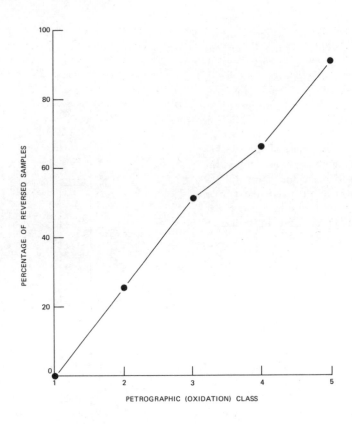

17. 3 The polarity-oxidation correlation in basalts from the Columbia Plateau, Oregon.

suppose that this does not apply to all of the world's reversed rocks of whatever age.

This is confirmed by other, distinct, evidence. Wherever an igneous rock has baked another rock, either igneous or sedimentary, both the baking and baked rocks are found to have the same magnetic polarity. This is just what would be expected from field reversal since both components of the baking-baked pair would have acquired their magnetizations at the same time. If self-reversal were predominant, however, we would expect that in some cases one of the two components would possess the self-reversal property and the other not. This never occurs. Further, it is possible to find sequences of rocks in which the directions of magnetization gradually change from one polarity to the other, the rocks between the extreme polarities having intermediate directions. This can only be interpreted in terms of field reversal, for self-reversal would produce, in addition to the fully normal magnetizations, only magnetizations directly opposed to normal.

POLARITY-OXIDATION CORRELATIONS

The evidence for field reversal is now overwhelming; and there can be no doubt that almost all reversely magnetized rocks were produced in a reversed Earth's magnetic field. This being the case, it follows that all rocks magnetized in a reversed (normal) field acquire a reversed (normal) magnetization irrespective of their chemical compositions or physical characteristics. There should, therefore, be no correlation at all between the physical, chemical or mineralogical properties of rocks and their magnetic polarities. Indeed, the presence of such correlations would be evidence in favour of self-reversal because self-reversal implies some difference between normally and reversely magnetized rocks which causes some to reverse and others not. But despite the conclusive evidence for field reversal, in recent years correlations have been discovered between the magnetic polarity and the state of oxidation of the magnetic minerals in certain rocks.

The mineral which accounts for the magnetic properties of most rocks, and especially basalts, is titanomagnetite. More correctly, titanomagnetite is not a single mineral but a series of mineral compounds having the general formula x Fe_2TiO_4 $(1 - x)$ Fe_3O_4, where x can take any value from 0 to 1. In basalts, x for titanomagnetites is usually about 0.7. When titanomagnetite grains are observed through the microscope, they are found to exist in various states of oxidation. Five fairly well-defined stages in the oxidation process (Classes 1 to 5) are recognized (Fig. 17. 3 and colour plate facing p. 201). Class 1 grains, which represent the lowest state of oxidation, are homogeneous when viewed under the microscope at magnifications up to about 1800. As oxidation proceeds (Classes 2 to 3), the titanomagnetite splits into two components, ilmenite ($FeTiO_3$) and new titanomagnetite which has a lower value of x than the original. Ultimately, in the highest state of oxidation (Class 5), the original titanomagnetite has been converted mainly to pseudobrookite (Fe_2TiO_5) and titanohaematite (Fe_2O_3) containing a little titanium. The oxidation we are talking about here is, incidentally, primary— that is, it took place at the time the rocks were formed and is not the result of subsequent chemical alteration.

Correlations between magnetic polarity and oxidation state become apparent when the oxidation state of some rocks (as represented by the class number for the relevant titanomagnetite grains) is plotted against the percentage of samples which are reversely magnetized. The correlation is particularly well illustrated by the Tertiary basalts from the Columbia Plateau, Oregon. Figure 17. 3 shows that for these particular rocks no samples in which Class 1 titanomagnetite grains predominate are reversely magnetized, that about 90 per cent of samples in which Class 5 grains predominate are reversely magnetized, and that between these extremes the correlation between polarity and oxidation is approximately linear. It is clear that there is no one-to-one correlation— that is to say, not *all* Class 5 samples are reversely magnetized, and generally (but not in the case of the Columbia Plateau basalts) not all Class 1 samples are normally magnetized. The correlation is, rather, statistical; but exists nonetheless. We can say that, *on average*, reversed samples are more highly oxidized than normal ones.

The same type of correlation also holds for Tertiary lava flows from the Isle of Mull, Scotland, and from Iceland. Furthermore, the reversely magnetized rocks from Iceland and the Columbia Plateau also exhibit independent evidence for field reversal in the form of baking-baked pairs of the same magnetic polarity and by the existence of samples magnetized during a polarity change. The dilemma posed by these results is clear. We must assume that either field reversal or self-reversal is predominant, for if both effects are common there would presumably be little order at all—that is to say, there would be neither well-defined polarity-oxidation correlations nor well-defined polarity-time sequences. But which is the dominant process? In the case of the Columbia Plateau basalts, for example, general experience together with the particular evidence from baked rocks suggest field reversal; the polarity-oxidation correlations suggest self-reversal.

Field reversal and self-reversal cannot *both* be responsible for the reversely magnetized rocks; and there must therefore be some way of reconciling the conflicting evidence. There seem to be, in fact, just three possibilities. The first is that the polarity-oxidation correlations are just coincidence. At first this seems quite likely, especially as some rock collections do not show any correlations at all. The problem with this is, however, that where correlations exist they are always in the same sense. There are no known correlations in which the normally magnetized samples are, on average, more highly oxidized than the reversed ones.

The second possibility is that field reversal has not occurred; but the evidence for field reversal is too strong to be denied.

The third explanation is that field reversal has occurred, but that the polarity-oxidation correlations are nevertheless real and significant. This can only mean that there is some sort of correlation between the activity in the Earth's core, where the magnetic field is produced, and processes occurring in the upper mantle, where the magma originates. In our present state of knowledge this seems unlikely, if not impossible. But if the correlations are valid, it can be the only solution.

FURTHER READING:

R. L. WILSON AND S. E. HAGGERTY 1966. Reversals of the Earth's Magnetic Field. *Endeavour, 25.*95.

D. W. STRANGWAY 1970. History of the Earth's Magnetic Field. *McGraw-Hill.*

It is now well established that the Earth's magnetic field has reversed many times in the past. But what exactly happens at the transition zone, the period during which the field is actually reversing from one polarity to the other? Does the dipole just rotate through 180° without losing any of its original strength? Does it rotate but lose strength as well? Or does it decrease in strength to zero and then build up to its original strength but with opposite polarity? Because the dipole must always lie roughly along the Earth's rotational axis, rotation of the dipole is highly improbable; and we are thus left with the third of these possibilities.

But what, if anything, happens to the Earth when the dipole field is removed —that is, at the point during a transition when the dipole is exactly zero? At this point the non-dipole field is still there, but with the dipole temporarily gone the total field is greatly reduced.

In 1963, Uffen suggested that the reduced fields in transition zones may explain the faunal extinctions whose occurrence had long been known. He argued that if the geomagnetic field was greatly reduced, the charged particles originally shielded from the Earth by the full field would fall to the surface and produce evolutionary changes in the Earth's fauna. During the 1960's several people examined this proposition; but in the end Dr. D. I. Black, then of Cambridge University, showed that it was just not tenable. Here Dr. Black explains why. It is an interesting example of how scientific hypotheses are put forward, subjected to rigorous analysis, and sometimes proved wrong. Meanwhile, the reasons for faunal extinctions remain obscure.

18 Polarity Reversal and Faunal Extinction

by D. I. BLACK

The Earth is partially shielded from radiation by its atmosphere and its magnetic field. Can a periodic decline in the field strength account for the catastrophic extinctions that have disrupted the evolution of species?

THE FOSSIL RECORD of life on Earth shows that many catastrophic extinctions have disrupted the evolution of species. For tens or even hundreds of millions of years, groups of animals and plants flourished, only to disappear abruptly. For example, at the close of the Cambrian period, about 500 million years ago, roughly two-thirds of the existing families of trilobites vanished. Such extinctions appear in the history of every major group of fossils, with particularly widespread extinctions occurring at the end of the Cambrian, Ordovician, Devonian, Permian, Triassic and Cretaceous periods. The cause of these extinctions is as yet an unsolved mystery.

Any hypothesis to explain these evolutionary upheavals must account for at least their following characteristics:

(i) generally, floral extinctions did not accompany mass faunal extinctions;

(ii) species disappeared all over the world simultaneously;

(iii) many mass extinctions have occurred, each lasting for a short period of geological time;

(iv) mass extinctions of land animals were sometimes accompanied by mass extinctions of marine organisms.

A satisfactory hypothesis must also be open to quantitative test, and of the many ideas put forward, two explain the above four features and are amenable to investigation.

Both theories invoke extra-terrestrial radiation which can account for the above observations naturally. Floral reproduction through long-lived, radiation-resistant spores implies that plants could survive an acute radiation dose that might kill a faunal species. Even if the mature plants were to die, spores could later perpetuate the flora. If the external radiation dose was spread over the Earth then world-wide effects might be explained. 'Short' bursts of radiation, separated by long time intervals, would lead to the temporal distribution of the extinctions. Land and marine organisms would be simultaneously affected if the radiation had sufficient energy to penetrate the sea to the depth at which the organisms live.

The first hypothesis is that supernovae explosions produce fatal radiation doses, from direct cosmic-ray bombardment, at sea level. The explosions, and hence the radiation doses, are short-lived and happen rarely. A quantitative study of the energies and time intervals involved has made this idea plausible, but there is no evidence to support the hypothesis.

Here we will examine the second hypothesis, for which *prima facie* evidence has been found. In 1963, Uffen predicted that evolutionary surges would occur at every polarity reversal of the Earth's magnetic field. At a reversal, the field intensity is much reduced for a 'short' period, during which time the Van Allen belts would spill on to the Earth's atmosphere and the solar wind would bathe the Earth. The hypothesis suggests that there would be high radiation dose rates, and hence high mutation rates and rapid evolution, at a polarity reversal.

The postulated mechanism must be altered slightly. A high mutation rate is unlikely to cause significant evolutionary changes, since there are ample natural mutations to permit response to any environmental selection pressure. If the stress of the hypothesis is changed to 'high radiation dose rates leading to faunal extinctions', then it plausibly predicts evolutionary changes at field reversals, and allows a quantitative examination of the suggested mechanism.

FAUNAL EXTINCTIONS AT RECENT POLARITY REVERSALS

Before surveying the evidence that has come to light to support this hypothesis, the recent history of the geomagnetic field must be described. Potassium-argon dating combined with the magnetic measurement of lava flows has determined the sequence of reversals of the field over the last 4 million years. The field shows long epochs of one polarity, of the order of 1 million years each; and in each epoch there may be short 'events' of the order of 10^5 years of the opposite polarity. The lengths and names of the more recent epochs are shown in the left-hand column of Figure 18. 1.

Ocean sediments are magnetized in the direction of the prevailing geomagnetic field at the time of deposition, so that sediment cores show a continuous profile of field changes. The sediments also contain fossils of marine micro-organisms, and several investigations have found correlations between faunal boundaries and magnetic reversals in ocean cores. Figure 18. 1 shows a striking example of such work. Four faunal zones, each consisting of 3–6 species of Radiolaria (single-celled micro-organisms with a siliceous skeleton of typical radius equal to 0.1 mm) had previously been defined, and two of the faunal boundaries roughly coincide with field reversals. Studies of Arctic, Pacific and Indian Ocean cores also show that faunal changes coincide with polarity reversals. Possible correlations between faunal boundaries and changes in temperature (i.e. climate), sedimentation rate and sediment type have been looked for but not found.

Ocean cores also provide the most detailed information on the behaviour of the geomagnetic field as it reverses. The intensity of the field decreases in about 10000 years to about 15% of normal, and in about a further 10000 years the original intensity is regained. The inclination of the field changes much more abruptly, in less than 1000 years.

THE RADIATION HAZARD FROM INCREASED PARTICLE FLUX AT ZERO FIELD

The unit of radiation used here is the *rad*. 1 rad is an absorbed dose of 10^{-2} J kg^{-1}. The more familiar unit, the *roentgen*, only applies to X-rays and γ rays, but for this class of radiation the rad and roentgen are roughly equal. Different types of radiation have different biological effects. The Relative Biological Efficiency (a measure of the biological damage) is 1 for X-rays, γ rays and electrons, 2.5 for slow neutrons (< 100 ev) and 10 for fast neutrons and protons.

Cosmic Rays

Primary cosmic rays are isotropic in space. Meteorite studies of the 'fossil' record of cosmic radiation show that the flux and energy spectra have been constant over millions of years. The Earth is partially shielded by its magnetic field from cosmic rays. There is no shielding at the poles, while at the equator only particles with energy over 15 Gev can reach the atmosphere. The equatorial flux is only about 10% less than at the poles (apart from the neutron component whose biological effect is negligible) and by latitudes 45° has risen to the polar value. At a reversal, the cosmic ray dose rate at the equator would rise from its present value of 23 mrad year^{-1} to 25 mrad year^{-1}. The rate of increase would be very slow compared to variations in dose rate with sunspot cycle, so that both the increase in dose rate and its rate of change are very small.

Solar Flares

The largest solar flare ever recorded at sea-level, in 1956, produced an integrated increase in radiation intensity in polar regions equivalent to doubling the cosmic-ray flux for $1\frac{1}{2}$ hours. The protective atmosphere ensures that there is no hazard from solar flares at sea-level either now or at a time of zero field.

The Solar Wind and the Van Allen Belts

Neither the protons in the solar wind, nor Van Allen belt particles, (both discussed in more detail below) are sufficiently energetic to penetrate through the entire atmosphere at zero field, so that there will be no direct radiation dose at sea-level from either source at a field reversal.

OTHER HAZARDS FROM THE SOLAR WIND AND VAN ALLEN BELTS

It is relatively easy to dismiss the hazard of an increase in direct-radiation dose rate at a field reversal; but there are other possible hazards that could arise from the exposure of the atmosphere to these particle fluxes.

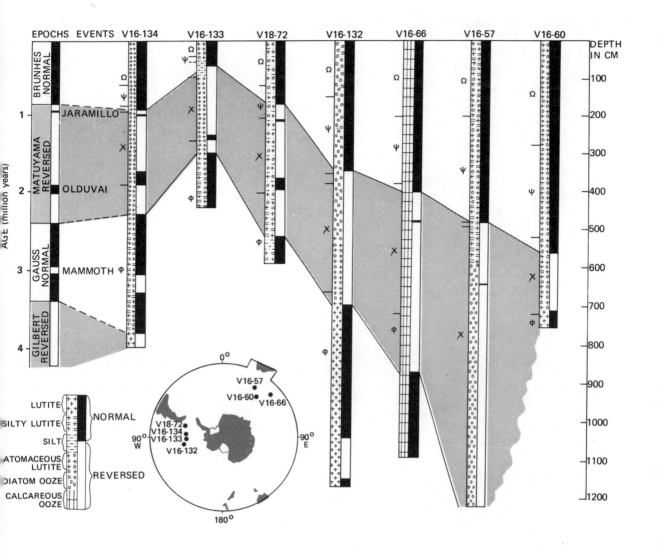

18. 1 Correlation of magnetic and faunal zones in seven Antarctic cores. Ω, Ψ, X and Φ are four Radiolarian zones.

The steady solar wind incident on the Earth has a typical velocity of 400 km s^{-1} and a proton flux of 10^{12} m^{-2} s^{-1}. The typical energy of each proton is 800 ev, and of the accompanying electrons 0.5 ev. The solar wind is supersonic, in the sense that its speed is ten times the Alfvèn wave speed, so that a detached shock front is formed in front of the magnetopause (see Fig. 18. 2) at a subsolar distance of about 12 Earth radii (R_E) from Earth. The solar wind protons are reflected at the magnetopause, and so never enter the magnetosphere. However, at a field reversal the atmosphere will be exposed to the solar wind if the minimum field is so small that the magnetopause comes down to $1R_E$. It is thought that the field never decreases below 15% of normal, so that exposure would not occur. If, however, this result is incorrect, and the solar wind enters the atmosphere, then the protons will be absorbed in its outermost extremes (about 150 km up).

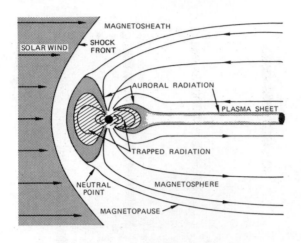

18. 2   **Magnetospheric domains.**

The protons will ionize the air, and create ozone (O_3). It is known that the ozone content of the atmosphere (concentrated between 15 and 40 km up) has a 'greenhouse' effect and retains warmth. However, the solar wind could not increase the equilibrium photochemical content of ozone by more than 1 part in 10^7, so that no climatic changes from ozone production could be caused by the solar wind.

A simple calculation shows that the solar wind could not literally blow away any appreciable quantity of the Earth's atmosphere in the time scale of a reversal.

Ultraviolet radiation at sea-level is controlled by both oxygen and ozone in the atmosphere, but as the solar wind can affect neither significantly there will be no change in ultraviolet dose rate.

The Van Allen belts are zones of stably trapped particles within the magnetosphere. The particles consist of protons with energies from 100 kev to 500 Mev and electrons with energies from 10 kev to 5 Mev. It was once thought that aurorae were produced by the Van Allen belts 'overflowing' and dumping electrons on the atmosphere. It is now known that the belts are inadequate for this rôle and that an external source both replenishes the belts and causes aurorae. If all the electrons in the belts were suddenly dumped on the atmosphere they would cause only a minor aurora. The proton flux in the belts represents a balance between replenishment from the magnetosphere and absorption by the atmosphere. This steady flux

into the atmosphere would be no more than doubled if the Van Allen belts were dumped over a period of 10 years.

At a polarity reversal, when the field decreases over a period of 10 000 years, the effects of the gradual absorption of the Van Allen belts into the atmosphere would be quite undetectable.

OTHER HAZARDS FROM COSMIC RAYS

Cosmic rays create many radio-isotopes in the atmosphere. Radioactive isotopes in the environment cause three radiation hazards to animal life. They contribute to the background (external) dose; ingested radio-isotopes give an internal radiation dose; and when incorporated in the genes may destroy the genes from the violence of their decay. At a polarity reversal, the production rate, and therefore the concentration and the hazard of these isotopes, will increase. It is known that of the radio-isotopes produced by cosmic rays (and solar flares) the production rate of ^3H, ^7Be, ^{10}Be, and ^{14}C will double at zero field, while the production rate of ^{22}Na, ^{32}P, ^{32}Si, ^{33}P, ^{35}S, ^{36}Cl and ^{39}Cl will increase by a factor of five. These isotopes emit between them α, β and γ rays of typical energies 4 Mev, 0.5 Mev and 1.5 Mev respectively and have corresponding ranges in sea water of 0.03 mm, 2 mm and 1 m.

The majority of radiation from dissolved radio-isotopes in the sea comes from ^{40}K, which gives

Radiolaria external and internal doses of 3.6 and 0.7 mrad year^{-1} respectively. None of the isotopes created by cosmic rays gives a significant external dose, but organisms concentrate carbon, and the internal dose from ^{14}C will be 0.8 mrad per year at normal field intensity and 1.6 mrad per year at zero field. At the poles the only increase in radiation at a polarity reversal will be from ^{14}C, and it will be a mere 3% rise. At the equator, the enhanced cosmic-ray flux plus the ^{14}C effect cause a total increase of 12%. It is inconceivable that this increase could cause the extinction of a species.

There remains the hazard of radio-active decay in the genes. Only ^{14}C, ^{32}P and ^{33}P need be considered—the half-lives of the other isotopes are too short for them to become incorporated in organic matter—and if every decay in a gene is lethal, then at zero field just 6×10^{-6} of the Radiolaria population will die every year.

organisms and sufficiently disturbed the Earth's core to reverse the field. The other hypothesis is that a shower of energetic magnetic monopoles from space could both reverse the geomagnetic field and cause the biological damage that cosmic rays fail to do. Both of these indirect mechanisms have been effectively dismissed in other papers.

Any new hypothesis will gain considerably in credence if it can explain, or encompass, the observations that most species survive each reversal, and that those species that do disappear have survived many previous reversals. We are left with a dilemma. On the one hand there is the evidence from independent investigations of widely scattered oceans; on the other, there is still no semblance of a satisfactory theory to explain a causal relation. We must, for the present, suspend judgement.

CONCLUSION

The many effects of extra-terrestrial radiation looked at above are collectively inadequate to explain the disappearance of species. However, the evidence continues to mount that improbably many faunal changes occurred close to magnetic boundaries, if chance alone were responsible.

Other causal mechanisms to relate polarity reversals and faunal extinctions have been sought. Climatic changes are frequently suggested, but not only is it hard to envisage how a polarity reversal could directly influence the climate, but studies of climatic changes in ocean cores, by oxygen isotope analysis or comparison of relative abundances of warm and cold water species, show no correlation with faunal changes. Nevertheless, 'volcanic maxima' have been put forward as agents of climatic change and hence of faunal change; it is suggested that such maxima occur at polarity reversals because of the postulated increase in upper mantle activity at such times.

The most direct possibility is that some species are sensitive to changes in the ambient magnetic field. In contrast, two indirect mechanisms have been proposed that consider extinctions and reversals to be mutual results of a third phenomenon. The Australasian tektite field was deposited 0.7 million years ago, coincident with the last polarity reversal, and it has been suggested that the tektites were created by a meteor impact which simultaneously killed off surface-living marine

FURTHER READING:

BULLARD, E. C. 1968. Reversals of the Earth's magnetic field. *Phil. Trans. Roy. Soc. A 263*, 481.

HAYS, J. D. AND OPDYKE, N. D. 1967. Antarctic Radiolaria, magnetic reversals and climatic change. *Science, 158*, 1001.

NEWELL, N. D. 1963. Crises in the history of life. *Scientific American, 208*, 77.

TERRY, K. D. AND TUCKER, W. H. 1968. The biologic effects of supernovae. *Science, 159*, 421.

UFFEN, R. J. 1963. Influence of the Earth's core on the origin and evolution of life. *Nature, 198*, 143.

WILLIAMS, D. J. AND MEAD, G. (ed.) 1969. International Symposium on the Physics of the Magnetosphere. *Rev. Geophysics*, 7, Nos. 1-2.

HAYS, J. D. 1971. Faunal extinctions and reversals of the Earth's magnetic field. *Geol. Soc. Am. Bull., 82*, 2433.

So continents drift and sea-floor spreads—or do they? You may think this all rather speculative and little more than gazing into a scientific crystal ball; if so, this chapter on plate tectonics will be of little interest to you. But if the arguments presented in Chapters 15 and 16 convince you and you believe, as Professor Vine says, that 'there is no serious objection to the acceptance of sea-floor spreading and continental drift as facts rather than theories', then there are other questions to answer: what are plates, how are they created, and what causes them to move?

First, let us look briefly at the development of the 'Plate Tectonics' theory which proposes that much of the present surface of the Earth is formed by six or seven internally rigid, but relatively thin, plates. The chain of reasoning, which only recently culminated in the plate theory, began in the early 1960's when the late Harry Hess of Princeton University, on the basis of circumstantial evidence and brilliant geological insight, proposed his sea-floor spreading hypothesis. A few years later, the interpretation of linear magnetic anomalies in the ocean basins by Vine and Matthews provided the quantitative break-through and showed that the spreading theory was not just a speculation, but must be treated seriously.

Up to this time 'plates' had not been mentioned. The first usage of the term seems to have been in 1967 by W. Jason Morgan of Princeton University. Morgan developed the ideas of J. Tuzo Wilson of Toronto University concerned with the relationship between oceanic ridges and transform faults, and suggested how the shape of plates could be determined and their movement predicted. In the meanwhile, D. P. McKenzie of Cambridge and R. L. Parker of the Scripps Oceanographic Institution arrived independently at similar concepts by the study of earthquake first motions in the north Pacific. Between them, the papers published by these three scientists ushered in a phase of previously unparalleled innovation and upheaval in the Earth sciences. The 'New Global Tectonics', now usually known by the less grandiose title of 'Plate Tectonics', had arrived. Virtually no branch of geology, geochemistry or geophysics remained unaffected by the revolution which followed and which continues to-day.

In this chapter, Dr. Oxburgh, of the University of Oxford, first describes how plates are surrounded by three different types of boundary—the constructive boundaries where new crust is created, the destructive boundaries or sinks, where plates are destroyed, and the conservative boundaries where plates neither gain nor lose surface area but just slip past each other. He then discusses how plates move over the surface of the Earth, and how movement in one plate affects the movement of others. Then, after a brief survey of the contributions of seismological observations to the understanding of plate tectonics, the more controversial topic of how plates are formed and what causes them to move is considered. Dr. Oxburgh concludes that, although several mechanisms are possible, convective movements within the Earth's mantle are, most probably, responsible for plate creation and movement.

19 Plate Tectonics

by E. R. OXBURGH

This great, unifying concept draws sea-floor spreading, continental drift, crustal structures and world patterns of seismic and volcanic activity together as aspects of one coherent picture.

IT IS PROPOSED that the entire surface of the Earth comprises a series of internally rigid, but relatively thin (100–150 km) plates. Although the size of the plates is variable, much of the Earth's present surface is occupied by half a dozen or so large plates (see Fig. 16. 13). These plates are continuously in motion both with respect to each other and to the Earth's axis of rotation. Virtually all seismicity, volcanicity and tectonic activity is localized *around plate margins* and is associated with differential motion between adjacent plates.

Boundaries between different plates are therefore regions of particular interest, and we distinguish between a *plate boundary*, the surface trace of the zone of motion between two plates, and a *plate margin*, the marginal part of a particular plate; two plate margins meet at a common plate boundary.

There are three types of plate margin. The first is known as a *constructive* margin. During the spreading process, which occurs at oceanic ridges, new crust is created and moves away from the ridge along with the underlying, uppermost mantle. The newly generated crust and its upper mantle is effectively welded to the plate's trailing edge. Thus a ridge represents a zone along which two plates are in motion *away* from each other; yet they do not separate, because new material is continuously added to the rear of each. Boundaries at which the net effect of motion is to generate surface area are here termed *sources*. The second type of margin is *destructive* and occurs at the deep ocean trenches. At trenches, two plates approach each other and one slips down under the margin of the other at an angle of about 45°. Plate boundaries at which the net effect of the motions is to destroy surface area are called *sinks*. Plates may also have *conservative* margins.

These are margins at which the plates neither gain nor lose surface area, but simply slip past each other.

One other important consideration is that, although plates may comprise either continental or oceanic crust or both, it seems that only those parts of plates which are capped by oceanic crust can participate in the main processes of plate growth and destruction. It is fairly clear that the spreading processes which we observe operating to-day at ridges produce only the thin, layered oceanic crust and not the thick and geochemically distinct crust of the continents. That continental crust is not destroyed at sinks is less easy to demonstrate. It seems that when the thick, low-density (ca. 2.85 g cm^{-3}) continental crust reaches a sink it may descend a short distance, but its buoyancy with respect to the mantle (density ca. 3.35 g cm^{-3}) is so great that the downward motion cannot continue: the continental part of the plate remains at the surface and the sink ceases to operate.

Let us now consider the important assumptions which underlie the general theory of Plate Tectonics—there are three which are more important than the others:

(1) that sea-floor spreading occurs, i.e. that new oceanic crust is continuously generated at irregular line 'sources' (active ocean ridges);

(2) that the Earth is of constant surface area, or, if not, the area changes at a rate which is small by comparison with the rate of generation of new surface area by spreading;

(3) that, once formed, new crust forms part of a rigid plate which may or may not incorporate continental material.

Taken together, assumptions (1) and (2) require that the sinks destroy crustal material at the same rate as it is generated. Assumption (3) implies that deformation tends to occur around rather than within plates.

Assumption (1) is the subject of Chapter 16. Assumption (2) you are asked to accept, as its treatment is beyond the scope of this volume; but there are good reasons to believe that the Earth's radius has not increased by more than 5% during the last 600 million years. Assumption (3) is best verified by consideration of the crustal structure of the oceans. Seismological studies have established that, in contrast to the relatively thick and unlayered, or irregularly layered, crust of the continents, oceanic crust is thin and layered with a high degree of uniformity. Three horizontal and virtually undisturbed layers are usually present above the M-discontinuity, which lies about 5 km below the ocean floor. The uppermost layer (Layer 1) consists of sediments in various degrees of lithifaction and ranges in thickness from zero at the ridge crest to 0.75 km. The composition of Layers 2 and 3 is more controversial and is briefly discussed in a later section; we note here only that they are of the order of 1 and 4 km thick respectively. If any major crustal deformation had taken place, such as characterizes the orogenic belts of the continents (see Chapter 20), the simple uniform layering of the oceanic crust would have been buckled, thickened and distorted—this is not found. The same inference can be drawn from the preservation of a simple and regular pattern of magnetic strip anomalies well away from the ridge at which they were formed (see Chapter 16).

So far the discussion has centred upon the oceans because their simple structure allows deformation to be readily recognized. Suitable deformation indicators are generally not available for the continental parts of plates, and because they form the only 'permanent' and therefore locally rather ancient part of the Earth's surface, their history and structure is much more complex. If, however, we consider the present world-wide distribution of active oceanic ridges and trenches (Fig. 19. 1) and then, where possible, extend these lines into the continents, we find (1) that all pass into zones of current continental tectonic activity (e.g. the Indonesian arc/trench system passes into the easternmost part of the Himalayan system); (2) that all zones of current continental activity either lie along the continuation of such

oceanic features *or* lie along plate boundaries identified as sources or sinks (e.g. the Andes lie parallel to the edge of the South American plate bounded to the west by a sink—the Peruvian trench). If the continental continuations of the oceanic sinks and sources are regarded as the continuations of the plate boundaries it becomes possible to recognize a world-wide system of plates encompassing both continents and oceans.

GEOMETRICAL ASPECTS OF PLATE MOTION

The plates which we have discussed in the previous section are continuously in motion with respect to each other. They also form the closed surface of a sphere. This means that the motions of all plates are to some extent interdependent, and a change in the velocity or direction of motion of one plate cannot occur without concomitant changes in motions elsewhere.

As shown by W. J. Morgan, a rigorous geometrical analysis of plate motions can be made if plates are regarded as totally rigid bodies moving on the surface of a spherical Earth. Euler's geometrical theorem shows that every displacement of a plate from one position to another on the surface of a sphere can be regarded as a simple rotation of the plate about a suitably chosen axis passing through the centre of the sphere (Fig. 19. 2). All points on the plate travel along small circle paths about the chosen axis in passing from their initial to final positions. It follows that any plate boundary which is conservative (i.e. involves neither plate growth nor destruction) *must be parallel to a small circle*, the axis of which is the axis of rotation for the relative motion of the plates on either side. Conversely, any plate boundaries which are not parallel to such small circles must be either constructive or destructive.

A corollary of Euler's theorem is that the velocity of relative motion across a constructive or destructive boundary is proportional both to angular velocity about the axis of rotation for the motion of the plates, and to the angular distance of the point on the boundary under consideration from the axis of rotation. This relationship is illustrated in Figure 19. 2. It implies that velocities vary continuously along all constructive and destructive boundaries, being smallest in 'high rotational latitudes' and greatest in 'low rotational latitudes.'

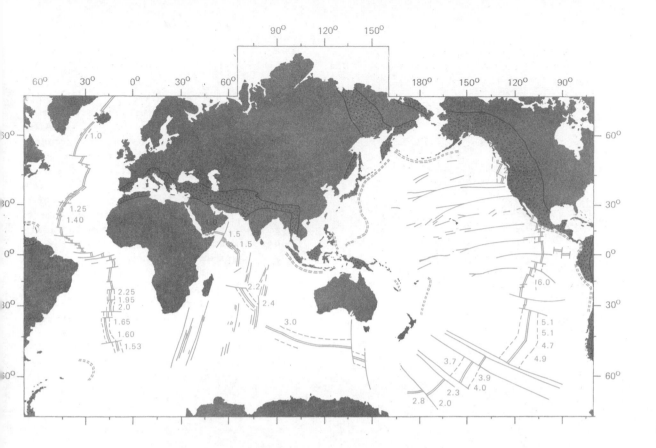

19. 1 **The main tectonic features of the Earth to-day. Colour bands mark Mesozoic and Tertiary orogenic belts and the larger island arcs; actively spreading ocean ridges are indicated by double lines with half spreading rates given in cm/yr, and the associated 10-million-year anomalies by a single dashed line; active trenches are shown by a double dashed line, and fracture zones by a single line.**

The importance of these geometrical considerations is that they afford a further test for the plate tectonics model. Let us take, as an example, W. J. Morgan's study of the Atlantic Ocean. Figure 19. 3 shows the coast lines and the mid-Atlantic ridge for the equatorial Atlantic. The ridge crest is displaced along numerous transform faults whose importance here is that they represent conservative sectors of plate boundary separating constructive sectors—the spreading parts of the ridge. All active transform faults on the same ridge ought to be segments of coaxial small circles if the plate model is valid. Figure 19. 4 shows the intersection of great circles constructed normal to the strike of the equatorial Atlantic transform faults. All the great circles should intersect at a single point on the surface, the apparent pole of rotation, if the model is

satisfied. All but one of the great circles pass through the area indicated by the heavy circle centred at 57.5° N and 36.5° W. The intersection of the great circles is such that the longitude control is good but the latitude control rather weak; bearing in mind the uncertainties in the mapping of the sea-floor topography, this result is about as good as could be expected.

Spreading rates have been established from the magnetic profiles at more than a dozen points on the mid-Atlantic ridge. In Figure 19. 5 the spreading half rate (Chapter 16) is plotted against latitude; strictly, the latitude should be latitude with respect to the axis of relative motion, but as the mid-Atlantic Ridge runs approximately north-south, and the pole of rotation is close to the ridge, it is satisfactory, and more convenient, to use geographic latitude. The variation of the observed

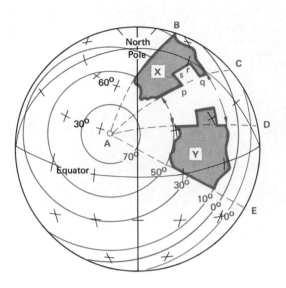

19. 2 The spherical geometry of Plate Tectonics. Consider an original segment of the Earth's surface ABE which splits into two pieces, ABC and ADE; they separate, remaining in contact only at point A. A is then a pole of rotation about which the separation occurs. Points originally in contact move away from each other along small circle paths which are 'lines of latitude' (in light colour) about the pole of rotation. Surface velocities of separation are zero at A and increase to a maximum 90° from A (0° latitude in colour). Exactly the same considerations apply if an original area XY (colour) splits, and the two parts separate; their motion can be described as a rotation about a pole of rotation (again A), and originally adjacent points must separate along small circle paths about A. If the sides (ps and rq) of the rentrant are parallel, they must also be parallel to lines of latitude about A. Geographical latitudes and longitudes are shown in black.

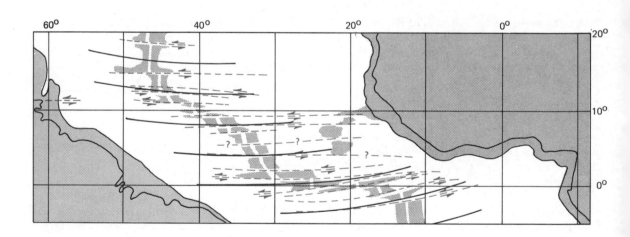

19. 3 The strike of transform faults (broken coloured lines) in the equatorial Atlantic compared with small circles (solid black lines) concentric about a pole 58° N, 36° W.

values of the spreading rate is in agreement with the theoretical model (solid line) for rotation about an axis 62° N and 36° W and a maximum spreading half rate of 1.8 cm/yr. This axis is 4° further north than the axis derived earlier and is preferred as it is compatible with the rate observations and with the fracture zone data, which alone give rather poor control on the latitude of the axis.

Two further points of general significance may now be mentioned. The direction of relative spreading is parallel to the transform faults across a ridge and not, other than by coincidence, at right angles to the trend of the ridge crest or to the magnetic anomaly lineations; it is vital that this be taken into account when computing spreading rates from anomaly patterns. Further, if an active transform fault passes laterally into a dead fracture zone following the trend of the same small circle, this is evidence that the relative directions of spreading have remained constant for some time. Conversely, abrupt changes in the direction of these inactive fracture zones within plates are evidence of past changes in spreading direction.

We return now to a more detailed discussion of plate margins and boundaries between plates. It may happen that across a particular boundary between two plates the nature of the adjacent plate margins may be different. For instance, a destructive margin for one plate may abut against a different kind of margin for the plate with which it is in contact; indeed this is what happens in active island arc regions (Figure 19. 6) where one plate is lost into the mantle along the line of the ocean trench while the edge of the adjacent plate is either conserved or, as discussed later, slowly grows. There is in general no *geometrical* constraint on the character of adjacent plate margins. There do, however, appear in practice to be various physical constraints. For example, it appears that the margins of *both* plates which meet at an active ridge are constructive, and spreading seems in all cases to take place away from the ridge in both directions. There is, however, no geometrical reason why spreading should be a symmetrical process.

The point where three plate boundaries meet is termed a *triple junction*; it is not intended to elaborate on the significance of such junctions, but the way that movement between three plates can occur is shown in Figure 19. 7.

Let us now consider the boundaries and margins of an individual plate. We find that, provided only one plate is considered, there are no geo-

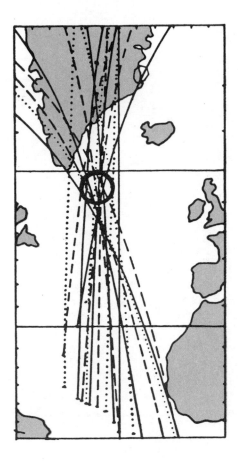

19. 4 **Great circles perpendicular to the strike of the offsets of the mid-Atlantic ridge; see text for discussion.**

metrical constraints and that a plate may be of any shape and may be surrounded by any combination of boundaries and margins, conservative, destructive or constructive. On a world-wide scale, however, the constraint remains that the overall pattern of boundaries and relative velocities must be such that the Earth's surface area remains the same. The boundaries of the main plates in existence to-day show that a variety of different boundary combinations exist (Fig. 16. 13). Some of these boundary combinations are very instructive. For example, the African plate is bounded on the west by the mid-Atlantic ridge and on the east by the Indian Ocean ridge. Both of these ridges are active centres of spreading, and although the transform faults on each are not

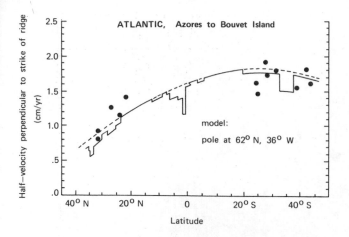

19.5 Variation of spreading rate with latitude. Dots indicate observed rates; dashed line, predicted rate parallel to the direction of spreading; solid line, predicted rate normal to the trend of the ridge.

exactly parallel, they run, more or less, E–W. There is no evidence whatsoever of the existence of any sink between these two ridges bounding the African plate (indeed, if there were, it would not be a single plate). This means that the plate has two opposed constructive margins and that it must therefore at present be expanding in an E–W direction and changing its shape. This conclusion has several important implications. If some plates are able to increase their area, others must become smaller; otherwise the constant surface area constraint would not be met. So, in theory, plates may be completely destroyed, and mid-ocean ridges and deep ocean trenches must be free to migrate with respect to each other and with respect to conservative features on the Earth's surface.

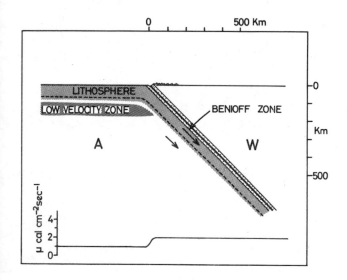

19. 6 Schematic section of relationships in an active island arc; the plate on the left is being destroyed and is sliding beneath that on the right. Heat is generated in the zone of slip between the two plates (coarse stippling) and gives rise to high surface heat-flow (profile below). The trench and arc are shown at the surface.

We have arrived at a picture of plates which are continuously in motion with respect to each other and which are normally either growing or diminishing in size unless, by chance, the rates of construction and destruction on opposite sides of a plate exactly balance each other, or, by chance, a plate is entirely surrounded by conservative boundaries. But physical arguments suggest that a plate which contains continental crust can never be completely destroyed. If this is so, whenever the continental part of a plate arrives at a sink there must be some change in the pattern of inter-plate motion. Either the boundary motion changes, or relative motion between the two plates ceases. This second possibility involves adjustments in the motions at the other boundaries of the two plates and the formation of a single plate from the two which existed previously. This appears to be the only possibility when two continental masses approach a sink from either side. It will be suggested (Chapter 20) that 'continental collisions' such as these have occurred along the lines of many of the Earth's fold mountain belts.

Conservation of constant surface area for the Earth requires that on any great circle round the globe the sum of the parallel components of constructive and destructive velocities at all sinks and sources crossed by that great circle be zero. There is no requirement that the number of sinks and sources along any particular great circle be equal —if ridges are more numerous than trenches, then on average the trenches must consume the plates faster than ridges generate them. These considerations are illustrated in Figure 19. 8.

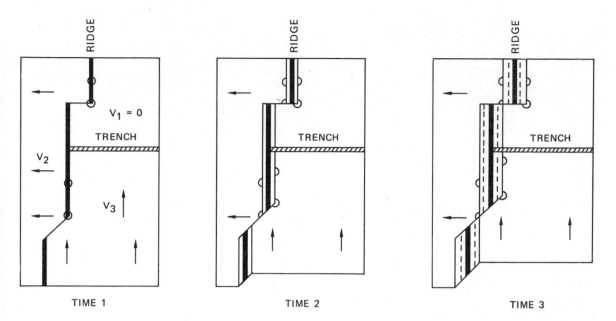

TIME 1   TIME 2   TIME 3

19. 7   Three crustal blocks bounded by a ridge, trench and faults are shown at three successive stages in spreading.   The upper right-hand block is regarded as fixed.   Note the displacements of the four circular markers shown at Time 1.   The strike of a transform fault is parallel to the difference of the velocities of the two sides.

SEISMOLOGY

Although seismology played a relatively minor rôle in the early development of ideas on Plate Tectonics, it was very soon realized by seismologists, notably those at the Lamont Geological Observatory, New York, that Plate Tectonics offered for the first time a possible all-embracing explanation for the distribution of the Earth's seismic zones which, although reasonably well mapped, had never been fitted together into any convincing, large-scale tectonic framework.   Furthermore, earthquake focal mechanism studies offered a possible means of testing an important aspect of Plate Tectonics theory.

But let us first consider the thickness of the plates. Clearly Plate Tectonics does not involve the movements of an infinitely thin skin; we may deduce that the moving plates must involve material at least as deep as Layer 2 of the oceanic crust, for this is the depth of origin of the pattern of magnetic lineations which record the plate motions (see Chapter 16).   In so far as continents are involved in plate motions and have their own deep structure extending down to 40 or more kilometres, plates must be at least this thick.   It is,

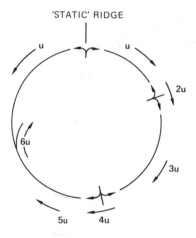

19. 8   A great-circle section through the Earth, traversing three ridges and one trench. All ridges assumed to have the same spreading half-velocity, u, and the ridge at the top to be 'static'. All other features migrate with respect to the static ridge, at the velocities shown, if surface area is to be conserved.

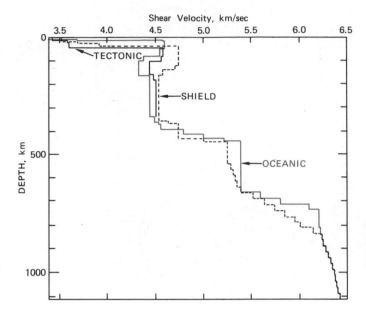

Shear Velocity, km/sec

DEPTH, km

19. 9 **Shear wave velocity distributions with depth for oceanic, shield and tectonically active areas, as derived from surface wave studies. Note that the step-like increases in velocity result from the computational technique employed, and in nature velocity would vary more smoothly.**

however, equally clear that the plates are of finite thickness and that at some depth they become decoupled from what lies beneath.

In 1948, the late Beno Gutenberg of the California Institute of Technology proposed the existence of a seismic 'Low Velocity Zone' within the Earth's upper mantle at a depth of about 80 km in order to explain certain anomalies in earthquake body-wave amplitudes. At the time, this suggestion did not meet with widespread acceptance, but in the last decade detailed analyses of the travel times of surface waves have generally confirmed Gutenberg's predictions, and curves are now available for the variation of both compressional and shear wave velocity with depth (Fig. 19. 9). These studies have yielded two important results: (1) there is in the upper mantle a world-wide zone at which the velocity of transmission of *shear* waves reaches a minimum; (2) the characteristics of this zone are different in regions which have different crustal characteristics. In general, the low velocity zone for shear waves (LVZS) occurs between 100 and 200 km depth.

The velocity of compressional (*P*) and shear (*S*) waves is dependent on the compressibility, rigidity and density of the transmitting medium. Furthermore, it is known that an increase in pressure increases, and an increase in temperature decreases,

the wave velocities. This suggests that within the low velocity zone there is an abnormally high thermal gradient, for pressure must be assumed to increase steadily with depth. Also, cold materials transmit a greater fraction of the elastic energy they receive than they would at high temperature. The low velocity layer absorbs an abnormally high amount of energy; it is therefore probable that this zone is abnormally hot.

At temperatures close to their melting temperatures, rocks are very weak. Furthermore, studies of body waves, velocities and absorption have led some workers to suggest that LVZS contains a small amount of liquid—that the mantle is just beginning to melt. If this were true, the rocks would certainly have a very low shear strength. It thus appears that the LVZS represents a zone which is mechanically weak, and at which plates, 100–150 km thick, are decoupled from the deeper mantle. This uppermost 100–150 km of the Earth which is relatively strong, and which appears to be bounded beneath by a relatively hot and weak zone, is commonly known as the *lithosphere*; the underlying weak zone is sometimes known as the *asthenosphere*.

It is clear that the majority of earthquakes have occurred along certain narrow, semi-continuous, well-defined zones and that there are extensive, irregularly shaped areas which are virtually free of seismicity. Comparison of Figure 16. 1 with Figure 16. 13 shows that the seismicity is concentrated along plate boundaries. This is not altogether surprising in that the seismicity was one of the factors taken into account when the boundaries of the plates were drawn. Seismicity is virtually restricted to ridges, trench-arc systems, major fracture zones and young orogenic belts (see Chapter 20), features which are regarded as diagnostic of plate boundaries.

If we now subdivide earthquakes by depth of focus, and consider only those with foci deeper than 200 km (intermediate and deep-focus earthquakes), a very different distribution pattern emerges. Comparison of the two figures shows that they are restricted to oceanic trench areas, the regions where plates descend into the mantle. If the earthquakes from a typical trench area are projected on to a single vertical plane and viewed in profile (Fig. 19. 10) it is seen that the broad band of epicentres at the surface, some hundreds of kilometres wide, corresponds to a rather narrow inclined zone of seismicity defined by the earthquake foci. The zone meets the surface close to the line of the deep ocean trench and dips away beneath the island-arc. These inclined zones of seismicity characterize *all* active island-arc systems and are known as *Benioff zones*, after the seismologist Hugo Benioff who drew attention to the importance of such zones around the margins of the Pacific. Benioff zones vary in their angle of inclination from about 30° to 80° but commonly are close to 45°. They may be somewhat curved both along their length and down dip. In some cases they can be recognized down to 300 or 400 km but in others the seismicity may extend down to about 700 km, the maximum depth at which seismicity has been recognized. In a general way, it seems that Benioff zone seismicity is associated with the downward motion of the lithosphere. At first it was thought that the earthquakes were generated along a slip-zone (Fig. 19. 6) between the descending slab and the mantle material immediately overlying it. Studies of fault plane solutions, however, make this interpretation improbable.

Fault plane solutions are based on analysis of earthquake first motions and have provided an important predictive test for the plate tectonics theory. It has been shown that all known earthquakes have characteristics which strongly suggest their generation by a double-couple mechanism, i.e. by slip on some kind of fault plane or shear

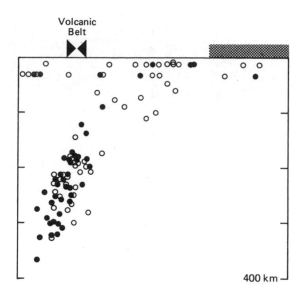

19. 10 Vertical profile through the seismic zone of N. Island, New Zealand. Vertical and horizontal scales equal. Earthquakes indicated by black symbols (open symbols for those which are less well located). Stippling of the surface shows the location of the oceanic trench.

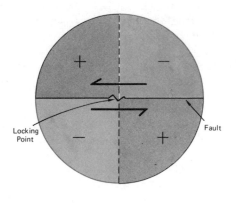

19. 11 The double-couple fault mechanism. The grey tone shows the quadrants within which the first motions are compressional, and the coloured tone those in which they are dilatational.

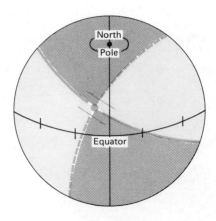

19. 12 The four quadrants defined by first motion analysis. The two nodal planes are shown in colour. The origin of the earthquake lies on their intersection. Without additional information it is not possible to tell which is the fault-plane (solid line in colour). Geological observation generally makes this possible.

zone (Fig. 19. 11). Typically, slip occurs on an existing plane when the stress difference across it is sufficient to produce rupture at the point where the two sides of the fault are locked. On failure at the locking point there is an explosive release of elastic energy in all directions; however, on either side of the fault, the first waves to leave the source are compressions in the direction of motion and rarefactions away from it. This gives the symmetrical distribution of compressional and rarefactional first motions shown in Figure 19. 11. When an earthquake occurs, the motion it produces may be recorded at many different and widely distributed seismological observatories. Some will record compression and others rarefaction as the first motion associated with any particular earthquake, those recording compressions being concentrated geographically into two broad quadrants separated by two other quadrants within which the first motions are rarefactions (Fig. 19. 12). It is possible to identify *two* mutually perpendicular, so-called *nodal planes*, either of which may have been the fault plane. Choice between them must be made on geological grounds. Making certain other reasonable assumptions about the shear fracture character of the fault, it is possible to derive (1) the direction of the maximum principal stress—at 45° to the fault plane, normal to the null direction (the line of intersection of the nodal planes) and within the sector of rarefactional first motion; (2) the direction of the least principal stress—as for (1) except that it lies within the sector of compressional first motion; (3) the direction of slip—normal to the null direction and parallel to the fault plane.

The ability to identify stress and slip directions

is clearly important for the examination of all kinds of plate boundary. Lynn R. Sykes of the Lamont Geological Observatory was responsible for the large-scale application of these methods. Focal mechanism studies on precisely located earthquakes have shown that conservative boundaries are characterized by horizontal movement in the direction predicted by plate theory. Earthquakes on ridge crests are found to have steeply inclined slip directions consistent with normal faulting and extension at right angles to the ridge.

In the Benioff zones maximum principal stresses are aligned uniformly parallel to the direction of dip of the zone (Fig. 19. 13). This suggests that the descending slab of lithosphere is under compression parallel to its length, and that the earthquakes take place *within* it.

The shallow earthquakes in these regions appear to be of two kinds: (1) those at or just outside the trench which indicate normal faulting parallel to the trench and probably reflect the extension of the upper surface of the lithosphere as it flexes downwards into the mantle; (2) those on the island-arc side of the trench, occurring slightly deeper than group (1), and which are consistent with a thrust faulting mechanism and appear to result from slip between the oceanic lithosphere and the rocks above it.

Seismicity therefore appears to be a characteristic of the lithosphere and results from failure within it, or slip along the margins of adjacent plates. Where the lithosphere is shallow, only shallow seismicity occurs; where it appears to descend into the deeper mantle, intermediate and deep seismicity is found. *The motions associated with earthquakes on constructive and conservative boundaries are as predicted by plate theory.*

PLATE FORMATION AND THE
CAUSES OF PLATE MOTION

We turn now to a topic which has proved to be highly controversial. Certainly it is one of considerable complexity, and one in which the line between legitimate inference and unwarranted speculation is particularly difficult to draw. The problems arise partly from ignorance of the physical properties of the interior of the Earth, their dependence upon variables such as pressure and temperature, from ignorance of the vertical and lateral variation of temperature within the Earth, and from uncertainties about the amount and distribution of the Earth's radioactivity. Given time and effort these and other important quantities could, at any rate in theory, be determined by observation and experiment. Further difficulty arises from the inadequacy of existing fluid mechanical theory to cope with situations as complex, and containing as many dependent and independent variables, as the Earth.

The problem of 'how the Earth works' may, in the face of these difficulties, be approached *analytically* or *numerically*. With both approaches, the first step is to attempt to identify those variables which are likely to have the most marked influence upon the behaviour of the system; as a first approximation the remaining variables in the system are assumed not to vary. The analytical approach is then to set up equations relating the important variables to properties of the system which it is desired to investigate (e.g. velocity distribution, heat transfer characteristics etc.).

The advantage of an analytical solution is that it can be derived without assigning numerical values to the variables themselves. Thus a solution might be an algebraic expression relating flow velocity to viscosity and temperature. In view of the observational uncertainties mentioned earlier, solutions in this form are very convenient and allow the effects of a whole range of numerical values for the prescribed variables to be readily examined. The difficulty with approaches of this kind is that the simplifying assumptions which have to be made in order that the equations may be solved can be so extreme as to render the solution irrelevant to the real problem under consideration.

The numerical approach involves the use of a computer and consists of prescribing precise numerical relationships between the various variables and arriving at a precise numerical answer. The advantage of this approach is that very many more variables may be handled simultaneously than by the analytical method, and in theory a highly realistic model may be investigated. The disadvantage is that the amount of computer time required for a single solution is very great and that this solution is unique only for the particular combination of numerical relationships chosen. It is not known how much influence each of these assumed relationships had on the final result unless one, and only one, is changed and the programme run again.

Both analytical and numerical methods have been used to study plate motions and the associated flows in the interior of the Earth. In the early

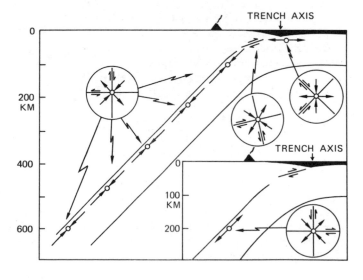

19. 13 **Vertical sections perpendicular to strike of an island arc showing typical orientations of double-couple mechanisms. Compressional and tensional axes are shown by converging and diverging arrows respectively; null axes are normal to plane of the section. The circles show the orientations of the two possible slip planes in each case. The inset shows a mechanism which would indicate tension parallel to the dip of the zone.**

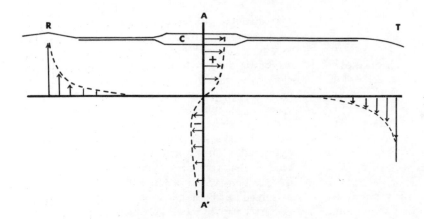

19. 14
For discussion see text.
C = rigid block
R = source
T = a sink

stages analytical methods have proved very helpful, but as Earth models become more sophisticated and the uncertainties about physical properties
become less, numerical methods will undoubtedly
play the major rôle.

We begin by considering the flow implications
of moving plates of finite thickness at the surface
of the Earth. In Figure 19. 14 a closed system is
represented in which a block of rigid material (C),
tapering off in both directions to a thin layer, is
shown floating in a viscous fluid. If the surface
layer and block move to the right, there is mass
transfer at the surface from left to right across the
reference axis A-A'. For the system both to conserve volume and to maintain hydrostatic equilibrium there must be some kind of compensating
counter flow at depth from right to left. We
can therefore construct velocity vector envelopes
as shown. At some depth below the block the
velocity will fall to zero and then increase once
more with opposite sign. The shape of the
vector envelope outside the block is arbitrary,
except that the area contained by the positive
part of the envelope must be equal to that contained by the negative part. If a horizontal
reference axis is drawn through the zero velocity
point on the vertical axis (Fig. 19. 14), it is seen
that in order for the counter flow to fill the
volume vacated by the surface flow there must be
an upwards flow across the horizontal axis on the
left and a complementary downflow on the right.
The same considerations apply to the vector
envelopes as previously.

We have therefore shown that any motion of
plates of finite thickness on the closed surface of a
sphere must involve a counterflow if hydrostatic

equilibrium (isostasy) is maintained and volume
conserved. Thus *mass transfer* circuits of some
kind must exist within the Earth. If the thickness
and velocity of the plates are known, constraints
may be set upon flow velocities etc. in other parts
of the flow circuit. We shall see later that the
velocities of such flows are of great importance for
the understanding of the Earth's thermal structure.
This is because conductive transfer of heat in
silicates is so slow that if flow of material takes
place the isotherms are displaced in the direction of
flow, i.e. they are carried along by the moving
material. The transport of heat by the movement of hot material in this way is known as
convective heat transfer. In the Earth, flow
velocities of the order of 0.1 cm/yr are sufficient
for flow effects to dominate over conduction and
for the isotherm distribution to be controlled by
the flow.

The Earth's crust and mantle are nearly everywhere entirely crystalline and below their melting
temperatures. Geochemical and petrological evidence indicate that the mantle has an ultramafic
composition and that the main minerals of the
upper mantle are olivine and pyroxene (Chapter
3). Yet the large-scale movements at the surface
require some kind of compensatory flow at depth.
We appear therefore to be dealing with the fluid-
like behaviour of a crystalline solid below its
melting temperature. Perhaps the most familiar
occurrence of solid-state, crystalline flow is that
of ice in a glacier. But there is other, geological
evidence that rocks have deformed in fluid
fashion while undergoing recrystallization at high
temperatures and pressures.

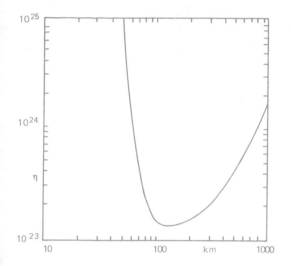

19. 15 Variation of viscosity with depth in the oceanic upper mantle at a point about 100 km from a ridge crest. The unit on the viscosity scale is the 'poise'.

There is therefore no doubt that crystalline solids can exhibit fluid behaviour, but it is important to know what conditions are required for this behaviour and what determines the value of the viscosity.   Experimental studies have shown that the rate of 'flow' is dependent upon (1) composition of the material, (2) the temperature, (3) the confining pressure, and (4) the stress difference (i.e. the difference between the maximum and minimum principal stresses affecting the body undergoing deformation).   In general, deformation by 'flow' is enhanced by high relative temperatures (the ratio of the actual temperature to the melting temperature), and low stress differences.   Flow remains possible at high stress differences only if the relative temperature is close to one (i.e. near the melting point); if the relative temperature is significantly lower, failure by fracture occurs.

Flow is thought to take place by the diffusion of atoms through crystals away from faces of higher stress to faces of lower stress; there is a complementary diffusion of lattice vacancies in the reverse direction.   This process is sometimes known as *diffusion creep*. However, diffusion creep requires that relative temperatures be high and stress differences low, and these conditions may not be present in the upper mantle.   If this is so,

other small-scale processes involving the migration of dislocation defects within the crystals may become important. The position is uncertain, so that difficulties in treating a stress-, temperature-, and pressure-dependent viscosity mathematically have so far deterred attempts to incorporate such a viscosity in a theoretical Earth model.

In Figure 19. 15, a depth/viscosity curve is presented.   Although a particular temperature/depth relationship has been assumed in order to construct this curve, any other reasonable relationship would give the same *kind* of result, i.e. a very high viscosity 'cap' which may be regarded as virtually rigid—the lithosphere—where both temperatures and pressures are low. Because the thermal gradient is steep, the effect of pressure increasing with depth is outweighed by increasing temperature, and viscosity gradually decreases downwards, reaching a minimum at about 100 km; below this the thermal gradient is less steep and viscosity slowly increases once more.

THE CAUSES OF PLATE MOTION

We turn now to a consideration of possible causes of plate motions and the concomitant mantle flow.   This is a controversial question and it is useful to begin our discussion of causes by enumerating the various constraints and conditions which any causal mechanism must meet.

(1) *Thermal*: all active oceanic ridges are characterized by a wide scatter of heat flow values, some of which are very high (up to 10 μ cal cm^{-2} s^{-1}). With increasing distance from the ridge crest the scatter diminishes and the mean heat flow falls until it reaches average level for the oceans (Fig. 19. 16).   Oceanic trenches have abnormally low heat flow (less than 1 μ cal cm^{-2} s^{-1}), but a short distance away, in the adjacent island-arc, the flow is high (2–3 μ cal cm^{-2} s^{-1}). (The SI unit of heat flow is Wm^{-2}, which equals 23.9 μ cal cm^{-2} s^{-1}; see Chapter 5, Table 1.)

(2) *Rates of motion*: spreading occurs symmetrically at ridges at rates ranging from less than 1 cm to 6 cm per year.   At oceanic trenches crust is consumed at rates from 5 to 15 cm per year.

(3) *Oceanic crust formation*: whatever the detailed process of crustal formation at mid-oceanic ridges, it seems that crust of approximately the same thickness is generated whatever the spreading rate.

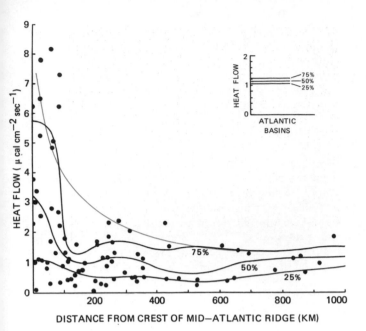

19. 16   **Heat-flow versus distance from the crest of the mid-Atlantic ridge; 75-, 50- and 25-percentile lines are given and (in colour) the heat flux predicted by a boundary layer convection model.**

(4) *Oceanic topography:* the mechanism must be consistent with the development of topographic ridges at centres of spreading; ridges rise 2 to 4 km above the level of the ocean floor and near the axes slope away more or less symmetrically from the crest.

(5) *Gravity:* observations show that ridges are close to isostatic equilibrium. Sinks are characterized by topographic trenches which may to a greater or lesser extent be filled with sediment; they are strongly out of isostatic equilibrium and show the largest negative gravity anomalies recorded on Earth.

(6) *Distance separating sinks and sources:* this is highly variable; in some cases newly formed lithosphere may travel only a few hundreds of kilometres before it is consumed. In other cases the sink and source may be two or three thousand kilometres apart.

(7) *Strength of the lithosphere:* even large lithospheric plates appear to be able to move great distances without undergoing significant in-

ternal deformation. In some cases the plates are twenty times as long as they are thick. With such a length-to-thickness ratio neither compressional nor tensional stresses could be transmitted from one end of the plate to the other unless the frictional resistance beneath the plate was very small.

The high heat flow associated with the ridges has suggested to many that some form of thermal convection within the solid Earth is the cause of the surface motions. Indeed, this was suggested by the late Arthur Holmes nearly 50 years ago without the wealth of observations available to-day. Various other suggestions have also been made for plate motion, but none has been shown to be both quantitatively acceptable and geologically realistic. We concentrate therefore upon thermal convection in its diverse forms.

One of the earliest investigators of thermal convection in fluids was Lord Rayleigh. He showed that when a layer of fluid was heated from below, the temperature difference between the bottom and the top of the layer could become so great that the fluid became unstable, and instead of heat being transferred from the bottom to the top by conduction, the fluid itself began to move; the hotter fluid moved upwards from the bottom, and the cooler surface fluid sank to take its place. This motion commonly occurred with the development of regular flow circuits (Fig. 19. 17) known as *convection cells.* Rayleigh showed that the stability of the situation was affected not only by the thermal gradient, but by the thickness of the fluid layer, its viscosity, its thermal diffusivity and its coefficient of thermal expansion. He used these parameters together to establish a stability criterion, now known as a Rayleigh number, R.

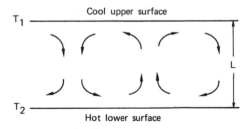

19. 17   **Simple convection cells; for explanation see text. T_1 and T_2 are temperatures, L is the depth of the fluid and β is the thermal gradient. $T_1 < T_2$, and $T_2 - T_1 = L\beta$.**

At a critical value of the Rayleigh number a system becomes unstable and convection begins. Generally, a system is said to be unstable if, when small perturbations are introduced, they tend to grow; it is stable if small perturbations are damped out and the system tends to return to its initial state.

Qualitatively, the Rayleigh number may be regarded as a ratio of those factors which tend to encourage convection against those which inhibit it. Large temperature differences and high coefficients of thermal expansion affecting a great depth of fluid enhance instability, whereas motions are damped by a high viscosity and are rendered unnecessary if heat transfer by conduction is efficient.

It should be emphasized that a Rayleigh instability as described above does not necessarily depend upon the existence of a density inversion, i.e. the material at the base of the layer being so hot that its density is less than that which overlies it. In situations where the depth of the layer is very large, the fluid at the bottom is compressed by the weight of the overlying fluid and, in many cases, an instability develops before the temperature is high enough to produce a density inversion.

Early analyses of mantle stability were made using the Rayleigh stability criterion and showed that instability was likely to be present. It also appeared from experimental work that the flow should have the form of extended rolls (Fig. 19. 18).

The velocities of surface flow, and the pattern of surface heat transfer at ridges are consistent with what would be expected from convection in the mantle. This approach led D. L. Turcotte of Cornell University and the present author to apply the concept of *thermal boundary layers* to mantle problems. Boundary layers are important for convective heat transfer in a fluid with high values of R such as the mantle. Once motion has begun, the flow pattern is as shown in Figure 19. 19b. As the fluid moves across the hot lower surface it is heated by conduction from below, but its conductivity is so low that by the time it reaches the beginning of its ascent conductive heating has extended only a small way into the flow—a warm thermal boundary layer has formed. The ascending flow is formed by the coalescence of two warm boundary layers to give a hot 'plume'. As it approaches the upper surface the plume flow diverges once more to give the symmetrical upper horizontal flow. The hot fluid now moves along a cold surface, and as it

19. 18 Finite-amplitude, steady-state laminar convection cells obtained by E. F. C. Somerscales and D. Dropkin. The view is of the upper surface of the convecting fluid. The long dimension of the view is about 100 cm. Note the irregular contortions of the cells around the margins of the experimental chamber.

flows laterally it loses heat by conduction upwards with the generation of a progressively thickening *cold* boundary layer. A descending cold plume is produced by the convergence of two such boundary layers. The interior of the cell plays no part in the heat transfer process, and within it there is an adiabatic temperature gradient. The heat is picked up, transported, and lost by a kind of thermal conveyor belt formed by the outside of the cell.

It turns out that as this 'constant-property' convection model is rendered progressively more realistic for the Earth (by such refinements as the introduction of a variable viscosity, heating by internal radioactivity and a gradational lower boundary to the flow), one aspect of the model remains unchanged—the upper, cold thermal boundary layer. The strong temperature dependence of mantle viscosity was discussed in an earlier section; when this is coupled with the boundary layer concept, the origin of the physical entities which we describe as 'plates' becomes clear: a plate represents the highly viscous and, in its upper part, quite rigid, cold thermal boundary layer. Figure 19. 20 shows the calculated thickness for the boundary layer associated with the

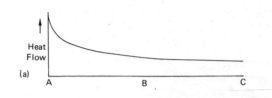

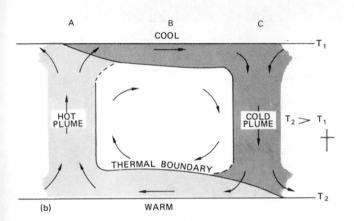

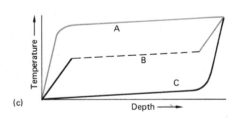

19.19 Boundary layer effects in laminar convection with high values of R. (a) shows the heat flow at the upper surface associated with a circulation as shown in (b); hot thermal boundary layer in colour; cold boundary layer, grey. The arrows show the general pattern of the flow. The vertical thermal gradients beneath points A, B and C are shown schematically in (c); note the influence of the boundary layers on the gradients.

East Pacific Rise. The boundary layer concept also shows why spreading is a symmetrical process; as the plates move apart they must continually split along their weakest line, which is where the most recent previous transfer of hot material to the surface occurred, and where, in consequence, the boundary layer is thinnest and weakest.

We now consider the refinement of simple models to represent mantle processes more closely. One important modification concerns the source of the thermal energy which gives rise to the instability. Up to now, we have assumed that our convecting system was heated from below. In the Earth, however, we are largely concerned with heating by the radioactive decay of unstable isotopes, the distribution and concentration of which within the Earth are not well known. We know from direct observation that both continental and oceanic crusts contain considerably greater (but unequal) concentrations of radioactive elements than the mantle from which they were formed. But we have no direct way of knowing whether the radioactivity which remains in the mantle is more or less homogeneously distributed or not. It was, at one time, thought that most of the mantle radioactivity must be concentrated within the crust and upper few hundred kilometres because any deeper distribution would give rise to general mantle melting. This argument is valid only if all heat transfer is by conduction and if large-scale mass movements within the mantle do not occur. We have, however, already seen that such movements must occur. Thus, although we are fairly sure that some internal heating occurs, we do not know how much, or where.

If only a part of the mantle is involved in movements, the convecting layer may be heated partly by internal sources and partly from below. It is possible for a fluid layer to become unstable even though it receives no heat from below and heating is entirely internal. It has been suggested by D. C. Tozer, on the basis of experimental work carried out at Newcastle University, that internal heating has the effect of changing the aspect ratio (ratio of horizontal to vertical dimensions) of the convection cells to give a pattern of wide shallow cells instead of the nearly square, box-like profile familiar in the laboratory. At the time this work was done it was regarded as removing a serious theoretical obstacle to mantle convection, because the known horizontal movements at the surface were on a scale of thousands of kilometres, implying the existence of cells extending nearly down to the core-mantle boundary. Yet observations on the figure of the Earth suggested that the viscosity of the lower mantle was so high that convection was impossible; cells with a shallow penetration but long horizontal dimensions offered a way out of this difficulty. However, the figure-of-the-Earth arguments are no longer accepted as providing any constraint on deep mantle viscosity.

The effects of combining a variable viscosity with a flow in a fluid layer without a clearly defined base (as would be the case in the mantle unless the flow extends down to the core boundary) have been investigated in a numerical computer model by K. E. Torrance and D. L. Turcotte of Cornell University. The resultant flow pattern is shown in Figure 19. 21. Note that there is a localized and accelerating (converging streamlines) ascending flow with a rapid horizontal surface flow and a very diffuse, slow, deep return.

The problem of the depth of the return flow is one which divides workers in this field at present. Some believe that there is virtually no flow below the Low Velocity Zone, and the Zone itself is regarded as the compensatory return flow associated with the surface motion of the plates (Fig. 19. 22a). Alternatively, the material in the Low Velocity Zone may (Fig. 19. 22b) be part of the upper horizontal flow and move in the same direction as the plate, the return flow being slow and diffuse (e.g. Fig. 19. 21). Advocates of the former model point out the rapidity with which ridge features appear to be able to migrate and regard this as possible only with a very shallow ridge structure, i.e. one not extending below the LVZ. However, consideration of the strength of the plates leads others to suggest that plates must be carried along passively on a flow beneath (Fig. 19. 22c). There is also seismological evidence that plates, or parts of descending plates, are recognizable at depths of 700 km. This means that at any rate the descending flow extends deeper than the LVZ and, from the conservation of volume arguments presented earlier, some kind of return flow must also be present at these depths. Finally, all fluid mechanical models which have employed a variable viscosity suggest the existence of a deep, slow, return flow.

The problem of flow continuity and its depth and direction becomes more complicated when the world-wide distribution of sinks and sources is taken into consideration (Fig. 19. 1). It is immediately apparent that there is not a one to one correlation between sinks and sources; there is

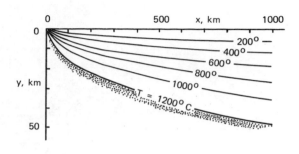

19. 20 **Temperature distribution below one half of the East Pacific Rise; y is depth, and x distance from the crest. The lower limit of the cold boundary layer is marked by the 1 200°C isotherm; note the thickening of the layer away from the ridge.**

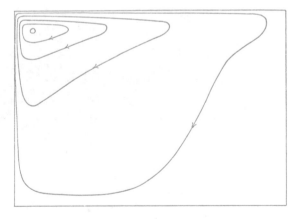

19. 21 **Convective flow in a fluid whose viscosity depends strongly on its temperature and pressure, heated from below; arrows on streamlines show the circulation.**

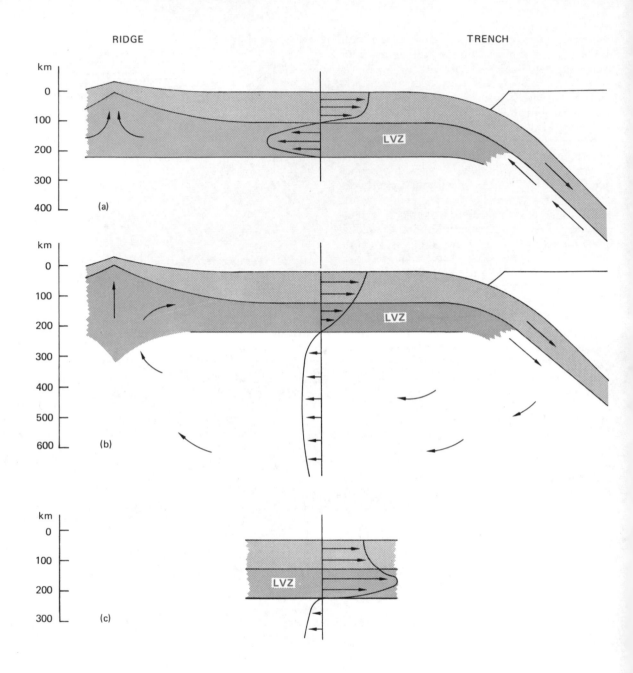

19. 22 Various models for Low Velocity Zone flow. (a): virtually no flow beneath LVZ, counter-flow within LVZ; (b) and (c): in both cases there is circulation within LVZ in the same direction as the plate, and there is a deep counter flow. In (b) the LVZ flow is slower than the plate and thus does not drive it; in (c) the LVZ flow is faster and carries the plate. LVZ is shown in colour and the plate (cold boundary layer) in grey.

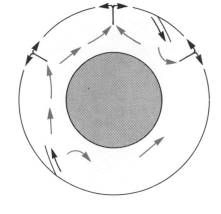

19. 23 **Worldwide distribution of sinks and sources may be irregular, and there is no simple one-to-one relationship. There can be no simple pattern of cells, and the return circulation (coloured arrows) cannot be simply related to the surface flow. The situation is further complicated by the continuous migration of the sinks and sources at the surface.**

a greater source length on the surface of the Earth than there is sink length. In some cases two or more sources may be adjacent to each other without intervening sinks, or two sinks may have no intervening source. This means that regular cells of the kind shown in Figures 19. 19 and 19. 21 are probably rather rare. In general, the return flows cannot be closely related to the surface motions and in some cases they transfer material more than halfway round the Earth. A crude representation of the kind of situation which may exist is shown in Figure 19. 23. Any localized deep flow (i.e. flow from below the LVZ) associated with ridges must migrate with them; laterally migrating vertical flows of this kind are known from laboratory convection experiments. The relative migration of sources and sinks which was mentioned earlier implies that the mantle flow is *unsteady*, i.e. the pattern of flow changes continuously, and it is therefore not possible to represent the flow by a set of streamlines.

We have so far dealt with the general stability problem, the possible flows associated with ridges, the horizontal flows at the surface, and the return; it now remains to consider the descending flow. At the sink, the plate has reached its maximum thickness, i.e. conductive cooling into the originally hot horizontal flow reaches its greatest depth —commonly 100 to 150 km. As in the situation shown in Figure 19. 19, the cold boundary is unstable because cooling has made it more dense than the material immediately below it. Ultimately it must sink to form a cold descending flow. In contrast to Figure 19. 19, however, the Earth's cold boundary layer is so viscous that the uppermost part is a rigid, elastic plate which grades

downwards into a viscous fluid. The plate has strength which tends to resist the downward body forces acting on it. Ultimately, when the plate descends, it bends as a flexured beam and goes down into the mantle at an angle between 30° and 70° with the horizontal. Thus, in contrast to sources, sinks are asymmetrical features with only one of the two plates at the destructive boundary contributing to the descending flow.

As the cold slab of lithosphere descends, it undergoes a series of phase changes to denser mineral assemblages; because the slab is cooler than its surroundings, the phase changes occur at shallower depths within the slab than in the surrounding mantle. This has the effect of increasing the density contrast between the slab and its surroundings and thus the effective downward body force. The most important of these phase changes is the breakdown of orthorhombic olivine to a cubic phase of the same composition with a spinel-like structure. The effect of this is most marked on the ultramafic part of the descending lithosphere.

As the lithosphere descends it carries its isotherms with it and forms an elongate, inclined, cold tongue equivalent to the cold plume of Figure 19. 19b, extending 700 km down into the mantle if we are justified in associating the deepest earthquakes with the existence of the slab. As the slab descends, however, it undergoes 'thermal erosion', i.e. it is warmed by conduction from its surroundings and is thus diminished in thickness. It has also been suggested from the distribution of earthquakes that descending slabs may undergo physical disruption and break up into a series of discrete blocks.

One of the most remarkable features of the sinks is the abnormally high heat flow and volcanicity which characterizes the peripheral part of the plate which is not contributing to the descending flow. Indeed the zone of high heat flow appears to overlie the inclined zone of seismicity associated with the descending slab. High heat flow in a zone of descending flow is not normally to be expected, because the flow depresses the isotherms and the diminished thermal gradient results in a reduced surface heat flux. These and other considerations led the writer and D. L. Turcotte to propose that frictional heating on the upper surface of the descending slab gives rise to the high surface heat flow. D. P. McKenzie and J. L. Sclater independently came to similar conclusions at about the same time.

The heat which is generated along the slip-zone between the descending slab and the relatively inert material against which it moves is partitioned two ways; part contributes to the heating of the cold slab and part is transferred upwards to the overlying mantle. The upward-moving contribution is rather small, however, until temperatures on the slip-zone are high enough for partial melting to occur. When this happens, heat is transferred upwards from the slip zone by rising magmas, a much more efficient process than conduction, and a high surface heat flow results. It appears that only a small proportion of the magmas generated on the slip-zone rise high enough to give surface volcanicity. Their volume, however, as judged from the surface heat flow, is very great and the rate of magmatic intrusion required to maintain the high heat flow behind, for instance, the Japanese arc in the sea of Japan, would result in behind-arc crustal extension at a rate of several centimetres per year.

There are other explanations of plate motion. It has been suggested that plates 'slide' off the flanks of ridges, using the very large amount of gravitational potential energy possessed by a plate on a ridge flank to drive the motion; or that plate motions are driven by the sinking of cold dense lithosphere at trenches which continuously drags the horizontal part of the plate after it, the resultant gap at the ridge being filled by a local passive upward flow. Some think that plate motion is driven by the injection of magmas at ridges which forces plates apart and, so that surface area is conserved, one plate must override

another (i.e. at trenches). To some extent these and other similar explanations are not incompatible with the fluid-mechanical analyses which have been discussed earlier; rather do they represent partial analyses of the problem. Certainly the potential energy associated with a ridge must play a part in driving the surface horizontal flow. Similarly the density excess associated with the descending plate must play its part in the system, just as the density of the cold boundary layer in the cold plume of Figure 19. 19 plays its part. However, it is not reasonable to attribute the plate motions solely to one, or to a combination, of these causes.

CONCLUSIONS

I favour, on the balance of present evidence, a mantle which is in a state of unsteady thermal convection. The mantle has a highly temperature- and pressure-dependent viscosity and, in zones of ascending flow, warm material from below rises at a rate which is high by comparison with the rate at which it can lose heat by conduction. It thus undergoes a pressure reduction but little change in temperature, and its viscosity falls as it rises; as its viscosity falls its velocity of upward flow increases (note the upward converging streamlines in Figure 19. 21). The localized flow beneath ridges is of uncertain depth and clearly can have no well-defined base, but it may not be recognizable at depths of more than 300–400 km. The ridge is the topographic expression of the thermal expansion of the relatively high temperature mantle beneath it.

The hot material reaching the surface at the ridge diverges to give horizontal surface flows which extend down to about 200 km. Within these horizontal flows the thermal gradient is very steep and is the main factor determining viscosity. In the upper 100 km or so, temperatures are so low that this part of the flow is best regarded as a rigid elastic body. It passes downwards into the lower part of the horizontal flow which has not been significantly affected by conductive cooling to the surface, and thus behaves as a viscous fluid. Below 200 km, counter-flow velocities are very low. The flow geometry is complex and related to the world-wide distribution of sinks and sources, their relative activities, and their directions of migration.

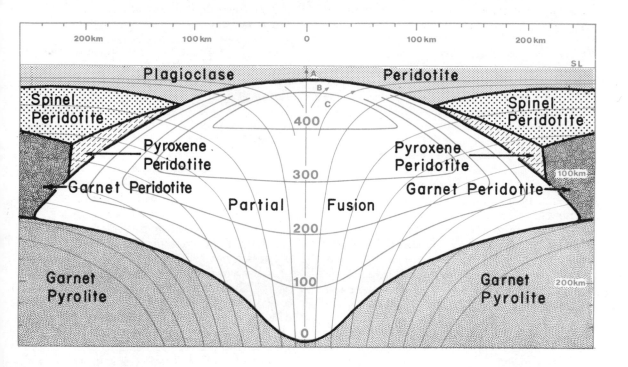

19. 24 A possible vertical profile through an oceanic ridge; paths of particle motion are shown by the upward and outward diverging fine lines. Computations of isotherm distributions allow the zone of partial melting to be delineated. The mineral facies of the mantle are controlled by the thermal structure and are shown. Within the partial fusion zone the amount of partial fusion is given in terms of arbitrary units, 100, 200, etc. If the thermal structure differed very slightly from that assumed here, the wings of the partial fusion zone would not terminate 200 km from the ridge, but would continue indefinitely as a very thin partially melted layer giving rise to the Low Velocity Zone.

At some distance from a source, the density excess associated with the cold boundary overcomes the ability of the plate to resist downward deflection, and localized asymmetrical downflow sets in. The sinking plate which is recognized only by its thermal structure is gradually destroyed by heating. Heated material which was previously part of the plate is indistinguishable from the surrounding mantle and participates in the general diffuse return flow. It is likely that in some cases there is positive overriding of the descending plate by the plate above through the interaction with other plate motions. This is one possible cause for the variation in angle of plate descent.

Virtually all of the Earth's magmatic activity occurs along plate margins, predominantly constructive and destructive margins (see Chapter 21). At ocean ridges the most important effect of magmatic activity is the building of the oceanic crust, largely by extrusion and intrusion of olivine tholeiite basaltic liquids. In the ascending plume, mantle material undergoes a decrease in pressure but almost no change in temperature and, in consequence, at shallow depth the melting curve may be intersected. Provided that both the shape of the convection-displaced mantle isotherms, and the temperature of the onset of melting as a function depth are known, it is possible to map a zone under a ridge within which partial melting is to be expected. The results of a computation of this kind are shown in Figure 19.24. The general diamond-shaped form of the melting zone is determined by the geometry of the flow; its dimensions, however, depend upon the detailed distribution of isotherms which is not well-known.

As a small unit volume of peridotitic upper mantle moves upwards in the ascending flow, it may or may not, depending upon its position, enter the zone of partial fusion and then, after changing direction to become part of the lateral flow, move out of it again. If melting exceeds 4 or 5%, the parent rock becomes permeable and the low temperature melting fraction will part company from its parent rock and move directly upwards at the relatively high speed of several kilometres per hour, to be extruded at, or very close to, the ridge crest, to form new oceanic crust.

Returning to a consideration of Figure 19.24, the elongate tapering 'wings' of the zone of melting pass laterally into the low velocity zone. Thus the oceanic low velocity zone may represent the highly attenuated continuation of the ridge melting zone. Occasionally, conservative plate boundaries have a small amount of volcanicity associated with them; magmas in these situations tend to be alkaline basalts which probably represent a rather small degree of source rock partial fusion, and may indicate a local tapping of the low velocity zone.

In zones of descending flow, two important but distinct processes occur. The first concerns the unconsolidated sedimentary material which forms the upper part of Layer 1 of the oceanic crust. This material, which has a low density ($2.3 \, \mathrm{g \, cm^{-3}}$) and virtually no strength, is not mechanically coupled to the lithosphere below. For these reasons it is not transported down into the mantle along the slip zone but accumulates at the surface and physically adheres to the margin of the 'static', non-descending plate, the crust of which steadily and slowly grows in an oceanward direction. Recent investigations using seismic profilers of the crust on the island-arc side of ocean trenches has revealed chaotic accumulations of strata which appear to have originated in this way. Thus, strictly speaking, at a destructive boundary between plates, one plate margin is destructive while the other is constructive; the rate of growth, however, is rather slow, about 1 millimetre per year. None the less, over long periods of geological time this is likely to be an important crust-forming process.

The second major process operating in sink regions is the partial fusion of the outermost lithosphere along the slip-zone. The outer lithosphere is composed of the basaltic rocks of the oceanic crust, overlain by any sedimentary material which was sufficiently indurated to become mechanically coupled to the descending flow. These rocks of the outer lithosphere are closest to the zone of heat generation and also have lower melting temperatures than the ultramafic part of the lithosphere. For these reasons they are the first material to melt. The liquids which are produced, however, are very different from those produced by partial melting of the ascending mantle material beneath the ridge; calc-alkaline liquids (Chapter 21) are generated which give rise to the andesitic extrusions and dioritic and granodioritic intrusions of the island-arcs and young orogenic belts (Chapter 20). Thus the crust in island-arc regions grows by two processes —the physical addition of oceanic sediment, the sea-floor 'scrapings'—and the addition of large volumes of calc-alkaline magmas. Basalts also occur in island-arc regions in subsidiary amounts; these may be generated by partial fusion within the ultramafic mantle rocks which overlie the slip-zone.

In the processes operating at plate margins we see possible mechanisms for generating the Earth's crust. By partial fusion of the mantle, it is possible to generate oceanic crust; by a second cycle of partial fusion, in which the oceanic crust is itself partially fused, it is possible to generate the more silicic rocks of the continental crust.

The theory of Plate Tectonics has therefore gone some way towards providing a unifying theory for understanding the wide variety of Earth phenomena which were previously neither understood in themselves nor in any wider context. We see some pattern in the distribution of world-wide seismicity and volcanicity, a mechanism for the generation of the two main crustal types, clear evidence of large-scale motions between crustal plates and between the continents which some of them contain. We see that, whilst oceanic crust is 'disposable', continental crust cannot return into the mantle. Is it too presumptuous to assert that the general outline of the processes we have discussed is sufficiently consistent internally to suggest that it is probably correct?

Drs. Vine and Oxburgh have elegantly described the concept of sea-floor spreading (Chapter 16) and the theory of plate tectonics (Chapter 19). One concerns itself with the evidence that oceanic crust is created at an oceanic rise, and the other with the rigid crustal plates, the way they move and what causes them to move. Neither of these articles considers what happened in the distant geological past. Indeed they cannot, for nowhere is the Earth's oceanic crust more than 200 million years old and it is, almost inevitably, consumed at the oceanic margins where it underrides the bounding continents, plunges deeper into the mantle, and is digested in the Earth's hot interior.

All this is very unfortunate for the land-bound geologist to whom the last 200 million years represent less than 6 per cent of geological time and just over 4 per cent of the time that has elapsed since the Earth came into existence as a separate planet. The quandary in which he finds himself is as follows: in sea-floor spreading and its 'offspring', plate tectonics, we have a unified, revolutionary and elegant hypothesis within which the Earth's main structural features, created during the last 200 million years, can be readily understood. But has sea-floor spreading operated throughout geological time, and have plates of lithosphere always been engaged in an *ad hoc* peregrination about the surface of the Earth? If so—how can we tell?

Obviously the place to look is on the continents, for whereas the oceanic crust is young and self-destructive the continents are old and preserve within them clues to their long and complex geological history.

In this chapter John Sutton, Professor of Geology at Imperial College, London, has accepted the rôle of our hypothetical geologist. Orogeny is the geologist's term for the process of mountain building and, as Professor Sutton points out, the major mountain chains that embellish the Earth to-day occur where moving plates collide and one underthrusts the other. By examining the history of structural and compositional changes in the mountainous margins of the continents created during these impacts, it is possible to re-examine the internal structure of the continents to see if the old, inactive mountain chains are 'fossil' plate margins. This, John Sutton suggests, is the case. And although it can never be enhanced by the geophysical perspicacity which supports the present cycle of ocean floor creation, it is nevertheless a vast step forward in geological thinking to realise that sea-floor spreading and plate tectonics could have operated in the distant past, and not just during the last 200 million years.

20 Orogeny

by JOHN SUTTON

The remarkable surface relief of the Earth is one of its most distinctive features. It reflects, to an extraordinary degree, the deep-seated processes which have determined the structural development of the Earth.

NOW THAT WE CAN COMPARE the relief of the Earth with the surface features of other parts of the solar system, we can better appreciate the extent to which the relief of the Earth is distinctive. The surfaces of Mars and the Moon are now sufficiently well known to show that highlands and lowlands are present just as on Earth, where the continental plains and plateaus stand several kilometres higher than the abyssal plains of the ocean floors. Moreover it is now known that the Moon's crust was built up by igneous activity, and that there are rocks on the Moon not very different from the

20. 1 Folded sedimentary rocks in the Alpine-Himalayan Chain.

igneous rocks on Earth. It may be that the highlands and lowlands on Mars and the Moon reflect differences in the densities of the underlying rocks in rather the same way that continental crust of Earth is constructed of lighter material than the deep ocean floor.

Despite these similarities there is one great difference between the relief of the Earth and that of the Moon or of Mars. Earth has, in addition to the extensive highlands and lowlands, another group of major structures which Mars and the Moon apparently lack. These are the great systems of linear features which cross the ocean floors and continents and which we now know originate from the movement of large plates of the Earth's crust at rates of a few centimetres a year. Mountain chains are probably the best known of these structures, and were certainly the first to be scientifically investigated.

The term *orogeny* was introduced by the American geologist G. K. Gilbert in 1890 to describe the process of mountain building. The mountain chains Gilbert had in mind are the familiar mountains such as the Rockies or the Alps which are often termed *fold mountain belts*, for they are built up of folded rocks which result from compression of the crust. There is another system of mountains constructed in a very different way, whose true extent was unknown in Gilbert's day because these mountains are largely submerged. These are the ocean ridges and rises which extend as a network 60 000 kilometres long across the ocean floors. Although they are just as much mountains as the world's fold belts, the term orogeny is not applied to them since they originate through processes quite distinct from those which produced chains such as the Rockies or the Alps.

The gradual discovery of the extent of the submerged mountain chains has, however, played a most important part in making clear the processes involved in orogeny. As exploration of the ocean floor continued a general picture of the surface relief of the whole Earth became available for the first time. We now know that several other varieties of relatively narrow linear structure, in addition to the two types of mountain chains, form world-wide features which are superimposed on the fundamental division of the crust into low-lying ocean floor and upstanding continent.

There are six types of structure that are particularly important: the *oceanic ridges*, the *rift valleys*, the *continental slopes* which form the boundary between continent and oceans, the *ocean trenches*, the *island arcs* and the *fold belts* or mountain chains of everyday speech. All originate from the displacement of crustal plates at velocities of a few centimetres a year (see Chapter 19). The presence of these structures shows that the Earth is a planet whose deeper parts move in a fashion which can break up and then transport the outer lithosphere as large slabs. In contrast, Mars and the Moon show no comparable systems of global structure, and have either developed a crust so strong as to resist such movements or have static interiors which are not in motion.

Mountain chains of folded rocks like island arcs and oceanic trenches originate where plates of crust converge. Ocean ridges and rift valleys mark regions of tension where blocks of crust are moving apart. The continental slopes around the younger oceans (the Atlantic, Indian, Arctic and Southern Oceans) originated at Mesozoic and Tertiary fractures formed when pre-existing larger continents were broken up and moved apart. They are accordingly tensional features; and it seems probable that the continental slopes around the Pacific, though now in a very different tectonic situation, may be traced back to a much older fracture system, perhaps of Precambrian age. These date from the time when the Pacific first formed and when continents, then on the site of a 'proto-Pacific', were separated.

The length of time structures may survive varies greatly as between continental and oceanic crust. The continental crust contains rocks and structures as much as 3500 million years old, whereas ocean crust is manufactured and consumed so rapidly that oceanic rocks over 150 million years in age are rare. Because of the long life of continental crust and because of the distinctive nature of the structures formed when a belt of continental crust has been compressed, many examples of the roots of old, inactive mountain chains have been recognised, several dating back to the Precambrian. Study of these old orogenic belts provides one way of attempting to reconstruct the large structural changes in the crust that occurred over the long period of time that elapsed before the formation of all the structures in the present-day ocean basins.

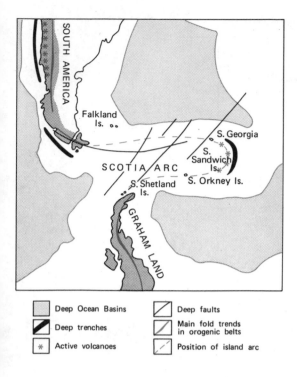

▨	Deep Ocean Basins		◪	Deep faults
◣	Deep trenches		◪	Main fold trends in orogenic belts
✳	Active volcanoes		◪	Position of island arc

20. 2 The Scotia Arc, the structural link between the fold mountain belts of Antarctica and South America.

THE LOCATION OF MOUNTAIN CHAINS

The active mountain chains—that is to say those where movements occur at the present, as indicated by the occurrence of earthquakes—lie within the Alpine and circum-Pacific belts. The position and structure of these mountains demonstrate that orogeny occurs where a block of continental crust is carried into collision with another crustal plate. The Himalayas provide an example of the effects of the collision of two continental plates, but at the present time most active mountain chains lie on continental plates in collision with oceanic crust, as is the general rule around the Pacific. This may not have been the rule in the geological past, for there are reasons for believing that the present global tectonic pattern is atypical.

The seismologist Benioff established that the foci of earthquakes which occur in a seismically active region near the edge of an ocean are located in a zone dipping underneath the adjoining conti-

nent and extending to depths of several hundred kilometres (see Fig. 19. 6). This asymmetric arrangement has developed around much of the Pacific where the seismic zone marking the convergence of the plates of Pacific crust with those of the Americas, Asia and Australia dips for the most part below those continents. There are exceptions due to complex local movements. As Chapters 16 and 19 show, one finds successive zones of shallow, intermediate and deep earthquakes as one goes outwards, away from the Pacific basin. The movements within the crust at the present day are consistent with those in the past, and how the direction of movement can be deduced from the geological structure of the crustal plates will now be discussed.

The asymmetry of a mobile belt near the boundary between ocean and continent has several important consequences. As we shall see, when considering the internal structure of orogenic belts, it is reflected in an asymmetry across the width of some orogenic regions. There is also an effect of quite a different kind which directly influences the distribution of orogenic belts. To appreciate this we need to consider the movements of continents once more. We know that on a global scale the rates at which crust is manufactured and consumed balance rather precisely (see Chapter 19), but this is far from true for individual plates. Where two plates converge, and the leading edge of one is made of oceanic and the other of continental crust, most of the consumption of crust will occur as the oceanic plate is thrust below the continental rock and so carried into the mantle and eventually melted. There is usually no such asymmetry in the regions where crust is generated and where both plates move outwards at comparable rates from the zones where new igneous rock is reaching the crust to form new lithosphere. Plates move as new crust is generated along the trailing edges. A moment's thought will show that a plate whose leading edge is of continental material will gradually increase in size, for new crust will be added where oceanic crust is generated at its trailing edge but little or no continental crust is being consumed at the leading edge. This growth is compensated for elsewhere as other plates decrease in size. The decreases occur when oceanic crust at the leading edge of a plate is destroyed at a rate greater than that at which it is being produced at the trailing margin of the plate. The consequence of this simple relationship, which arises from the difference in density of

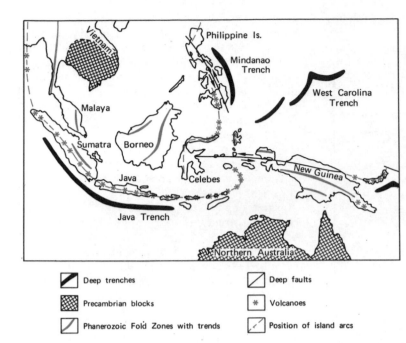

20. 3 The complex pattern of
ocean trenches and island arcs
between south-east Asia and
Australia.

◪ Deep trenches   ◪ Deep faults

▦ Precambrian blocks   ✳ Volcanoes

◪ Phanerozoic Fold Zones with trends   ◪ Position of island arcs

continental and oceanic crust, is that the location
of the zones where these plates converge is constantly changing relative to the zones where crust
is forming. In other words, as the continents are
displaced they carry with them, near their leading
edges, zones of compression along which mountain
chains are produced. In this way, the orogenic
belts are being transported across the face of the
Earth and moved relative to rifts.

In an active belt of compression which extends
over more than one continent the continental
segments are linked across the ocean floor by
trenches and island arcs which lengthen or shorten
(usually by folding) to maintain the continuity of
the mobile belt as the distance between the continents varies. For instance, the Scotia arc, linking
the mountains of the Antarctic Peninsula and of
South America, is an example of a mobile belt
which has been extended in this way as the two
continents of Antarctica and South America
separated during the break-up of the former
Gondwanaland (see Chapter 15). The complex
pattern of ocean trenches and islands between
south-east Asia and Australia reflects the very
complicated history of movement in that region

since the Mesozoic. During that time the mobile
belts on which plates converged, near the margins
of both the Pacific and the Indian oceans, were
themselves displaced as the relative positions of
Australasia and South East Asia varied. While the
details of such movements of mobile belts are not
yet understood, one can readily see that before the
break-up of Gondwanaland and Laurasia, the
Mesozoic Pacific, together with the arm of that
ocean which, as the Tethys Sea, extended between
Southern Asia and Gondwanaland, were larger
than the present Pacific (Fig. 15. 19). The circum-
Pacific and Alpine belts have been themselves
transported across the outer parts of the diminishing
Pacific and Tethys.

So, when we are considering horizontal displacements during orogeny, we are dealing with a
complex situation. While crustal blocks converge
upon active orogenic belts, the site of the mobile
belt itself may shift as movement continues.
Before considering the rates of these horizontal
displacements it may be helpful to look briefly at
some of the structures which are typical of orogenic belts on continents.

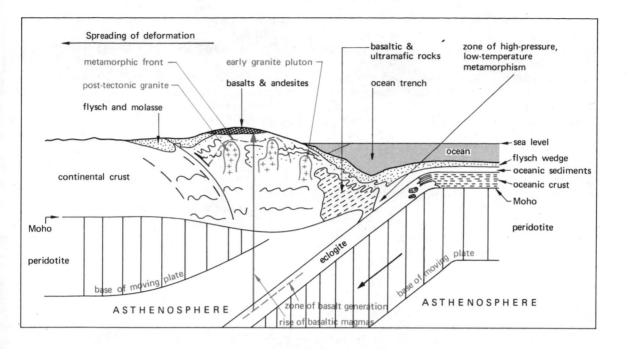

20. 4 Schematic model of orogenic belt structures created when an ocean plate underrides a continental plate moving in the opposite direction.

THE INTERNAL STRUCTURE OF MOUNTAIN CHAINS

The geological structure of a mountain chain may differ from that of surrounding areas in any one of a number of ways. By studying such distinctive features one can establish the geologic history of an orogenic region and so find out the steps by which a mountain chain was formed. Knowledge gained from the investigation of the internal structure of recent mountain chains can also be used when examining the numerous older orogenic belts where the mountains have been removed by erosion. In these latter cases it is only by the recognition of their deeply eroded internal structures that they can be correctly identified.

It is hardly possible to generalise about mountain structure. An orogenic belt may be distinguished by any of the following features, though it is unusual for all of them to be developed in any one section of a chain. Furthermore, some are not diagnostic taken in isolation, and it is usually through study of several characteristics that the older, eroded orogenic belts can be identified.

Sedimentation in Orogenic Belts

Some orogenic belts have formed along the site of a pre-existing elongated system of sedimentary basins. These basins contained much thicker deposits than those laid down over the same time interval outside the region where the mountain chain was to form. Many of the late Precambrian and Lower Palaeozoic mountain chains show such a relationship to late Precambrian deposits. Precambrian deposits up to 10–15 km thick are known in several sedimentary basins in Spitzbergen, East Greenland, and Scotland which were developed in the region where the Caledonian fold mountain chain formed several hundred million years later. On the other hand, so many examples are now known where no such relationship can be established that one has to conclude that, while mountain chains may develop over unusually thick sedimentary or volcanic sequences, this is by no means inevitable. A more consistent relationship between sedimentation and mountain building emerges from the study of sediments formed later, either after some deformation of the crust has occurred, or still later at the end of mountain building when the newly elevated chain could be subjected to rapid erosion. These two groups of deposits, the *synorogenic* sediments that accompany mountain building and the *post-orogenic* that follow it, were first identified as such in the latter half of the last century, particularly by Swiss geologists. As a consequence, two Swiss terms have been widely employed in this connection. The first term, *flysch*, was introduced to describe marine Lower Cretaceous to Oligocene (130–40 million years ago) impure sandstones and argillaceous rocks that formed thick successions in basins on the north side of the rising Alpine chain between west Switzerland and Vienna. The second term, *molasse*, describes the largely continental argillaceous marls and sandstones which formed in troughs north of the Alps at and immediately after the uplift which completed the development of these mountains in the Upper Oligocene and Miocene(40–20 million years ago). Whereas flysch is involved in the later episodes of folding and thrusting, the molasse is essentially later than all but the very last of the horizontal displacements. While it may be best not to apply these terms to other mountain chains, they do bring out very clearly the fact that certain types of sedimentation are closely linked with mountain building. Many greywackes (see Fig. 1. 21), a type of impure sandstone containing fragments of varying size set in an argillaceous matrix, have formed in orogenic areas after mountain building had begun. These rocks are characteristically laid down in long troughs into which the coarse greywackes have been transported in suspension by turbidity currents. The direction of these currents can be established by studying the sedimentary structures the rocks contain; and this may assist in the determination of the palaeogeography of the times when sedimentation occurred (Chapter 13). In this way it has been possible to show the relation of the many lower Palaeozoic greywackes in southern Scotland, the Lake District and in Wales to ridges and basins formed early in the history of the Caledonian mountain chain in those districts. The greywackes in question range in age from late Precambrian to Silurian (600 to 400

million years ago) and so date some of the early movements.

The Old Red Sandstone of north-west Europe and the Arctic, a younger deposit formed from 400 million years onwards in the Devonian period, is a post-orogenic formation which lies above and on the flanks of the folded rocks of the Caledonian mountain chain. It is composed of coarse continental sandstone and conglomerates which locally are inter-bedded with andesites (see Chapter 21). This typical post-orogenic deposit accumulated after the main folding and metamorphism of the orogenic belt. It results from the rapid erosion of the newly uplifted mountain chain, and as many of the sedimentary basins are fault-bounded it demonstrates that subsidence of blocks of crust was taking place as the Old Red Sandstone was laid down.

Rock Deformation, Folding and Thrusting during Orogeny

Perhaps the most characteristic feature of the present-day mountain chains is the deformed condition of the rocks. The fact that similar deformation can be seen in some regions of older rocks is one of the most convincing arguments for supposing that such regions represent the roots of older, now eroded mountains. All mountain chains contain either folded or thrust rocks, usually both. Thrusts are gently inclined planes along which rocks have been moved so that they come to overlie rock which originally lay some distance away from them. Thrusting of this nature suggests horizontal shortening of the crust. It does not *prove* shortening because it is often difficult to demonstrate that the overlying rocks have not slid down the thrust plane as a result of purely superficial movement such as the collapse of an elevated region under the influence of gravity. Many thrusts involve deep-lying rock, and it is not uncommon to find slices of crystalline basement rocks which originally underlay a sedimentary succession thrust in such a way that one or more vertical repetitions of the basement and cover are found above the sedimentary succession.

The French word *nappe* (literally 'a sheet') can describe either a thrust mass or a folded body of rock in which the fold is recumbent, that is to say it is a fold whose limbs are almost parallel and roughly horizontal. Nappes, first identified in the European Alps, have been found in many mountain chains. The rocks involved may be highly deformed, so that what was a cube of rock before deformation is converted to a body with its greatest dimension as much as thirty times longer than the shortest. Originally spherical pebbles can be deformed into elongate rods like walking-sticks. In the deeper parts of a mountain chain rock flowage through creep becomes important, and many examples of strongly distorted objects such as pebbles or fossils indicate the very great changes in shape which may occur. Probably through flow of this kind the entire thickness of the crust below a mountain chain may be altered, so that in place of the 25 to 35 kilometres of crust common outside recent mountains, thicknesses as great as 70 km are reached.

The deformed rocks in most mountains originated as part of the continental crust, for they are either crystalline rocks with a larger proportion of granitic types than would be found on ocean floor, or they are sedimentary and volcanic rocks which, though now modified, can be recognised as having been laid down on a basement of such continental crystalline rock. Such a basement can be seen in its original situation in the more northerly parts of the Alps. The overall picture in this area is of a crystalline crust with a relatively thin sedimentary and volcanic cover which has been broken up into a series of folds and nappes during the convergence of the African and European continents. But if this is a correct picture, as it appears to be, one is left with a major uncertainty. Was there oceanic crust between the two converging continents, and if so, is any part of this oceanic crust to be found in the Alpine chain? This is not an easy question to answer, for it appears that to-day ocean floor is carried downwards where plates converge, so that any Alpine ocean floor which may have been present is likely to have been destroyed in a similar fashion. Although the Alps are predominantly made of continental crust, to which magmatic rock has been added from the mantle, one cannot be certain that continental masses involved in the mountain building may once have been separated by an ocean whose floor has been almost wholly destroyed during orogeny. Fortunately there are some clues remaining as to the possible existence of such an ocean, as for instance on the island of Cyprus, where within the Alpine fold belt an association of basic and ultrabasic rocks occurs. This Mesozoic rock assemblage includes pillow lavas, ultramafic rocks and abundant basic dykes, and in many ways resembles the association exposed on Iceland on the mid-Atlantic ridge; it

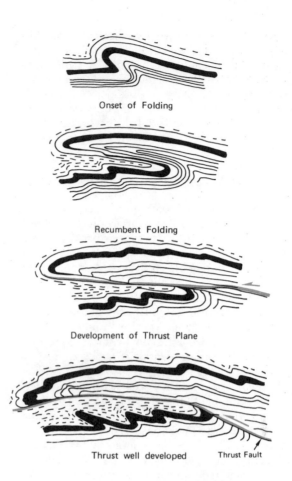

Onset of Folding

Recumbent Folding

Development of Thrust Plane

Thrust well developed   Thrust Fault

20. 5 Stages in the development of thrusts.

may represent fossil Mesozoic ocean floor now largely destroyed by mountain building.

Some striking examples of possibly deformed ocean floor occur in the circum-Pacific fold belts. In Japan, for example, where mountain building occurred both in the Trias and during the Tertiary, the resulting structures formed on each occasion include pairs of metamorphic belts. One member of each pair consists of rock metamorphosed at unusually high temperature relative to pressure, and includes numerous granitic bodies. It appears to represent metamorphosed and deformed continental crust. The other metamorphic belt in each pair shows evidence of metamorphism under unusually high pressures and is composed largely of altered ultrabasic and basic rock. It has been suggested that these belts represent portions of the pre-Triassic and pre-Tertiary Pacific which had been thrust below the Japanese arcs of those times to form two successive orogenic belts. Many other fold chains around the Pacific contain rather similar occurrences of basic- and ultrabasic-rich belts metamorphosed under high pressure, low temperature conditions. Much information about mountain building processes can be obtained by the study of the crystalline rocks within orogenic belts, and we can now turn to this.

Igneous and Metamorphic Activity during Orogeny

At the present time nearly all the magmatic rock which reaches the Earth's surface does so along one or other of the two types of mobile belt, that is to say either along the rifts and ridges between separating blocks of crust, or along island arcs, and mountain chains formed where blocks converge (see Chapter 21). We are here concerned only with the compressional belts and principally with mountain chains. Both compressional and tensional belts are intruded by basaltic magma probably formed at depths of 60 kilometres or more and therefore originating in the mantle. Ultramafic rock similarly occurs both in widening rifts and closing mountain chains and is likely to be found in the crust above regions where the mantle has been tapped. In contrast, granitic rocks, while they certainly form in minor amounts along widening belts of oceanic crust, as for instance in Iceland, are predominantly associated with mountain building. This association arises because the light continental crust is crumpled and heated and so may be partly melted to produce granitic magma. In the deeper parts of many mountain chains, extensive granitic migmatite complexes occur. (Migmatite is a term applied to composite rocks, such as gneisses, produced by the injection of granitic magma between sheets or *folii* of a schistose or gneissose rock.) These originate in regions where temperatures and pressures have been such that partial melting has begun and a granitic melt has formed from which ultimately large granite intrusions may develop. Migmatite complexes develop largely from the pre-existing rocks in their immediate vicinity and may not have been greatly displaced from the sites where they formed.

The production of granite within mountain belts is of particular interest because it provides a model from which to understand the production of higher temperature magmas which originate at greater depths and are never later exposed by uplift and erosion. Essentially granite magma evolves in deformed mountain belts first at large numbers of points where rock begins to melt and migmatites form. Subsequently, the new magma collects and rises through the crust as molten bodies to produce the intrusive granites which can consolidate high in the crust. This granite series, which proceeds from migmatites to granite plutons, can be linked with the actual process of mountain building. The migmatites are syntectonic as are many of the early granites, but the latest granites are characteristically intruded after tectonic activity has ceased and immediately after the uplift which produces the mountain ranges. These late granites cut the molasse-like deposits with which they are almost contemporaneous. Acid volcanic rocks are frequently linked with such late granites which may rise high in the crust and approach within a few kilometres of the Earth's surface. In these cases some of the 'magma' escapes at the surface through volcanic activity. Thus eruptive and intrusive rocks are closely related in time and space.

Acid volcanics can also appear much earlier in the history of a mountain chain, as for instance in the Caledonian mountains of Britain where extensive Ordovician rhyolites were erupted about 100 million years before the Devonian uplift which ended that orogeny.

Only as a very broad generalisation is the conventional view—of basic activity early in the evolution of a chain followed subsequently by more acidic eruptions—a correct hypothesis. It may be better to look at the igneous activity in terms of place as well as time. Early basic rocks may be present as dyke swarms or sills or in lava sequences, particularly where oceanic crust is preserved in a chain. Intermediate volcanic rocks, especially andesites, can arise where partial melting of ocean crust during down-warping has taken place and are thus of particular importance in belts which have formed through the compression of island arcs or other structures at the continent-ocean margins. Acid rocks, as we have seen, are characteristically formed during and after the deformation of continental crust.

Just as igneous rocks form along the length of mountain chains, so are metamorphic rocks arranged in zones along orogenic belts. The metamorphic minerals in each zone provide an indication of the temperatures and pressures reached. The degree of metamorphism increases in general downwards and towards the central axis of a chain. There are many exceptions. Some chains may be virtually unmetamorphosed, just as some mountain ranges such as the Himalayas have no active volcanoes along their length. The type of metamorphism can vary across a chain as in the paired metamorphic belts of Japan. In one belt heat flow was unusually high and in the other unusually low with respect to pressure. Abnormally high heat flow at a certain point within a mountain belt can produce a concentric system of metamorphic zones grouped about the thermal culmination. Such an unusually high heat source may continue to exist for many tens of millions of years. It is of considerable interest that the sources of unusually high heat flow indicated by such metamorphic changes and also by magmatic activity must migrate with the orogenic belts when mountain chains are displaced as plates of crust move across the Earth. This relationship provides one indication that the upper mantle is displaced at the same time as the overlying crust.

THE TIMING OF OROGENY

The process which leads ultimately to the uplift of mountains is a long one, which can be shown in many chains to have continued for several hundred millions of years. Despite this, the actual uplift which ends the process is geologically speaking rather rapid. Vertical movements of several kilometres are known to occur in active mountain chains in a few million years. It would appear that during orogeny an elongate belt in the outer parts of the Earth responds to compression in a variety of ways. Uplift is simply the final stage in a varied sequence of events.

If the prime cause of mountain building is accepted to be the compression of the crust where large plates of crust converge as new crust is generated elsewhere, we have a working hypothesis for orogeny. We can test this hypothesis in a variety of ways, including an examination of the history of typical mountain chains.

Broadly speaking there are four kinds of geological phenomenon which may have their origin in such converging movements and which can leave a permanent record in the crust. Firstly there is the possibility of forming open downwarps in the crust, a few tens or hundreds of kilometres wide and up to a few kilometres in depth. Although such structures are insignificant when compared with the radius of the Earth, they have marked effects on the accumulation of sediments.

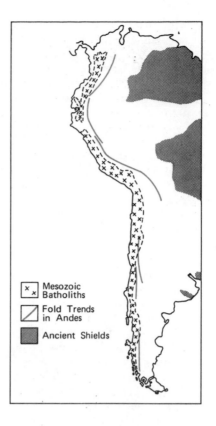

20. 6 **The distribution of the vast South American batholiths intruded into the Earth's crust some 100 million years ago.**

Sediments may be deposited in such depressions or geosynclines, and thus a chain of elongated basins may come into existence, filled with detritus from higher regions. Such a subsidence may be the earliest indication of deeper movements below the crust. There is certainly some correspondence between the regions where, for example, late Precambrian deposits over 5 kilometres in thickness accumulated, and those regions which several hundred million years later became mountain belts during the Palaeozoic.

A second effect of such crustal movements is the deformation of the crust resulting from compression. This leaves a permanent record in the shape of distorted rocks, such as slates which have been flattened and have developed a wholly new structure as the result of pressure. The cleavage which allows slates to form thin slabs is the result of the parallel arrangement of metamorphic minerals which have grown during the deformation of the rock (see Chapter 1). Other permanent records are the thrusts and folds found in mountain chains. It is possible to date many of these structures and in that way to obtain a deformation timetable of an orogenic belt.

A third effect is the metamorphic and magmatic activity which results from an unusual accession of heat. It is true that the sources of this heat are not known, and to this extent it may be premature to identify an abnormally high heat flow with unusual mobility of the crust and upper mantle. There are, however, several lines of evidence pointing to such a conclusion. Deformation and metamorphism often affect the same area of crust and die out together. There is a close link in time between deformation as demonstrated by the ages of folds, thrusts and of new fabrics in rocks, and the metamorphism as indicated by the periods when new assemblages of minerals formed.

There are several links between magmatic activity and mountain building. Many chains contain masses of granitic rock which are restricted to the orogenic region along which they may extend for several hundreds of kilometres. In contrast, they do not extend laterally outside the belt of strongly deformed crust. These granites, as has been mentioned, are both syntectonic and post-tectonic, but in the latter situation are known to have been emplaced very shortly after the ending of orogeny. There is thus a good temporal and spatial link between certain types of magmatic activity, regional metamorphism and mountain building. The source of heat must lie deep within the mantle, for the thermal structure of mountain belts is remarkably simple and is not greatly affected by near surface features. For example the boundaries of metamorphic zones outline large simple structures which cut through the detailed tectonic structures indicated by the folds and thrusts. A map of the regional metamorphism in the Scottish Caledonide mountain chain illustrates this point. The simple thermal structure is also brought out, by the distribution of igneous rocks such as the Mesozoic granites of the Andes which lie in a narrow belt along the chain. Thus, although it is difficult to establish a direct connection, there is much evidence that areas affected by unusually high heat flow are also at the same time areas which lie on the converging margins of plates.

The last effect is the uplift itself which produces the mountain chain as it stands at the end of a period of mountain building. A permanent record of uplift may be preserved in several ways. The sudden change in the nature of sedimentation which takes place when the first coarse detritus begins to accumulate is one indication, particularly where the earliest deposits lie on the deeply dissected crystalline rocks. The appearance of post-tectonic granites and the ending of all kinds of plutonic activity over an orogenic belt provides another line of evidence. In recently formed mountains which have not yet been greatly eroded it may be possible to date land surfaces. These, where faulted or distorted by uplift, can provide an upper limit to the time span in which mountain building must have occurred. Land surfaces cut across uplifted structures may set a lower limit on the possible age of mountain building.

The four kinds of evidence (the nature and distribution of any pre-orogenic basins, the nature and age of the deformation, the timing and nature of the plutonic activity and the time of the final uplift), together provide means of establishing the history of mountain belts.

We may now turn to this aspect of orogeny, being careful to distinguish between (i) the length of life of individual fold belts, (ii) the distribution in time of successive fold belts, and (iii) the distribution of mountain building in space.

The Life Span of Individual Chains

The earliest mountains to be studied were naturally the youngest, as they are both the most conspicuous and also contain much fossiliferous rock which could be dated by the classical stratigraphical methods which at that time provided the only way of determining the age of a geological structure. One of the most accessible parts of the Tertiary fold system lies in southern Europe, and much of the basic knowledge of mountain building comes from study of the European Alps. It is from this work, and from study of the somewhat older Nevadan fold belt of Western America and of the Palaeozoic belts of Eastern America and Europe, that a picture of Phanerozoic mountain building first evolved in the closing decades of the nineteenth century. The model established by this early work indicated that a period of some 200 million years was required for the whole sequence of mountain building operations. The cycle includes pre-orogenic sedimentation, deformation, syntectonic sedimentation, plutonism, uplift and finally the start of erosion of the newly completed chain.

It was appreciated that the preliminary stages took up most of this time span, and that the uplift or true orogeny was confined to a few million years at the end of the sequence. It was thus possible to speak of a long evolutionary sequence culminating in a revolutionary period of mountain formation. It was also appreciated that compression was required to form mountains, and it was suggested that orogeny occurred as the cooling Earth contracted. Whatever the cause, it was clearly appreciated, particularly by the Alpine geologists of the second half of the last century, that mountain building involved the convergence of continental blocks. At that time it was not possible to get a global view of the active mountain chains for many had not yet been geologically investigated; nor was it possible to establish the chronology of the unfossiliferous parts of any mountain belt, whether young or old. The growth of knowledge which has led to the current hypotheses on global tectonics has largely, as far as mountain building is concerned, taken two forms. First was the exploration of the geological structures of South America, Asia, Africa and Australia which, at the time of the classic Alpine work, were relatively little known. Together with studies of the sea floor the results provided an essential framework of fact about the Earth's crustal structure. Secondly, the ability to date, through radiometric measurements, unfossiliferous rocks and minerals made it possible to assign ages to virtually any structure and so to investigate Precambrian orogeny.

When study of the Precambrian fold belts began, it became clear that the active life of an orogenic belt might greatly exceed the 200 million years conventionally attributed to an orogenic cycle. As we have already seen, some mountain chains whose development began with late Precambrian sedimentation were not completed until 500 million years later in Devonian times, as for instance in the Caledonian chains of north west Europe and the Arctic.

It also became apparent that, in addition to these late Precambrian events which had contributed to the evolution of mountain belts completed in the Palaeozoic, there were many much older Precambrian orogenic belts whose active life had been completed within the Precambrian. This became clear first through structural analysis of the Precambrian shields, which are extensive outcrops of Precambrian rock, in some instances, such as the Canadian Shield, several million square

kilometres in extent. It was then realised that the shields were made up of systems of orogenic belts similar in many respects to parts of the Phanerozoic mountain chains. Although now so deeply eroded that no visible chain of mountains survives, the strong resemblances to the deeper parts of geologically young mountains demonstrate that the shields were once the sites of Precambrian mountain belts.

Just as with the younger belts, those of the Precambrian provide evidence that fold belts remain active for somewhere within the limits of 800 to 200 million years. In the longer-lived Precambrian belts a complex history of deformation and metamorphism can be established, marked by a succession of plutonic events interspersed with renewed sedimentation, volcanism and erosion from time to time. The sediments in question are largely syntectonic and can be compared with, say, the Palaeozoic greywackes of the British Isles. A remarkable feature of the long-lived belts is that there is no evidence of regional uplift and the accumulation of the post-tectonic deposits of molasse type during any period within the life of the chain prior to the final uplift.

We thus arrive at the conclusion that the process which ends orogeny is in some way different from any of the earlier phenomena which accompany orogeny. While the latter may be repeated, the uplift and consequent accumulation of molasse is an event which occurs only once in the life of a mountain chain. At that point the life of a mobile belt appears to end. The rigid stable crust that is then developed may later be involved in vertical movements, but if so it moves as a block. Arrangements of Precambrian fold belts bounded by more stable rigid plates have formed since about 2 800 million years ago. In still earlier Precambrian time—between 4 000 and 2 800 million years ago—the whole crust was probably so mobile that a distinction between fold belts and stable blocks can hardly be made.

The time relations of the deformation and metamorphism which occur during the development of an orogenic belt have been studied in considerable detail. It is possible to map successive sets of structures such as sets of cleavage, or axial structures such as lineations and folds, and to determine how these were distributed in space and time. If the age of the minerals which define these structures can be established, a precise chronology can be developed.

The net result of such work has been to show that, at least in the upper parts of mountain chains accessible to study, rocks often deform in the following way during mountain building. Large thrust planes extending along the strike for many hundreds of kilometres, and involving movements of some tens of kilometres, form and are then refolded several times, during which time metamorphism may begin. In other words temperature rises in the upper parts of mountain chains after the initial deformation has occurred. This sequence of thrusting followed by folding on a smaller scale may be repeated several times with or without further metamorphism. Long-lived chains show several such repetitions, whereas in a short-lived belt, such as the Alps, the metamorphic history is limited to a few million years.

Many if not most mountain belts are poly-metamorphic; in other words, they provide evidence that the temperatures within the crust have varied by several hundred degrees during the history of the chain. The type of metamorphism—that is, the rate at which temperature varies with pressure—may change from one metamorphic event to the next, but a certain range of temperature-to-pressure ratio appears to prevail over a certain stretch of fold belt throughout its history. The explanation may lie in the rate at which blocks of crust converge on the belt, for it is known that contemporary belts characterised by high pressure–low temperature metamorphism overlie regions of unusually rapid crustal convergence. The consistency in metamorphic facies may indicate a consistent rate of crustal movement during the time when metamorphism took place.

The uplift that ends mountain building, and which, within the orogenic zone, is shown by the ending of plutonism and the accumulation of molasse type sediments, can affect a very much larger part of crust than is underlain by the fold belt itself. An excellent example is provided by conditions during the Devonian in Europe and North America. Early in the Devonian the Caledonian orogeny ended and Old Red Sandstone sediments formed along the region where plutonism had come to an end along the Caledonian mountain chains which run from New-foundland north-eastward through north-western Europe and the Arctic. At the same time continental Old Red Sandstone deposits began to form over a much more extensive region. This region, the so-called Old Red Sandstone continent, extended from the region of Leningrad to the Canadian plains with a southern coastline passing through the south of the British Isles and the vicinity of New York. Marine deposits formed in Devonian times around this continent, particularly to the south. The emergence of the Old Red Sandstone continent is of considerable geological interest, for it shows that uplift occurred far outside the orogenic belt and that a much wider region of the crust than the mountain chain itself was affected in some way. Moreover, since that time there has been no more orogeny in the region, whereas in contrast the area to the south, in which marine conditions prevailed, was affected by mountain building until the end of the Palaeozoic.

The conclusion we can draw is that the long series of events within a mountain chain provides a record of the convergence of blocks of moving crust which compress narrow mobile zones along which the mountains ultimately emerge. The uplift itself marks the end of the converging movements and the welding together of the two moving plates. From that point in time these behave as a single stable block, and may remain in that condition for many hundreds of millions of years. When ultimately orogeny once more affects such a stable block, a wholly new system of mountain chains will evolve. To investigate this phenomenon we need to look at the way in which orogenic belts follow one another in time.

The Succession of Orogenies

In 1960 Gastil showed that plutonic rocks and minerals had not been formed at constant rates throughout geological time. Strictly speaking, what was established was that minerals cooled to the temperatures at which the products of radio-active breakdown began to be trapped in the crystal lattices, at a non-uniform rate through geological time. The explanation appears to be that large areas of crust have, at various times, been through the process which produced the Old Red Sandstone continent when the Caledonian orogeny ended. This phenomenon ended pluton-ism over a large area of crust—several million square kilometres in extent—although mobile belts continued to be active elsewhere in the crust. Such a change in the behaviour of the crust will be recorded as a peak of activity in any record of radiometric age determinations on plutonic rocks. For such a record establishes the times when large quantities of igneous and metamorphic rocks were cooled at the same period. These periods are the occasions when large areas of crust were uplifted, stabilised and cooled. The connection between cooling, the ending of plutonism and uplift is critical, for it marks the change in any part of the crust from orogenic to non-orogenic conditions.

What is shown by a plot of plutonic activity against geological time is the way in which continental crust progressively develops larger areas of stabilised crust. Each time this occurs the network of mobile belts is reduced in size and the area of stabilised crust is increased. One can find a record of many such changes over the last 3000 million years. If we take as an example the amount of orogenic activity since late Precambrian time, we find that many orogenic belts in Africa became inactive in the Cambrian when uplift occurred over large parts of the centre of that continent. The Devonian uplift then gave rise to the Old Red Sandstone continent in the northern hemisphere and was followed by a Triassic uplift which ended the Hercynian (Upper Palaeozoic) mountain building. As a result of these successive stabilisations, active mountain belts became restricted in Tertiary times to those which bound the Pacific, and which extend as the Alpine chain from Indonesia to western Europe and North Africa. Somewhat similar changes leading to a restriction in mountain building occurred between about 2800 million years ago and 2000 million years ago when, as to-day, there were large areas of stable crust and restricted mobile belts. Between about 1800 and 1200 million years ago another such restriction in orogeny occurred. In contrast, new and extensive arrangements of orogenic belts appear to have been established at the start of three long periods of geological time beginning about 2800, 1900 and 1100 million years ago. On this view the history of orogeny is one of an alternation of rather short periods when an extensive orogenic belt system was established with very much longer periods over which they decayed. Short-lived belts are belts which become inactive early in one of these cycles or which for some reason became established only for a short period at some later stage in a cycle. The long-lived belts are those which remained active throughout most of a cycle. The outstanding current examples of long-lived belts are the circum-Pacific belts still active at the present day, but which include within their structure rocks dating from as far back as the late Precambrian.

Looking at mountain building in relation to other large-scale geological phenomena we see a very simple relationship. When orogeny ends, uplift occurs and a stable continental crust is formed in which several smaller plates of crust become linked by the newly stabilised mobile belts to form a new and more extensive block of rigid crust. The extent of rigid crust gradually increases in size as other stable blocks form. At some stage rifts form within the stable blocks which are then fragmented as oceanic crust disrupts the once continuous continental blocks. During this disruption mountain chains continue to evolve on the leading edges of the separating continents, but any older mountain chains within the blocks have by this stage ended their active life. Eventually, since the Earth is a sphere, dispersion of continental blocks brings them together again in new combinations to form either one or two super-continents which are now compressed by the growth of new crust in the surrounding oceans. It is possibly in this way that a new system of fold belts begins to evolve in the interior of the new groupings of super-continents. In time, stabilisation and rifting will break up these continental masses yet again, so starting a further period of continental dispersion in which mountain chains are confined to the leading edges of the separating continents.

Such a view of geological history, which is admittedly speculative, suggests that the proportion of continental crust involved in mountain building has varied through geological time in long cycles some 800 million years or more in length.

21.0 The classic volcano shape is beautifully displayed by this great Alaskan cone. The steep sides are due to the comparatively high viscosity of the lava, a variety called andesite, and the abundance of interbedded pyroclastic material. Volcanoes built of less viscous basalts are more dome-shaped.

Volcanicity is the process by which matter is transferred from the Earth's interior and erupted on to its surface. The oldest rocks exposed on the surface of the Earth were produced by volcanic activity. So, volcanoes were active 3 500 million years ago and, although the intensity of activity has waxed and waned through geological time, their products form, at the present time, most of the oceanic and part of the continental crust.

Volcanoes are undoubtedly one of the greatest natural hazards to life on this planet. It is therefore a strange natural quirk that, without them, there would be no life on Earth, for the water brought from the Earth's interior by volcanic activity has allowed the creation of the hydrosphere and the atmosphere and—if we take the evolutionary process a bit further—us!

There are many problems connected with volcanic activity. We might well ask how a volcanic liquid, a magma, is formed in the Earth's interior or, once formed, how it gets to the surface. Although these are fundamental problems, answered in part by Dr. Oxburgh and Professor Sutton in Chapters 19 and 20 respectively, we are here concerned with the rôle of volcanism in the formation of the Earth's crust; obviously it is a long story going back all the way through the Earth's history. John Wright, the author of this chapter, takes a good look at the geologically recent volcanic rocks of the Earth's crust before speculating backwards in time.

21 Volcanism and the Earth's Crust

by J. B. Wright

This contribution discusses the volcanism of the ocean floors, mountain chains, island arcs and continents. It concludes with some approximate calculations of the rate of crust formation.

IGNEOUS ACTIVITY TRANSFERS molten rock (magma) from the upper mantle to the crust, or from lower to higher levels within the crust itself. In the special case of volcanic activity, magma is transferred from below or within the crust to its outer surface, either as flowing lava (Fig. 21. 1) or as explosive pyroclastic deposits (ash, tuff, ignimbrite, Fig. 21. 2). Minor manifestations of volcanic activity include steam emission and hot mud pools (Fig. 21. 3).

The magma may be conveyed to the surface along essentially tubular conduits to build conical or dome-shaped volcanoes (Figs. 21. 0 & 21. 4) whose form and size depend on the abundance, viscosity and composition of the magma. Alternatively, the magma may rise through crustal fissures to form plateau accumulations of horizontal lava sheets (Fig. 21. 5). The amount of material added to the Earth's surface by volcanism in any particular region may be quite small when the activity consists only of small cinder cones and thin lava flows (Fig. 21. 6). At the other extreme it may be prodigious, great volcanic domes such as Hawaii and Kilimanjaro (Fig. 21. 4) involving thousands of cubic kilometres of lava; elsewhere, as in north-western peninsular India, *hundreds* of thousands of cubic kilometres have been erupted from fissures to form great continental basalt plateaux. Eruptions from fissures are also responsible, as we have seen in Chapter 16, for the ocean floors, which cover about 70% of the Earth's surface.

In assessing the rôle of volcanism in crustal development, we should first consider the distribution of volcanoes in relation to major features of the Earth's surface, the ocean basins, mountain ranges, island arcs, continental margins, and rift valleys. It is this relationship, the *geotectonic* setting of volcanism, which determines both the formation and composition of magma, and hence controls the amounts and proportions of elements being added to the crust in a given region.

There is such good correlation between earthquake activity and volcanic activity that Figure 16. 1 provides an indication of how the world's active volcanoes are distributed. The main concentration lies around the Pacific margin, where lithosphere destruction is actively in progress. There are smaller concentrations about the Atlantic and Pacific ocean ridge systems, where new lithosphere is being formed. Volcanism is also associated with other regions of crustal disturbance, for example the Mediterranean and the East African rift valleys.

VOLCANISM OF THE OCEAN FLOORS

Ocean Ridges:

Oceanic volcanism mainly takes the form of non-explosive fissure eruptions from active, spreading ridge systems. The rate of lava emission must be fairly uniform when averaged over long stretches of ridge (tens of hundreds of kilometres) and substantial periods of time (millions of years) to produce the well-documented parallel magnetic strip patterns flanking the ridge systems (see Chapter 16).

It is this continuous and, because it is largely invisible, unspectacular kind of volcanism which is responsible for almost all the crust beneath the world's oceans. The lavas are of the basalt variety known as *tholeiite* (Table 1, col. 2). They are believed to form by partial melting of the upper mantle at relatively shallow depths, under conditions of comparatively low pressure and high temperature, such as would be expected in the ascending limbs of a convective system. Essentially similar lavas form some of the largest plateaux of the continental regions.

21. 1 Lava flow, Tristan da Cunha. Eruption of this lava in 1961-62 all but overwhelmed the small settlement on this South Atlantic island, which had to be evacuated. A small volcanic vent can be seen in the centre foreground of the picture, near the origin of the flow.

21. 2 Mount Tarawera, New Zealand. In 1886 it literally split open with a titanic explosion, erupting clouds of volcanic blocks and ash (pyroclastic materials), which overwhelmed several nearby villages, killing over 100 people.

However, it must not be supposed that the ocean floors necessarily resemble the continental plateaux in having a uniformly layered structure. In the first place, being erupted under water, the lava flows tend to disaggregate and form accumulations of sub-cylindrical bodies called *pillows* (although the presence of such pillow lavas in ancient rock sequences (Fig. 21. 7) is unequivocal evidence of sub-aqueous eruption, it cannot be taken as proof of eruption from a spreading ocean ridge system). Secondly, the high heat flow and the wet environment encourages chemical alteration of the ocean floor basalts. They may even become metamorphosed to green schistose rocks, in which the minerals (albite, epidote, chlorite and hornblende) are different from those forming the original basalt (calcic plagioclase, pyroxene and olivine), although the overall chemical composition of the rocks remains generally the same. The mechanism of Plate Tectonics (Chapter 19) requires that fissure-filling dyke systems underlie the basaltic pillow lavas, and that they in turn pass downwards into a more slowly cooled and coarser-grained equivalent of basalt, called *gabbro*.

Volcanic Islands:

The central volcanoes of oceanic regions are more spectacular and better known to us because many of them form island groups—older ones such as St. Helena, younger ones such as the Azores, and brand new ones like Surtsey, south of Iceland, which was not in existence before 1963. (See also Fig. 21. 15.) Such volcanoes are mostly found along the ridge systems, particularly in the Atlantic, but they are also scattered throughout the ocean basins, some lying quite close to island arc systems and continental margins. The latter are still regarded as of oceanic type, however, either because they lie on the outer, usually convex sides of oceanic trench systems, as in the Pacific, or because the continental margins border expanding ocean basins, as in the Atlantic.

We must realise that the islands formed by these volcanoes are merely the summit regions of great mountains rising from the ocean floors, which may exceed 10 000 metres in total height. It is generally held that the lower parts of such volcanic edifices consist of tholeiitic basalt, similar to that of the ocean floors from which they rise. This is not confirmed, but whether true or not, it is certain that the higher parts, including the actual islands, are mostly characterised by a somewhat different type, the *alkali basalts*, which have comparatively lower silica and higher alkali contents than the tholeiites (Table 1, column 3). Because these volcanoes are large and their volcanic history is long, the magma reservoirs beneath them, whether initially of tholeiite or alkali basalt composition, have fractionated, giving rise to diversified lava assemblages. These include rocks of still higher silica and alkali content, with corresponding lesser amounts of calcium, magnesium and iron, rocks such as trachytes, phonolites and rhyolites (Table 1, cols. 4, 5 & 6).

Composition of Magma Source:

Volcanoes actually on the ocean ridges erupt from a region of high heat flow, since they are situated on a rising convective current. In contrast, those in the flanking ocean basins rise from comparatively cooler ocean floors. In 1967, Dr. Alexander McBirney of the Oregon Center for Volcanology, and Professor Ian Gass (then at Leeds and now at the Open University) attempted a correlation to see whether these contrasted oceanographic and thermal environments yielded any contrast in volcanic products. They plotted the distance of the volcanic islands from the ridge systems against the geochemical character of the islands based on the proportion of silica in the rocks. This was then compared with the heat flow profiles across oceans at right angles to ridge axes, and a sympathetic variation was evident (Fig. 21. 9). Those basalts with proportionately higher silica contents, that is, those nearest to tholeiites in character, build volcanoes on or near the high heat flow regions of ridge axes. As the heat flow falls off on either side of the ridge axis, the silica proportion in the lavas diminishes. Clearly if the heat flow is less in the ocean basins, the lavas erupting from them must have come from deeper in the mantle. We do not yet know for certain whether this is the result of actual changes in mantle composition with depth, or signifies merely different selective fusion products under higher pressure conditions. The results of numerous laboratory experiments in many countries suggest that the latter interpretation is probably the right one.

This relationship is of interest, for it poses the question of whether these volcanoes in the flanking basins originated on ridge crests and remained active while they were carried away upon the spreading sea-floor, progressively changing the nature of their lavas as they moved into regions of lower heat flow, or whether they originated well

21. 3 (a) Hot mud pools, like those above in the North Island of New Zealand, are common in volcanic regions. They accompany geyser activity, the expression of vast reserves of near-surface energy, which can be utilised for industry. The plumes of steam in the lower picture (b) are artificial geysers, the safety valves of the geothermal power station at Wairakei, New Zealand.

21. 4 The top of the great dome of Mount Kilimanjaro, northern Tanzania, with its ice capping.

away from the ridges, perhaps on a lateral fracture system. We would have to drill down into such islands to find out, and this would be an expensive business (see Chapter 25).

Crustal Contribution:

The importance of the volcanic islands to the romantic novelist substantially exceeds their contribution to oceanic crust, for they are merely excrescences or pimples upon the ocean floor. Even allowing for the thousands of eroded volcanoes present as seamounts (*guyots*) hidden below sea level, the total volume of lava involved is not great by comparison with that which forms the crustal layer of the actual ocean floors.

Whatever the relative crustal contributions of oceanic ridge fissures and island volcanoes may be, both are ephemeral in terms of geological time. None of the ocean floors seem to be older than Jurassic, while the South Atlantic did not open until well into Cretaceous times. That is to say, any given portion of oceanic crust has a life span of not more than about 200 million years beneath the world's oceans, before it disappears down a Benioff Zone (see below), taking our romantic volcanic islands with it.

Oceanic crust is therefore being cyclically created and destroyed, and this raises the interesting question of whether its composition changes with successive cycles. The answer seems to be negative, for tholeiites erupted among ancient submarine sediments appear to be chemically almost identical to those erupting to-day. Since oceanic crust is a partial fusion product of the upper mantle, the inference is that upper mantle composition has not greatly changed since the beginning of *recorded* geological time, some 3 400 million years ago.

The special case of Iceland:

Iceland, the largest single piece of land which is entirely of volcanic origin, is built of fissure-erupted lava plateaux and large conical volcanoes. Because of its position athwart the actively spreading mid-Atlantic ridge, Iceland is overwhelmingly of basalt close to tholeiite in composition, with lesser volumes of more silica and alkali-enriched fractions, such as rhyolite (Table 1, col. 6). For the same reason Iceland is continually being stretched, its two halves being pulled apart by the spreading sea-floor beneath. The tension causes many cracks to develop in the crust, parallel to the ridge axis, and these become fissures for volcanic eruptions. With subsequent waning of volcanic activity, magma solidifies in the fissures to form swarms of parallel dykes, each dyke representing an increment of crustal extension (Fig. 21. 8). The aggregate width of these dykes has been estimated to correspond to a total crustal extension

21.5 Part of the immense Karoo basalt plateau in Natal, South Africa. An approximately 1 000 m succession of virtually horizontal flows is shown—a small proportion of the total volume of basalts erupted.

of some 400 km since the island began to form approximately fifteen million years ago. The oldest rocks in Iceland are therefore in the extreme east and west, and present-day active volcanism is almost entirely confined to the central zone directly over the mid-Atlantic ridge (Fig. 21. 10). This model of crustal spreading for Iceland may well apply also to the ocean floors, in which case Iceland offers us a unique laboratory for the study of physical mechanisms at growing lithosphere plate margins.

VOLCANISM OF MOUNTAIN CHAINS AND ISLAND ARCS

Mountain chains and many island arcs result from compression, metamorphism and isostatic uplift of great thicknesses (up to 20 000 m is not unusual) of marine sediments, eroded from continents and accumulated in *geosynclines*, elongate depressions which develop along continental margins.

In such depressions two main stages of volcanism can be noted. The first occurs during the ocean opening phase, when mainly basalt lavas are extruded on the ocean floor, often as pillow-form accumulations. Here they may be subsequently affected by introduction of sodium, possibly from the sea water in wet geosynclinal sedi-

ments, and transformed into rocks called *spilites* (Table 1, col. 7), rich in the sodium feldspar called *albite*.

Later, as ocean-closing movements begin, volcanic island arcs rise from the geosynclinal floor, adding lava flows, ash and eroded debris to the still growing sediment pile. The kind of lava erupted from such volcanoes is principally of a magma type intermediate in composition between basalt and rhyolite, called *andesite* (Table 1, col. 8). Both spilitic and andesitic lavas are well represented among the folded geosynclinal sediments of, for example, North Wales and the Lake District in Great Britain and the mountain chains of North and South America.

The second major stage of volcanism occurs after the pile of sediments and lavas has filled the trough, and has been compressed, heated and metamorphosed into new continental crust, which becomes isostatically elevated to form mountain ranges or major island arcs such as New Zealand and Japan. Within these mountainous belts rise the great andesite cones of classical volcano shape, such as Fujiyama in Japan, Egmont in New Zealand, or Paricutin in Mexico which did not start erupting until 1943. Andesitic volcanism is here accompanied by smaller amounts of basaltic lava, and by voluminous fissure eruptions of an incandescent pyroclastic froth called *ignimbrite*, which builds substantial plateaux and has the composition of rhyolite (Table 1, col. 6).

21. 6 In many volcanic regions the activity is confined to small cinder cones and thin flows, but cinder cones can also develop as parasitic features on the flanks of large volcanoes, as in this picture of Mount Longonot, Kenya.

21. 7 Ancient pillow lavas, largely metamorphosed to greenstones, exposed on the coast of Anglesey, North Wales. The lava pile has been tilted to a near vertical position, as can be seen from the alignment of the long axes of pillow cross-sections.

Origin of Andesite:

Oceanic trench systems are the sites of descending lithosphere plates where the oceanic crustal veneer on top of the lithosphere is carried diagonally downwards along gigantic thrust planes of the Benioff Zones, which are marked by belts of earthquake foci deepening behind the arcs (see Chapter 19).

Andesitic volcanism is closely related to this process, and Drs. W. R. Dickinson and Trevor Hatherton, of Stanford University and the New Zealand Department of Scientific and Industrial Research (Geophysical Division) respectively, have shown how magma composition varies with depth of generation along the Benioff Zone (Fig. 21. 12).

Relationships of this kind cast doubt on the traditional view of andesitic magma as a mixture between basalt and geosynclinal sediments or granitic continental crust. Furthermore, some of the larger island arcs continue along their length into exclusively volcanic ones, such as the Tonga-Kermadec chain in the southwest Pacific. Along these minor arcs, andesite volcanoes rise directly from the ocean floor, on the inner side of trench systems, without evidence of thick sediments or continental crust.

On the basis of laboratory experiments and geochemical considerations, Dr. T. H. Green and Professors A. E. Ringwood and S. R. Taylor, of Canberra University, have concluded that andesite results from partial fusion of the upper mantle under the special pressure and temperature conditions associated with Benioff zones. In particular, the water in the sediments overlying the oceanic plate is thought to provide a hydrous zone in the mantle, in which andesitic melts are produced.

There is, however, the problem of generating sufficient heat to induce melting in the upper mantle along these zones. Drs. E. R. Oxburgh and D. L. Turcotte, of Oxford and Cornell Universities respectively, investigated this problem and calculated that the main source of heat would be friction along the actual thrust plane itself, i.e. along the upper surface of the oceanic crustal veneer on the descending lithosphere plate. They suggest that this oceanic basalt veneer (mainly tholeiite) is partly melted to provide the abundant andesitic volcanism of orogenic belts. If their view is correct, then it is clear that the contribution of ocean volcanism to crustal growth is by no means ephemeral, for a good proportion will be remelted and incorporated permanently into the continental crust (see Chapter 19).

An interesting minority view of the origin of andesites (and spilites) whose principal protagonists are Professor E. F. Osborn of Pennsylvania State University, and Drs. G. A. Challis and W. F. Lauder of Victoria University, New Zealand, holds that andesite is not a fusion product at all, but arises through fractional crystallisation of basaltic magma within great chambers deep in the crust beneath orogenic belts. Such fractionation processes, concentrating lighter elements like Si, Al and the alkalis leave *ultrabasic* residues specially rich in Mg and Fe. These residues are identified as the enormous bodies of rock known as *peridotite* (Table 1, col. 9), which are emplaced as great slices into many mountain chains and island arcs, as in the western U.S.A. and New Zealand.

Peridotites are regarded by most authorities however, as sub-oceanic lithosphere thrust upwards into the crust during the intense deformation which accompanies transformation of a geosyncline into a mountain system. Whatever their origin, these bodies are important constituents of the continental crust because, although comparatively minor in amount, their composition is so different from that of average granitic crust (Table 1, col. 1 & 9).

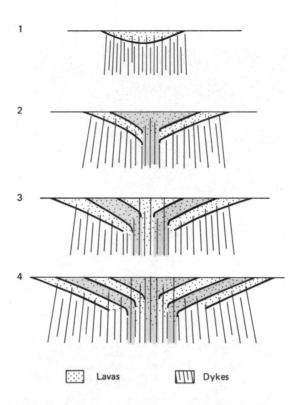

Lavas   Dykes

21. 8 Crustal extension by dyke injection, a simple model for accretion at lithosphere margins. Colour indicates alternate phases of lava eruption and dyke injection.

Origin of Ignimbrite:

Partly because its composition approaches the granitic composition of the upper continental crust (Table 1, col. 1 & 6), this incandescent pyroclastic froth is thought to originate at comparatively shallow levels within the crust itself. During the intense heating and deformation endured by geosyclinal sediments during mountain building movements, some selective melting undoubtedly occurs. The molten products will be of granitic composition and must rise towards the surface because of their lower density. During the earlier stages of orogenic movements, the confining pressures remain great and the liquids cannot rise far, so they crystallise at depth to form the immense granite batholiths, abundantly visible in every eroded mountain chain. However, in the waning stages of mountain building movements isostatic uplift occurs, and as confining pressures are reduced, later melts find their way to the surface and erupt as ignimbrites, rhyolites and ashes.

This process of partial fusion and rise of silica-rich melts in orogenic regions is probably the chief process by which the upper parts of continents have acquired, through geological time, their average granitic composition. The lower parts, enriched in more refractory residues, thus tend towards a complementary basic composition.

Continental Growth:

Since no part of the ocean floor is permanent, it follows that both major and minor island arcs, which are surrounded by oceanic crust, must ultimately become part of the larger continental blocks. The process of continental growth is thus a consequence of ocean closing movements, whereby continental plates ultimately collide together, thrusting and squeezing up between them any intervening island arcs and sedimentary accumulations, and creating giant mountain ranges flanked on either side by continental crust, as in the Alps and Himalayas (see Chapter 20).

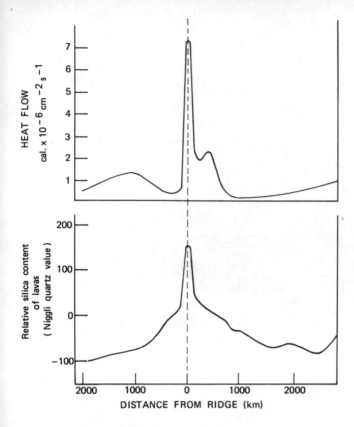

21. 9 Sympathetic variation of heat flow and chemical composition at varying distances from the crest of the Mid-Atlantic Ridge.

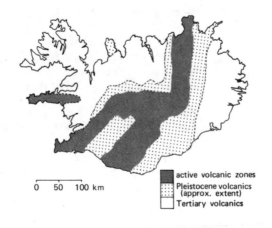

active volcanic zones
Pleistocene volcanics (approx. extent)
Tertiary volcanics

0   50   100 km

21. 10 Outline geological map of Iceland, showing how the age of volcanism decreases inwards towards the central belt.

The re-cycling aspect of continental crust development should not be overlooked: although a proportion of continental land is lost to the oceans each year by erosion, such a loss is only temporary, for the erosion products cannot accumulate anywhere but on the sea floor, whence ocean-closing movements will eventually ensure their return to the continents.

VOLCANISM OF THE CONTINENTS

Both oceanic and continental volcanism occur in regions where tensional forces predominate, with consequent similarities in eruptive patterns and magma types.

Basalt Plateau Volcanism:

On continents, as on the ocean floors, tholeiitic basalts erupt mainly from fissures, to build huge lava plateaux, which may contain several hundred thousand cubic kilometers of magma, scarcely varying in composition from the base to the top of horizontally layered lava sequences. Moreover, activity is monotonously basaltic and, though there may be a little rhyolite developed, pyroclastics are virtually absent. Some overlap to alkali basalt and associated lavas sometimes occurs, as in the Scottish Hebrides.

We have seen on p. 302 that oceanic tholeiites erupt where the heat flow is high and pressures are comparatively low. Since these are the conditions required for tholeiite eruption, they must also have obtained beneath the continental areas flooded by tholeiitic basalts.

The map, Figure 21. 11, shows a pre-continental drift (i.e. early Mesozoic) assembly of the southern continents (see p. 224), with most of the larger Mesozoic and younger tholeiitic basalt plateau regions marked. The latter lie on or across present day continental margins. It is evident that continental disruption must be initiated by growth of a spreading ridge system beneath the continents. Where such a system was locally more active, tholeiitic volcanism broke out upon the continents before actual disruption took place. In the special case of Iceland, this locally more intense activity persisted over the ridge axis and built a basaltic land mass on the ocean floor.

In regions where these basalt plateaux occur, we therefore have a manifestation of oceanic-type volcanism which perforce effects direct and permanent additions to the continental crust.

TABLE 1*

APPROXIMATE AVERAGE COMPOSITIONS (WEIGHT %) OF IMPORTANT IGNEOUS ROCK TYPES

	Upper Crust Granite	Tholeiite	Alkali Basalt	Trachyte	Phonolite	Rhyolite	Spilite	Andesite	Peridotite
SiO_2	67	50	46	63	55	73	51	60	43.5
Al_2O_3	16	16	15	18	20	13	14	17	4
Fe_2O_3	1	2	4	2.5	2.5	0.5	3	2	2.5
FeO	2.5	7	8	1.5	3.5	1.5	9	4	10
MgO	1.5	8	9	0.5	1.5	0.5	4.5	3.5	34
CaO	3.5	12	9	1	2.5	1.5	7	7	3.5
Na_2O	4	2.5	3.5	7	8	4	5	3.5	0.5
K_2O	3	0.5	1.5	5	5	4	1	1.5	0.3
TiO_2	0.5	1.5	3.5	0.5	0.5	0.5	3.5	0.5	1
H_2O	1	0.5	0.5	1	1.5	1.5	2	1	0.7

* The compositions differ slightly from those given in Table 3, Chapter 1 because they represent averages of figures from different sources with different distributions over lavas and plutonic equivalents.

Rift Valley Volcanism:

On most maps, the great rift valleys of Eastern Africa (Fig. 21. 11) are shown as a continuation of the global oceanic ridge systems, because they are thought to represent the sites of future opening ocean basins. Professor B. C. King of Bedford College, London has cited grounds for questioning this idea. One good reason is that the volcanism is non–tholeiitic and highly variable as regards both eruptive mechanisms and lava compositions. Even the great fissure flood basalts of Ethiopia are in fact alkaline basalts, while in Kenya especially, great volumes of phonolite and trachyte were erupted both as lava and pyroclastic deposits. Volcanic eruptions from central volcanoes of vast size like Mount Kenya and Kilimanjaro are as important as fissure activity, and are also related to the rift faulting.

Relationships similar to those displayed in Figure 21. 9 have been demonstrated for rift valley volcanics in both Kenya and Ethiopia by myself and by Drs. M. J. Le Bas and P. A. Mohr, respectively of Leicester University and the Smithsonian Institution. It may therefore be more reasonable to regard rift valleys as zones of earlier unsuccessful 'attempts' at ocean opening.

It was previously stated that phonolites and trachytes can develop by fractional crystallisation from alkali basaltic magmas. This process yields, however, only about 1 volume of such end products for every 5 to 10 volumes of basalt. In many rift valleys the proportion of phonolite and trachyte to basaltic lavas is much too high (between 1:1 and 1:2 for East Africa) for a fractionation relationship to be plausible, and it is becoming increasingly clear that these more silicic and alkaline lavas may also have their origin in the upper mantle, along with the associated basaltic varieties.

Rhyolite-granite Ring Complexes:

The possibility that trachytic and phonolitic magmas rising from the upper mantle may be modified by reaction with granitic crust to produce lavas of rhyolitic composition, provided the rise is slow enough, has been mooted by Drs. D. K. Bailey and J. F. Schairer, of Reading University and the Geophysical Laboratory of the Carnegie Institution of Washington respectively. This provides an excellent explanation for the igneous assemblages dominated by rhyolites and granites such as are found in several parts of the world, not directly or obviously connected with contemporary rift valleys. One of the best documented of

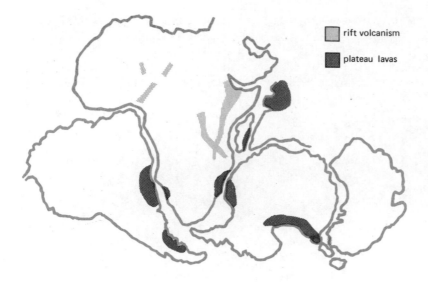

21. 11   Reconstruction of southern continents, showing distribution of regions affected by Mesozoic and younger tholeiitic plateau basalt volcanism, and some rift valley features characterized by alkali basalt volcanism.

these districts, in Nigeria, is famous for its associated tin and niobium ores.   In such regions both rhyolite lavas and ignimbrite froth were erupted from arcuate to circular fissures which subsequently became filled with granite.   Superficially the ignimbrites resemble those of mountain chains and island arcs, but their tectonic setting is totally different, and their magmas have a different origin.

SUB-VOLCANIC ACTIVITY

By definition, the products of active volcanism accumulate on the Earth's surface, thickening the crust from above.   When volcanism ceases, the magma in feeder channels and sub-surface reservoirs must also solidify, adding its quota to crustal growth, and that quota may not be negligible. The volume of magma involved in sill sequences and dyke swarms may rival that of some basalt plateaux.

Sills (Fig. 21. 13) are merely sub-surface lava flows, emplaced along interfaces in sub-horizontal sequences of sediments, usually when these are too light to support large thicknesses of surface basalt effusions.

When fissure eruptions cease, magma solidifies to form vertical *dykes* (Fig. 21. 8), although it must be emphasised that not all dykes are necessarily feeders to surface volcanism.   Extensive dyke swarms represent considerable crustal dilation, as for example in Iceland (Fig. 21. 10) and the Scottish Hebrides.

Magma in the feeder pipes of central volcanoes solidifies to form cylindrical *plugs* (Fig. 21. 14).

Magma Chambers:

In the crust beneath many volcanoes lie large reservoirs of magma, filled from below, in which crystal fractionation provides a variety of volcanic products at the surface.   If the chamber is large enough and the magma in it is of basaltic composition, the heat from it may be sufficient to melt the surrounding crust and generate granite magmas which further contribute to the local volcanism.   It is widely held that this happened in parts of the Scottish Hebrides.   At the close of volcanism, the magma in these chambers will also solidify to coarse-grained rocks having the same composition as the lavas which previously were being erupted.

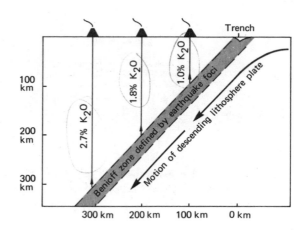

21. 12   Diagram showing how the composition parameter for andesite (K_2O content in lavas with 55–60% SiO_2) changes regularly with distance from the foredeep trench of a volcanic arc, and hence with depth along the Benioff zone.

CONTINENTAL COMPOSITION

The average composition of continental crust, according to Professor S. R. Taylor of Canberra University, is likely to be andesite.   This can be understood if we consider the processes involved in continental growth.   Erosion products from the continents must include granite from the older crust and mainly basaltic volcanics from the older plateau and rift valley volcanism.   The net composition of this material should therefore be somewhere between granite and basalt.   To this is added spilite and abundant andesite, the principal volcanic product of orogenic belts, whose composition also lies between granite and basalt. When all this material is swept up and plastered on to a continental margin it might be expected to have an overall andesitic composition.

Although not strictly within the scope of this chapter, we should not forget that volcanic processes are also responsible for all the water on the Earth's surface and for its atmosphere.   Expulsion of volatile materials from the Earth's interior has been going on throughout geological time, and volatile differentiation is to a great measure responsible for many of the important deposits of heavy metals, such as gold, silver, copper, lead, zinc and tin.   The principal reason is that such elements do not fit readily into the structures of common rock-forming minerals (see Chapter 1); this explains why they do not appear in Table 1 of average rock compositions.   The atoms of these metals are commonly transported in solution by the residual volatile (hydrothermal) fractions of crystallising magmas, from which they are concentrated into veins and pockets, often in volcanic rocks.   The major gold fields of the world, for example, are practically all associated with great thicknesses of orogenic andesites and basalts which have been affected by these hydrothermal processes.

APPENDIX

It is of interest to form some idea of the rate of addition of volcanic material to the Earth's crust. In the following very simple-minded calculations, the number of approximations and assumptions is so great that the answers can only be very gross approximations.

Oceanic Crust:

Icelandic volcanologists have estimated an annual volcanic output for Iceland of about 0.05 km^3, which is largely confined to a 350 km long strip directly overlying the mid-Atlantic ridge (Fig. 21 10).   A single volcanic eruption whether of lava or ash or both, probably involves an output of not more than 10^{-2} to 10^{-3} km^3 of magma.   Even the devastating explosion which demolished Krakatoa in 1883 may only have added about 0.1 km^3 of new material to the crust, even though something like 25 km^3 of the original volcano was involved in the explosion, whose main result was to achieve a re-distribution of pre-existing crustal matter.

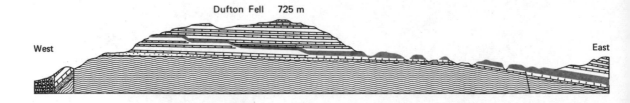

West

Dufton Fell   725 m

East

21. 13 Geological cross-section of the Whin Sill, a typical subterranean magma flow (colour). It is emplaced between beds of Carboniferous age, which overlie folded Silurian strata.

Iceland is a plateau on the mid–Atlantic ridge, so the rate of lava production there must be higher than in other parts, otherwise the whole ridge would be above sea-level. There is also a greater frequency of earthquakes beneath Iceland, suggesting a higher level of volcanic activity— 0.05 km³ per year is therefore an upper limit for the rate of lava production.   A figure of 0.02 km³ per year might be a reasonable estimate for the amount of lava added to the crust along a 350 km (length of central strip of Iceland) segment of an active ocean ridge system, the total globe length of which has been estimated at 60 000 km.

Then: if a 350 km length produces 0.02 km³/yr of lava, 60 000 km will produce

$$60\,000 \times 0.02/350 \ = \ 3.4 \ \text{km}^3/\text{yr}.$$

Note that (1) this figure would approximate the present rate of addition to oceanic crust; (2) clearly it must approximate the rate of destruction of oceanic crust at trench systems, otherwise the Earth would be expanding or contracting; (3) in the geological past, at times when no sea-floor spreading occurred, probably the only contribution would have been from isolated volcanoes in the ocean basins, and therefore very much less.

Continental Crust:

The continents are aggregates of mountain belts and island arc systems, progressively accreted in an approximately concentric pattern for most of recorded geological time. Figure 21. 11a suggests that the contribution from basalt plateau and rift valley volcanism, though locally immense, is globally negligible for a first approximation exercise such as this.

21. 14   Plug of basalt, the solidified pipe of a small satellite vent on the slopes of a much larger volcano.   It forms a positive feature because of its greater resistance to erosion. Auckland Islands, New Zealand.

If we consider width and thickness of the world's orogenic belts, the Andean range might represent a suitable average, somewhere between the gigantic size of the Himalaya and the narrow island arcs. For this average we shall take a width of 400 km and a thickness of 30 km over the 60 000 km length. This gives a volume for the world's orogenic belts of 720×10^6 km^3.

Continental mountain ranges are, however, principally composed of original continental crust (in great part remobilised), plus geosynclinal sediments derived from that crust further inland. The amount of volcanics in these belts is the contribution from geosynclinal, island arc and mountain range eruptions, amounting to between perhaps one-fifth and one-tenth of the total volume of the orogenic system, as estimated from the anatomy of ancient orogenic belts, i.e. between 70 and 150×10^6 km^3.

Growth of the present global orogenic belt system began perhaps 200 million years ago, so that the above volume of new crust has been added at an overall rate of 0.3–0.7 km^3 per year.

The present area of continental crust is about 150×10^6 km^2, and its average thickness is about 35 km. The total continental crust volume is therefore about $5\,200 \times 10^6$ km^3.

Note that (1) there is far more uncertainty about any of these figures than about those concerning oceanic crust; (2) any averages derived by this general kind of calculation take no account of the episodic nature of orogenic activity which is indicated by the considerable periods during which there apparently was no sea-floor spreading; (3) rough though these figures are, they do suggest that the amount of new crust added each year to the continent is less than that added to the oceans. This might perhaps be expected, particularly if andesite, the dominant volcanic product of the orogenic belts, is indeed derived from partial melting of basaltic ocean crust along Benioff Zones; calculations by Dr. Green and Professor Ringwood from the Australian National University suggest that one volume of basalt will yield only about $\frac{1}{4}$ to $\frac{1}{3}$ its volume of andesite.

More recent expert estimates of crustal growth lie between 10 and 60 km^3/yr. for the oceans (based on crustal spreading rates), and 3 km^3/yr. for the continents. Although these figures are about an order of magnitude greater than mine, they do agree on the *relative* rate of growth for oceanic and continental crust.

FURTHER READING:

Relevant portions of:-

A. HOLMES 1965. Principles of Physical Geology. *Nelson.*

A. RITTMANN 1965. Volcanoes and their Activity. *Wiley. Interscience.*

21. 15 White Island—a volcanic island in the Bay of Plenty, New Zealand.

So far we have concerned ourselves with Earth processes and phenomena that are primarily of academic interest. But the Earth must also supply our practical needs. Food, fuel and metals are the raw materials of our civilization; and their discovery and exploitation are essential to its maintenance and progress. Equally critical to man is the fact that our planet is a mobile, dynamic body. Hurricanes, floods, landslides and avalanches occur every day somewhere on Earth. More disastrous, when they happen in densely populated areas, are earthquakes and volcanic activity.

In a volume of this kind it is not possible to review, even briefly, all the natural resources that are of use to man, nor all the Earth's natural phenomena that affect us. Nevertheless we held it essential to include instances of how the Earth scientist serves man. We therefore invited three specialists to write short articles illustrating how their studies can be used for man's benefit.

In the first of these articles, Tom N. Clifford, Professor of Geology at the University of the Witwatersrand, shows, using Africa as an example, that certain metallic elements, such as copper, tin, lead and zinc, occur primarily within rocks that have been deformed during the last 1200 million years of the Earth's history. Other products—diamonds, chromite and gold, among others—are present in abundance only in Africa's most ancient rocks. So, although the old adage 'gold is where you find it' is undeniably true, it would have taken less time to find it if prospectors had had access to our present knowledge. Nevertheless, although many of the world's obvious mineral deposits are now known, more sophisticated, and hence expensive, geophysical techniques are needed to find hidden deposits. If the area of search can be limited by studies such as those of Professor Clifford, then not only will the exploration expense be reduced but other areas worthy of detailed study will be indicated.

22 Location of Mineral Deposits

by Tom N. Clifford

If the right correlation between geological setting and deposits of a mineral can be found, the search for further deposits can be more effectively directed. Africa provides a case in point.

MINERAL DEPOSITS are local accumulations or concentrations of rocks and minerals that can be recovered at a profit. Almost all of the concentration processes involve movement of fluids (liquids and gases) which, for the most part, fall into three categories: (1) fluids of igneous origin; (2) waters derived from the atmosphere; and (3) fluids associated with metamorphic processes—that is fluids, present in rocks, that are set in motion by changes in temperature and pressure. It will emphasize the importance of these concentration processes if we first consider the *average* concentrations of some economic elements in the Earth's crust; these are listed in Table 1. Clearly, many of these elements are present in low concentrations. More abundant elements, such as iron and aluminium, are generally locked in the crystal lattice of silicate minerals which precludes their easy separation and economic recovery; they are only workable when they occur as simple oxides, hydroxides or, in the case of iron, carbonates.

Two examples will serve to illustrate what is meant by economic concentrations:

(1) *Gold*, normally averaging 0.004 parts per million (ppm) of crustal rocks, forms 7 ppm of the rock in the great Witwatersrand Goldfield of South Africa; this represents a concentration of 1 750 times the crustal average.

(2) *Copper*, *Lead* and *Zinc* have averages of 55, 13 and 70 ppm respectively in the Earth's crust, but in one mining area in South-West Africa the rock mined has averaged 15% Pb, 7% Zn and 6% Cu—compared with crustal averages this represents concentrations of × 1 000 for Cu and Zn, and × 10 000 for Pb.

Pioneer work by de Launay (1913) and Spurr (1923) clearly indicated that there is a pattern of distribution of mineral deposits. In particular, there are geographic regions characterized by the

Element	Crustal Average	Element	Crustal Average	Element	Crustal Average
Aluminium	81 300	Lead	13	Silver	0.07
Antimony	0.2	Lithium	20	Tantalum	2
Beryllium	2.8	Manganese	950	Tin	2
Chromium	100	Mercury	0.08	Uranium	1.8
Cobalt	25	Molybdenum	1.5	Vanadium	135
Copper	55	Nickel	75	Tungsten	1.5
Gold	0.004	Niobium	20	Zinc	70
Iron	50 000	Platinum	0.01		

TABLE 1: AVERAGE AMOUNTS OF SOME ECONOMIC ELEMENTS IN THE EARTH'S CRUST IN PARTS PER MILLION (OR GRAMS PER TONNE)

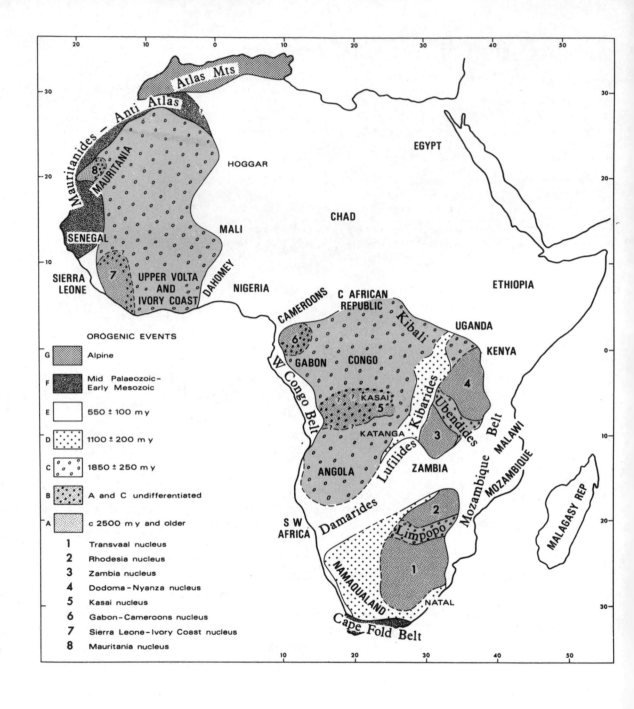

OROGENIC EVENTS

G   Alpine

F   Mid Palaeozoic–
     Early Mesozoic

E   550 ± 100 m y

D   1100 ± 200 m y

C   1850 ± 250 m y

B   A and C undifferentiated

A   c 2500 m y and older

1   Transvaal nucleus
2   Rhodesia nucleus
3   Zambia nucleus
4   Dodoma–Nyanza nucleus
5   Kasai nucleus
6   Gabon–Cameroons nucleus
7   Sierra Leone–Ivory Coast nucleus
8   Mauritania nucleus

22. 1   Generalized map of the major structural units of Africa.

occurrence of numerous mineral deposits, all with the same character; such regions are termed *metallogenic provinces*. Some areas such as Zambia and the Lake Superior area in North America are rich in copper; South Africa is rich in gold. Other areas are apparently barren of mineral wealth. The process by which the mineral deposits are formed may occur more than once in a particular region. When this happens the term *metallographic province* is used. It is quite evident that the fundamental control of regional patterns of mineral wealth is within the Earth; individual deposits are then enhanced by local concentration factors.

While there is no doubt that the ocean basins represent a vast reservoir for certain types of mineral wealth, most of our presently workable mineral deposits are found on the continents or on the continental shelves. We must therefore examine the relationship between the distribution of mineral deposits and the events in the geological history of the continents to see if there is any major control involved in the distribution of the deposits. A particularly good example of this study is Africa, for it represents 20% of the land area of the Earth and its rocks span a period of 3500 million years.

In Africa, a number of attempts have been made to delineate regions of similar metallogenic character. In the absence of detailed knowledge of the structure of the metallogenic regions, early classifications were based on the age of the mineral deposits or even just on their geographical pattern. More recently the distribution of deposits has been considered in relation to the structure of the continent. In this regard, at least seven major mountain-building events (periods of orogenesis) are recorded in Africa and are, from oldest to youngest (see Fig. 22. 1): (i) 3000 million years ago (Transvaalian orogeny); (ii) 2500–2800 million years ago (Shamvaian orogeny); (iii) 1850 ± 250 million years ago (Eburnian and Huabian orogenic episodes); (iv) 1100 ± 200 million years ago (Kibaran orogeny); (v) 550 ± 100 million years ago (Damaran-Katangan or Pan-African orogeny); (vi) middle Palaeozoic-early Mesozoic orogenies (Acadian and Hercynian) of north-western Africa and the Cape Fold Belt of South Africa; and (vii) the Alpine orogeny of the Atlas Mountains.

We can subdivide these seven episodes into two major groups. Events (i) to (iii) occurred in the distant geological past. Some of the areas deformed during this early period in the Earth's history have not been disturbed by orogenesis during the last 1500 million years of geological time, and as they have remained stable for this long period we invoke the use of the term *craton* (meaning stable) and call such areas the *older cratons*. The areas which have been affected by the geologically younger mountain-building events—(iv) to (vii)—are called the *younger orogens* and they have been deformed from time to time during the last 1200 million years.

When the major mineral deposits of Africa are considered in relation to these structural régimes, a clear relationship emerges. The younger orogens contain Africa's principal copper, lead, zinc, cobalt, beryllium, tin, tungsten and niobium-tantalum occurrences. In the older cratons the principal deposits of gold, diamonds, chromium, asbestos and iron are found. These two fundamental structural units have been called *tectono-metallogenic units*. Metallogenic *provinces* are then defined as more localised provinces of mineral concentration within the major units. We are now going to concern ourselves with metallogenic *units* and *provinces*.

TECTONO-METALLOGENIC UNITS

Younger Orogens

About 96% of Africa's *copper* production comes from these regions; in addition, almost all of her *lead* and *zinc* come from the younger orogens, as does *cobalt* for which Africa is the world's principal source. The greatest concentration of *niobium* and *tantalum* is also within the younger orogens, almost 70% of the ore reserves occurring in the eastern Congo.

The deposits of *tin* within the younger orogenic belts yield approximately 95% of Africa's annual production, the eastern Congo and Nigeria together providing 80% of this output. The principal exception is the tin mineralization associated with the Bushveldt complex (2000 million years) in the Rhodesia-Transvaal older craton which provides less than 5% of Africa's overall annual production. More than 95% of African *tungsten* production has been obtained from the younger orogens, largely from Uganda and the eastern Congo (Fig. 22. 3), whilst about 5% comes from the older cratons. *Beryllium*, although less distinctive than the other elements, is mainly won from the younger orogens, 90–94% coming from pegmatites within these zones.

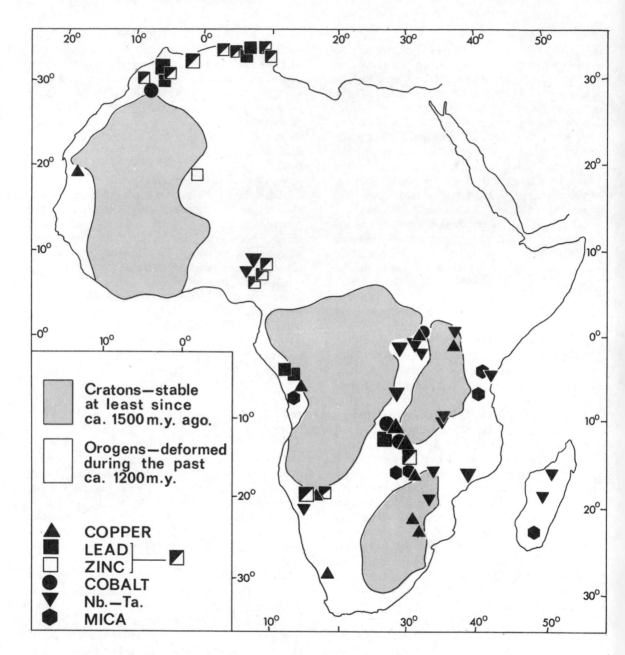

22. 2 Distribution of principal African deposits of copper, lead, zinc, cobalt, niobium-tantalum and mica. Large and small symbols indicate major and minor deposits respectively in this and the following figures.

Older Cratons

In contrast to the orogenic belts which have suffered tectonism within the past 1200 million years, the regions of ancient stability, at least during the last 1500 million years, are the *source* regions for most of Africa's gold, chromium,

asbestos and diamond. In addition, these regions contain very important iron concentrations.

The confinement of major deposits of these ore products to the older cratons is, however, not as clear-cut as the concentration of deposits that characterize the younger orogens. In particular, elements characteristic of the older cratons occur

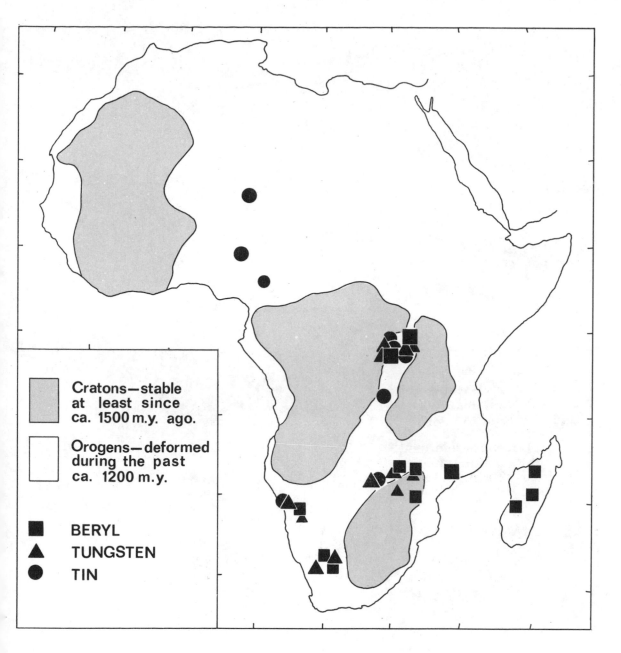

22. 3 Distribution of principal African deposits of beryl, tin and tungsten.

Legend:

Cratons—stable at least since ca. 1500 m.y. ago.

Orogens—deformed during the past ca. 1200 m.y.

■ BERYL
▲ TUNGSTEN
● TIN

where large segments of the cratons have been involved in the younger orogenic episodes; a number of gold, iron and asbestos deposits occurring near the margins of the younger orogens probably owe their distribution to this effect.

Something like 97% of Africa's *gold* is won from these ancient stable regions, or from the marginal

portions of the cratons where they have been rejuvenated by younger orogenesis (Fig. 22. 4). South Africa, with 90% of Africa's total output, is the principal producer; other regions are Ghana, the Congo and Rhodesia.

Diamonds present a difficult economic product in any consideration of the tectonic control of

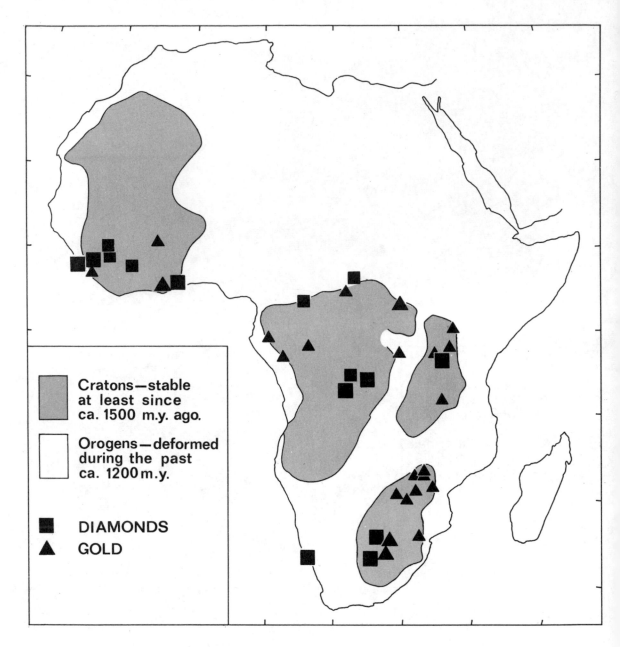

22. 4 **Distribution of principal African deposits of diamonds and gold.**

mineralization, since they are not only concentrated in the original source rocks—kimberlite—but are also present as placers (alluvial deposits) in subsequent sedimentary sequences. In the analysis of their genetic relationship to structural units it is, therefore, germane to discuss the source regions rather than the location of the derived deposits.

In many cases, major diamond deposits are directly worked in their kimberlite source, or in alluvial deposits very close to that source. The principal African deposits of this kind are in South Africa, northeastern Angola and the Congo. All of these areas lie within the older cratonic regions of Africa (Fig. 22. 4); indeed, although there are many kimberlite intrusions within parts of the

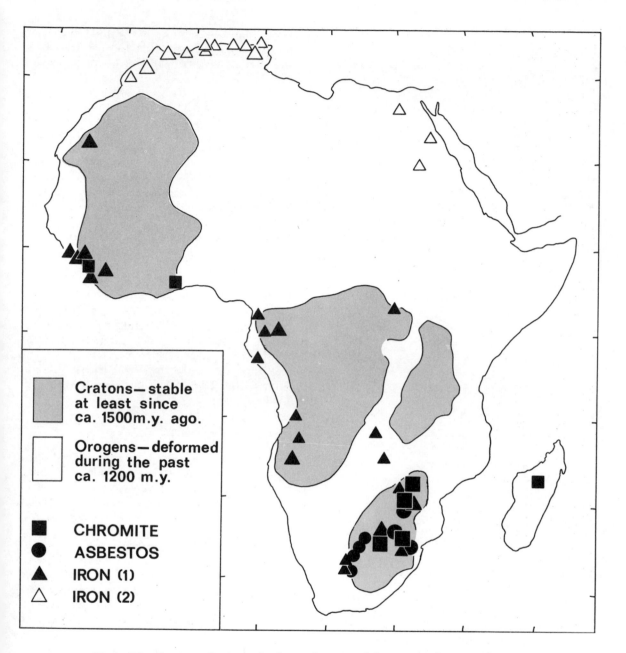

22. 5 Distribution of principal African deposits of chromite, asbestos and iron.

younger orogens, they are largely barren or have a low diamond concentration.

In addition, there are a number of regions, often a considerable distance from any kimberlite pipe, where alluvial diamonds are mined. In almost all of these cases, the source of the alluvial diamonds can be readily related to kimberlites in the older cratonic regions. In this regard especial mention must be given to the very important deposits of the west coast of southern Africa, to the north and south of the mouth of the Orange River, where diamondiferous gravels occur in Tertiary to Recent sediments. Diamond-bearing sediments occur all along this coast, from the Cape of Good Hope to Angola, but the stones decrease in size northwards and southwards from the mouth of

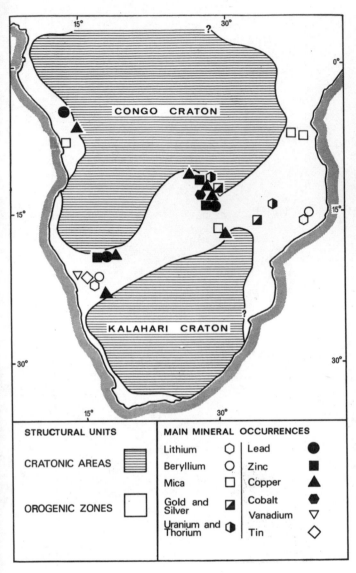

the Orange River. This has been taken to indicate an origin by way of the Orange River from
the interior cratonic region of South Africa (see
Fig. 22. 4).

Over 97% of Africa's *chromium* production is
from basic and ultrabasic igneous rocks in the
older cratonic regions in South Africa, Rhodesia
and Sierra Leone. Over 95% of Africa's production of *asbestos*, also commonly associated with
igneous rocks, comes from the older cratons in
South Africa, Rhodesia and Swaziland or from the
extensions of the rocks of the cratons that have
been deformed within the marginal zones of the
younger orogens.

Outside the northern and northeastern region,
Africa's principal *iron* ore production is from the
older cratons or from the portions of the older
cratonic sequences which have been involved in
the younger orogenic belts (Fig. 22. 5); the main
undeveloped reserves are also concentrated in
these stable regions. The principal producers are
South Africa, Sierra Leone, Liberia and Guinea;
South Africa produces approximately 20–25% of
the African total; other important deposits occur in
Angola, Rhodesia, Mauritania, Gabon, Swaziland
and the Congo.

METALLOGENIC PROVINCES

Metallogenic provinces are more restricted
regions, within the tectono-metallogenic units,
that are characterized by concentrations of certain
ore products. Although they can be identified in
both the younger orogens and the older cratons,
we will now, as an example, discuss some of the
features they exhibit in the former.

As in the case of the larger metallogenic *units*, an
important aspect of the concept of metallogenic
provinces is the influence of the structural environment. This is particularly well illustrated with
reference to the late Precambrian-early Palaeozoic
structural pattern of southern Africa, which consisted of a zone of crustal instability—an orogenic
zone—and two extensive stable areas, the Congo
and Kalahari Cratons.

Mineralization related to this period of African
history is concentrated in the mobile zones; in
particular, a large percentage of southern Africa's
base-metal mineral wealth is associated with the
unmetamorphosed (or weakly metamorphosed)
and folded late Precambrian rocks along the
southern and western margins of the *Congo Craton*
(Fig. 22. 6). In contrast, with the exception of

**22. 6 Late Precambrian-early Palaeo-
zoic structural units of southern Africa,
and the principal related mineral
occurrences.**

copper mineralization in Rhodesia and South-West Africa, similar rocks in the orogenic zone adjacent to the *Kalahari Craton* have not, to date, produced really major deposits of base metals. Equally, major deposits of these metals are poorly represented in the interior highly metamorphosed parts of the orogenic zone; however, the latter are characterized by significant pegmatite and other deposits of lithium, beryllium, tin and tungsten in South-West Africa, Central Africa and elsewhere (Fig. 22. 6).

Three metallogenic provinces associated with the late Precambrian-early Palaeozoic orogenic zone of southern Africa can be recognized from these data: (a) *province of base metal concentration*—in the unmetamorphosed (or weakly metamorphosed) folded rocks directly adjacent to the Congo Craton; (b) *province of lithium-beryllium-tin-tungsten-(uranium-thorium) concentration*—in the interior high grade metamorphic zones; and (c) *province of relatively low concentration of base and other metals*—in the unmetamorphosed (or weakly metamorphosed) folded rocks adjacent to the Kalahari Craton.

The principal distribution pattern in this specific example is widely applicable within the younger orogens. A selection of some other provinces is listed in Table 2. The Namaqualand and Kibara belts (see Fig. 22. 1), for instance, are comparable to province (b) above, whilst the North African region is similar to province (a) above. In addition to this evidence of other provinces, Table 2 illustrates a number of other important features: (i) mineralization took place at widely spaced intervals within the younger orogens tectono-metallogenic unit; (ii) similar mineralization recurred at different periods of geological time in Nigeria (Nb-Ta and Pb) and North Africa (Pb); and (iii) *iron* is a definitive metal for the tectono-metallogenic units of the older cratons in southern Africa, but in North Africa is concentrated in a younger orogens metallogenic unit (Fig. 22. 5).

Spurr, writing in 1923, considered that the unequal distribution of rare metals in the world, and the selection of certain regions for the concentration of significant amounts of economic elements, must mean that there exists in the portion of the Earth which is below the crust 'a highly individualized distribution of metals' and that this 'suggests a heterogeneous *underearth* which has remained stably heterogeneous during the whole of our geologic historical record'. A fundamental control of this kind is clearly suggested by the con-centration of diamond source-rocks in the older cratons. That the emplacement of these source rocks occurred at a number of periods in geological time is shown by the presence of detrital diamonds in the ancient sediments (older than 2000 million years) of the Ivory Coast and South Africa, the antiquity of the diamondiferous Premier kimberlite of South Africa (between 1100 and 1800 million years old), and the widespread emplacement of kimberlite pipes in South Africa about 100–250 million years ago. However, although diamondiferous source-rocks of various ages all lie within the older cratons, prolonged structural stability does not seem to be a prerequisite for their emplacement. Indeed, as the ancient diamondiferous sediments of the Ivory Coast were orogenically deformed some 2000 million years ago, it seems that the pattern of diamond concentration in that area was fixed even before it became fully cratonic. This suggests a mantle source for diamondiferous kimberlites; a view which is entirely consistent with the high pressure of formation of diamond.

In contrast, the present concentrations of gold and iron in sedimentary rock sequences in the older cratons apparently reflect concentration processes in the Earth's crust itself. Since it is probable that large segments of the older cratonic terrains have been affected by younger orogenesis, it seems likely that the relative paucity of *major* deposits of these ores in the younger orogens is due to the modifying and/or dissipating influences of younger tectono-thermal activity. However, it is also clear that the conditions (notably source-material and accumulation processes) required for the *widespread* concentration of these elements have not been favourable in the major part of the African crust during the past 1200 million years.

Lead, zinc, copper, niobium-tantalum, tungsten, tin and beryllium show their most important concentrations in the younger orogens; the older cratons show no comparable accumulations of these elements. It seems therefore that these elements either require repeated superimposed crustal orogenic activity for their concentration in major quantities in the crust, or reflect a chemically heterogeneous mantle which is only tapped under certain (more youthful?) orogenic conditions. That the mineralization in the younger orogens stems from one or other of these controls is illustrated by the repetition of certain types of mineralization at different periods of geological time (see Table 2). For example, tin mineralization

TABLE 2

METALLOGENIC PROVINCE	MINERALIZATION		EMPLACEMENT AGE
NORTH AFRICA	Pb*–Zn–Fe		Palaeozoic-Mesozoic-Tertiary
NIGERIA	Pb–Zn–Sn–Nb–Ta	Pb*–Zn	Younger than Middle Cretaceous
		Nb–(Ta)–Sn	Jurassic
		Ta–(Nb)	Lower Palaeozoic
NAMAQUA-LAND BELT	Sn–W–Be–Ta		*ca.* 1100 ± 200 m.y.
KIBARA BELT	Sn–W–Be–Ta–Nb		*ca.* 1100 ± 200 m.y.
DAMARA-KATANGA BELT (Central and South-West Africa and the Lower Congo)	Cu–Pb–Zn–Co		*ca.* 550 ± 100 m.y.

*Lead age data indicate a remobilisation of Upper Precambrian-Lower (and Middle) Palaeozoic leads (Pb) into much younger host rocks.

TABLE 3: AFRICAN PRODUCTION OF CERTAIN MINERALS

	Annual African production	African production as per cent of the World †
Asbestos	300 000 tons	20
Chromite	650 000 tons	44
Cobalt	11 000 tons	73
Copper	1 000 000 tons	27
Diamonds	26 000 000 carats	94
Gold	700 tons	69
Iron ore	9 000 000 tons	4
Lead	200 000 tons	3
Tin	19 000 tons	13
Tungsten	1 500 tons*	5
Zinc	260 000 tons	3

* 1957 production; presently the output is much less.
† Not including the Soviet countries.

occurred in the Kibaran belt 1000 to 1200 million years ago and in Nigeria some 200 million years ago. Lead mineralization took place 500 to 600 million years ago in southern Africa and at various times between 100 and 500 million years ago in North and West Africa; copper and niobium-tantalum mineralization show the same repetitive pattern.

The fundamental problem is to decide whether, as Spurr suggested, mineralization processes are related to properties of, and mechanisms within, the Earth's mantle or to mechanisms within the crust. In the case of diamonds there is overwhelming evidence that they originate in the mantle. Nevertheless, as we have seen, other elements appear to be concentrated in parts of the crust which have different structural characteristics and show repeated mineralization of the same type but of different ages. This implies that the crust itself has played an important role in the formation of certain mineral deposits. We need to know if the regions in which these deposits occur are themselves characterized by particular chemical traits. For instance, are there, in Africa, *geochemical* provinces whose disposition is related to the occurrence of mineral deposits, particularly in the younger orogens, and is the chemical heterogeneity of the crust a major controlling influence?

Even now we do not fully understand the fundamental origin of many mineral deposits.

However, we can with a fair degree of confidence state that economic concentrations of certain elements are most likely to occur in specific regions of the Earth's crust. This pattern applies not only to Africa but to other continental areas such as North America and Russia. Although, as expressed here, this correlation is mainly of academic interest, it is also of use to mining companies in their continuing search for new ore deposits. For, by using the principles outlined in the article, exploration geologists are able to concentrate their efforts in the most likely areas and thereby avoid considerable expense in the search of probably barren regions.

FURTHER READING:

W. LINDGREN 1933. Mineral Deposits. *McGraw-Hill, New York.*

B. MASON 1966. Principles of Geochemistry. *J. Wiley & Sons, New York.*

C. F. PARK AND R. A. MACDIARMID 1964. Ore Deposits. *Freeman & Co., San Francisco.*

J. E. SPURR 1923. The Ore Magmas, Vol. 2. *McGraw-Hill, New York.*

23.0 Vertical view of the San Andreas Fault, 8 km west of Taft, California. Note right-lateral displacement of river channels.

Dr. Kovach is a Professor of Geophysics at Stanford University in California. He is interested in earthquake prediction and modification, and the huge San Andreas fault runs conveniently (ominously?) close to his campus. Here Professor Kovach shows that earthquake prediction is by no means simple but that, almost by accident, techniques have been discovered which could well lead to processes whereby the strain in the Earth is released in small packages rather than in a huge, disastrous, natural earthquake. Such techniques may be of little interest to those of us who live in the aseismic areas such as western Europe and the Eastern United States, but those living in the crowded cities around the Pacific may one day be thankful for this seismic research.

23 Earthquake Prediction and Modification

by ROBERT L. KOVACH

Our understanding of earthquakes is still incomplete, but empirical evidence shows us that human activities can affect their incidence. This may help us to learn how to modify and, later, how to predict them.

BY THEIR SUDDENNESS of occurrence and absence of warning earthquakes can arouse fear in the multitude and stimulate the interest of scientists, philosophers, astrologers and religious fanatics. Earthquakes are among the most destructive of natural phenomena, and scarcely a year goes by without some portion of our globe being subjected to the effects of a major shock. Statistics reveal that earthquakes take an average yearly total of 14 000 lives and cause extensive property damage.

On 29 February 1960, the city of Agadir, Morocco, suffered major devastation from an earthquake and 12 000 people were killed—more than one-third of its inhabitants. 35 000 buildings were destroyed in the city of Tashkent, in Russia's Uzbek colony, from the earthquake which occurred on 26 April 1966. Six hundred people lost their lives in the well-known San Francisco earthquake of 1906 and the fires that followed. Damage to the city of San Francisco was severe. Because of the damaging earthquakes which have occurred in their countries, the United States, Japan and the Soviet Union have intensified their efforts in earthquake research. To-day only a few doubt that California, basking in the sunshine of prosperity, will again be subjected to a major earthquake, and this raises an interesting question: can earthquakes be predicted, or if not, what are the possibilities of minimizing the potential hazard?

Earthquakes cannot yet be predicted with complete certainty because we do not fully comprehend the mechanism that produces them. General zones or belts where earthquakes occur can be defined and viewed in the broader context of sea floor spreading and plate tectonics. For the most part, the major earthquake belts lie along island arcs and continental margins and represent the places where large rigid plates of the Earth's crust plunge downward. However, earthquakes are abrupt and sporadic in occurrence, and this seems to be a basic difficulty for prediction.

It may seem then somewhat surprising to discuss earthquake modification and control when we do not yet understand earthquakes with complete certainty. It is clear, however, that man has inadvertently caused earthquakes through his intervention in geologic processes. Increased earthquake activity has followed underground nuclear explosions in Nevada, the subsequent filling with water behind large dams and fluid injection into deep wells. Thus, even though we may not be able to predict specifically *individual* earthquakes, we may still be able to *alter* the seismic regime in zones where major earthquakes occur.

EARTHQUAKE MECHANISMS

Whatever the mechanism, an earthquake is thought to represent the termination of a series of events which have taken place in the Earth's interior. Scientists know that the Earth's crust is being distorted continuously by forces within the Earth. Take, for example, the San Andreas fault in California. The San Andreas is called a strike-slip fault. This means that there are two large land masses moving relative to each other; one is moving to the south-east, and the other to the north-west. (We might envisage that the San Andreas fault represents a major transform fault system associated with sea-floor spreading from the East Pacific Rise off the coast of South America,

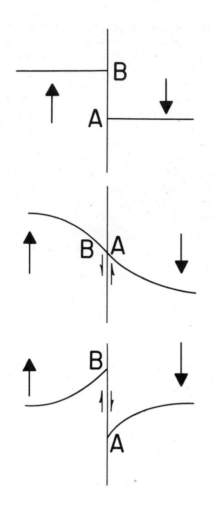

23. 1   Illustration of Reid's elastic rebound theory.

formulated by Harry F. Reid of Johns Hopkins University in 1910, some four years after the San Francisco earthquake.

The monitoring of movement along faults, such as the San Andreas fault, using geodetic surveying techniques is particularly illuminating in the context of the elastic rebound theory.   After the 1906 San Francisco shock a comparison of the relative position of survey markers across the fault prior to the shock and their positions shortly thereafter resulted in the pattern of displacements shown in the bottom of Figure 23. 1.

The curvature of the lines (grossly exaggerated) shows that points nearer the fault were displaced more than points farther away.   These measured displacements, of course, included any which occurred suddenly at the time of the earthquake together with any which may have accumulated slowly between the surveys.

Reid imagined that if any imaginary straight line were drawn from the fault at right angles to its surface trace after a major earthquake (in the presumably unstrained state) this line would become elastically distorted as shown in the middle of Figure 23. 1 until the strain is released by rupture.   Notice that if a horizontal reference line is drawn across the fault in the strain build-up state, then subtracting these displacements from the unstrained state gives the displacement pattern on the bottom of Figure 23. 1, as was actually observed.   We could also visualize the strain build-up state by superimposing a virtual left lateral dislocation on the unstrained state.   Adopting the elastic rebound theory as a working hypothesis allows us to examine geodetic data along major earthquake zones and study patterns of strain accumulation.

and that we have north-westward motion of a large rigid plate moving towards the Aleutian Islands where it descends into the Earth's mantle.)

The rocks near the fault can withstand this shearing force only so long, before snapping to release the accumulated strain. If an earthquake is caused by a sudden release of accumulated strain it seems reasonable to expect that one could monitor the strain accumulation by some appropriate measurements.   The postulation of a slow accumulation of strain until rupture occurs forms the basis of the *elastic rebound theory*.   This theory was

PHYSICAL CHANGES PRIOR
TO AN EARTHQUAKE

Observations of physical effects which have occurred prior to earthquakes have been reported, but to be a useful prediction tool these effects must precede large earthquakes regularly.   In general, these premonitory phenomena fall into the following categories: (1) strains and tilts in the epicentral zone; (2) variations in the level of background seismic activity; and (3) changes in the physical properties of the rock prior to the earthquake.

During the Matsushiro (Japan) earthquake swarm which began in August 1965, investigators from the Earthquake Research Institute in Tokyo observed that changes in the inclination of the ground strongly correlated with seismic activity, and anomalous changes in tilting took place before earthquakes of about magnitude 5 on the Richter scale. Anomalous Earth tilts prior to earthquakes have also been reported by Soviet investigators.

Changes in the pattern of occurrence of small earthquakes (microearthquakes) have sometimes preceded a major shock, but unfortunately the number of observations is limited and it is difficult to place a measure of reliability on the reported observations. In fact, all studies of premonitory phenomena require a large number of case histories and continuous monitoring and observation over a period of several years.

Changes in the Earth's magnetic field prior to earthquakes, believed to be responses to changes in the magnetic susceptibility and electrical conductivity of the underlying rocks, have also been noted. Both of these physical properties are known from laboratory experiments on rocks to be sensitive to variations in stress. Observations of the variations in the electrical resistivity of rocks (the resistance to an applied electric current) and the local geomagnetic field in earthquake belts are showing some promise and may prove to be extremely valuable field techniques for monitoring local stress changes.

MAN-MADE EARTHQUAKES

Now let us turn away from a discussion of premonitory phenomena and examine man's influence on earthquakes. Man-made earthquakes have occurred as a result of the accumulation of water behind large earth dams. After Lake Mead (Arizona) was filled behind the Hoover Dam in 1935, more than 600 local tremors occurred over the subsequent 10 year period. Since that time there have been other notable examples of earthquakes associated with dams, some with devastating effects. Major earthquakes were associated with the filling of a large artificial lake at Kremasta, Greece, and the filling of the Kariba Dam in Rhodesia and the Monteynard Dam in the French Alps.

One example of interest was the disastrous Koyna (India) earthquake of 10 December 1967. The Koyna Dam was constructed in a very non-seismic area, that is, no earthquakes were noted

prior to the impounding of the reservoir in 1962. After filling, a definite correlation was observed between the height of the water in the reservoir (which varied because of seasonal fluctuations), the duration for which the water was retained, and the seismic activity. The highest water level retained for the longest period of time occurred from August 1967 to December 1967, terminating in the major earthquake on 10 December 1967.

Before attempting an explanation of these man-made earthquakes, let us look at another example. A disposal well for waste fluids was drilled to a depth of 3 671 metres into Precambrian crystalline rocks in the vicinity of Denver, Colorado, for use by the U.S. Army Corps of Engineers. Injection of fluids began on a routine basis on 8 March 1962. A very striking relationship (Fig. 23. 2) was observed between the volume of fluid injected and numbers of earthquakes.

These earthquakes were extensively studied by members of the U.S. Geological Survey, and the strong correlation between fluid pressure at the bottom of the well and the amount of seismic activity from 1962 to 1966 is apparent from the graph; but the increase in seismic activity in 1967, after fluid injection was terminated, seems to rule out a simple correlation (Fig. 23. 3).

A physical explanation for the relation between fluid pressure and the triggering of seismic activity can be formulated as follows. In the Mohr-Coulomb empirical criterion for shear failure we have

$$\tau = \tau_0 + S_n \tan \phi \quad ,$$

where

$\tau = $ shear stress at failure

$S_n = $ normal stress across the fracture plane

$\tan \phi = $ coefficient of friction

$S_n \tan \phi = $ frictional term, resistance to fracture

$\tau_0 = $ cohesive strength.

Now in the presence of a pore fluid (fluid in the spaces between rock crystals or grains) at pressure p we have

$$\tau = \tau_0 + (S_n - p) \tan \phi \quad ,$$

that is, the result of an increase in pore pressure is to diminish the frictional resistance to fracturing by producing a *decrease* in the effective normal stress across the fault plane.

With this theory as a basis, a conceptual model was formulated by J. H. Healy and his co-workers

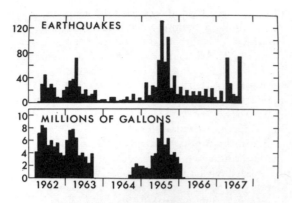

23. 2 Volume of fluid injected at Rocky Mountain disposal well versus numbers of earthquakes.

at the United States Geological Survey. Fractures pre-existed in the Precambrian basement rocks, and the lengths of these cracks are governed by the distance between stronger 'lock points'. The strength of these lock points is represented by the cohesive strength τ_o in the above failure criteria. When fluid is injected into the well there is a rapid increase in pressure near the well. The extension of existing cracks can then take place until the mean pore pressure over the fault surface is below the value for continued extension or until a stronger lock point is reached.

Now when injection ceases there is a rapid reduction of pressure near the well but a continued advance of the pressure front at a greater distance away from the well. This has the net effect of reactivating the larger cracks which had extended out beyond the pressure front during injection, but the shorter cracks near the well, in which the fluid pressure is decreasing, become inactive.

Increased earthquake activity has also been observed near the Rangely oil field in north-western Colorado. Oil is obtained at this site by a technique of secondary recovery which involves the injection of water under pressure. Earthquakes occur where the fluid injection process generates the largest increase in fluid pressure.

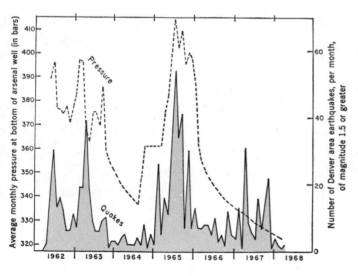

23. 3 Correlation between fluid pressure at the bottom of the disposal well and occurrence of earthquakes.

EARTHQUAKE MODIFICATION

The concept of fluid injection to lubricate fault planes can now be thought of in the light of the elastic rebound theory. Rather than letting strain accumulate to store energy to be subsequently released in a major destructive earthquake, it may be possible by fluid injection to reduce the frictional resistance to faulting and release the strain accumulation with a number of smaller shocks.

Some scientists have argued that, because of the logarithmic relation between the magnitude of an earthquake and the amount of seismic energy released, a huge number of shocks of smaller magnitude would have to be triggered to release the equivalent amount of energy generated by a destructive earthquake of larger magnitude. However, there is now some theoretical basis for believing that the amount of energy which is released at the focal point of an earthquake, in the form of radiated seismic wave energy, *decreases* with decreasing earthquake magnitude and actually approaches zero for fault creep. Therefore, the number of shocks required to release a given strain accumulation may be much smaller than one would expect on the basis of magnitude alone. The possibility that fluid injection could be used to reduce potentially dangerous crustal stresses by the initiation of non-seismic fault creep is tantalizing!

Underground nuclear explosions in Nevada have been extensively studied and there is ample evidence to indicate that natural tectonic energy, in addition to the explosion energy, has been released. We know this from an examination of the resulting seismic wave radiation patterns, the surface geological effects, the pattern of aftershocks and the permanent adjustments in the strain field, which are observed after the explosion.

The observation that nuclear explosions can change the strain field to distances of the order of tens of kilometres leads to the obvious suggestion that they might also be used to modify patterns of potentially dangerous strain accumulation. However, underground nuclear explosions in Nevada have not had a yield much larger than 1 megaton (equivalent to about a magnitude 6 earthquake) and one cannot overlook the possible hazards associated with higher yield explosions, particular-ly at the Amchitka Test Site in the Aleutian Islands where the level of natural seismic activity exceeds that of Nevada by at least an order of magnitude.

CONCLUSION

We see that the prediction of individual earthquakes is not yet possible, but we should seriously consider the possibilities of earthquake modification and control. Over the years much progress has been made in our ability to predict the weather, ocean tides and volcanic eruptions. A substantial reduction in hurricane deaths has been achieved through timely warnings given to coastal residents. Research towards earthquake prediction and control seems well worth the effort.

FURTHER READING:

L. C. PAKISER, J. P. EATON, J. H. HEALY AND C. B. RALEIGH 1969. Earthquake Prediction and Control. *Science*, *166*, 1467.

T. HAGIWARA AND T. RIKITAKE 1967. Japanese Program on Earthquake Prediction. *Science*, *157*, 761.

23. 4 The city of Chimbote after the Peruvian earthquake of May 1970, which directly and indirectly caused over 50 000 deaths.

Since 1939, several countries have undertaken research into the development of atomic weapons. In the 1950's it became very evident that if indiscriminate testing of these weapons continued it would, sooner or later, become a major menace to life on Earth. In the early 1960's a test ban treaty was signed by the major atomic powers, which covered tests in space, the atmosphere, and on the Earth's surface. Underground tests were excluded, primarily because of the difficulties of detection; some thought it impossible to tell a natural earthquake from one caused by a nuclear device. In the following article, Dr. D. Davies, formerly of the Department of Geodesy and Geophysics, Cambridge, and now at the Lincoln Laboratory, Massachusetts Institute of Technology, describes how man-made tremors can now be identified—with about 90% success—among the ever-present background of natural earthquakes.

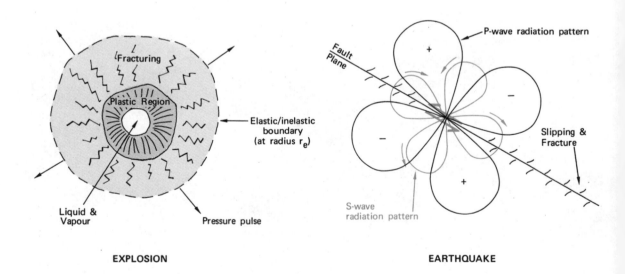

EXPLOSION

EARTHQUAKE

24. 1 Characteristics of explosions and earthquakes. Whereas the radiation pattern for an explosion has spherical symmetry, an earthquake releases shear (S-wave) and compressional (P-wave) energy in a nodal pattern. The P-wave pattern has alternate lobes of compression (+) and dilatation (−). The S-wave pattern has alternate lobes of reversed polarization.

24 Nuclear Explosions and Earthquakes

by D. DAVIES

The political need to distinguish underground nuclear explosions from natural earthquakes gave a fresh stimulus to research. This led to rapid developments in seismology, from which we have gained a clear idea of the extent to which a test ban treaty could be adequately monitored without territorial intrusion.

SEISMOLOGY HAS GROWN IN RECENT YEARS at a remarkable rate. In 1960 it was a small science, with friendly international cooperation between observatories with odd instrumentation attached to universities able to afford the luxury. In the early 1970's it is a big science with a large pool of talent, excellent unified instrumentation and still, fortunately, friendly international cooperation. This change came very largely through a political requirement—that it should be possible to identify underground nuclear explosions.

In 1958, an East-West conference of experts at Geneva looked into the problems of policing a ban on nuclear weapons tests. In 1963, agreement was reached between the Powers (but not including France and China) forbidding nuclear weapon tests in the atmosphere, in outer space and under water. *Underground* tests were not included, partly because of disagreement on the establishment of monitoring procedures. Moreover, the difficulties of discriminating between small earthquakes and tests of a few kilotons were seen to be formidable. The question arose: what characterizes these two types of disturbance?

TECHNICAL BACKGROUND

Let us consider the problem in three parts: the *source* of the disturbance, the *propagation* of the waves created, and the type of instrument required as a *receiver*.

Source

Since 1958 a large number of underground explosions have been fired. Table 1 gives some idea of the activity. As a result of this the pheno-menology of buried explosions is rather well understood from yields of a kiloton to yields greater than a megaton. Figure 24.1 shows the main characteristics of an explosion, and from the seismologist's point of view the elastic radius r_e of an explosion is important, as to all intents and purposes an outside observer can regard an explosion as a pressure pulse on a sphere of this radius. To a rough approximation

$$r_e = 140 Y^{1/3} \text{ metres,}$$

where Y = yield in kilotons; thus a kiloton and a megaton explosion have elastic radii of 140 and 1400 m. Obviously an explosion will not be buried very much deeper than is deemed safe from the possibility of eruption into the atmosphere. This means that the depth of burial is between 0.1 and 2.0 km, depending on yield. The sign of the pressure pulse is everywhere positive on this notional sphere, so all observers at a distance should record the first motion from a nuclear blast as outwards. Since the explosion is buried, elastic energy transmitted upwards will be totally reflected at the Earth's surface, and so the first signal from an explosion will be followed within less than one second by a second pulse.

An earthquake is a much more complicated affair, and we still have few acceptable generalizations. No earthquake has ever been reported from a depth greater than 700 km, and the great majority have a depth of focus less than 100 km. On very rare occasions an earthquake breaks surface in a region where there has been adequate surveying beforehand which allows a more detailed analysis than seismological observations at far-flung stations permit.

TABLE I

REPORTED NUCLEAR TEST EXPLOSIONS, 1960–1968*

YEAR	USA	USSR	UK	FRANCE	CHINA
1960	—	—	—	0 + 3	—
1961	9	0 + 32	—	1	—
1962	50 + 39	1 + 41	2	2	—
1963	26	—	—	2	—
1964	29	6	1	—	0 + 1
1965	27	9	1	2	0 + 1
1966	40	12	—	0 + 5	0 + 3
1967	28	13	—	0 + 3	0 + 2
1968	37	9	—	0 + 5	0 + 1

*These are the explosions announced by some agency. Undoubtedly many more have occurred but have not been disclosed. Where two numbers are given, the first is the number of underground explosions, the second the number of atmospheric explosions. Otherwise, the figures are for underground explosions.

The major feature associated with earthquakes is a fault plane across which the adjacent material slips. The length of the fault along which dislocation occurs may vary from a few tens of metres to several hundred km, its width is commonly less than 10 km, and the orientation of the fault plane can be anywhere between the vertical and the horizontal, depending on the local stress distribution, as movement along the fault plane is always such as to reduce the stored strain energy in the medium. Recent studies of activity along the San Andreas fault (where the fault plane is vertical and the strain energy released is that of 'stuck' portions of the Pacific and North American plates as they move past each other) show that slip movement never appears to exceed 5 metres, and is often as small as a centimetre. The drop in stress is less than 100 bars; stress drops as small as 0.5 bars have been reported. It appears that the velocity of rupture is 2 to 4 km s^{-1}, so an earthquake on a long fault will last for many seconds.

An earthquake generates shear and compressional kinetic energy, both of which are radiated away from the source as elastic waves. The source mechanism—orientation of the fault plane and direction of slip—will reveal itself in a nodal pattern of radiation of this energy (Fig. 24. 1 and Fig. 24. 2).

What distinguishing features can we then hope to observe at a distance? Clearly an event under water, or at a depth greater than 5 km, or one with a nodal radiation pattern is an earthquake. The presence of any seismic phenomenon which is related to a large source region or a long duration of the excitation will indicate an earthquake. The release of a substantial amount of shear energy will suggest an earthquake. Unfortunately the converse of none of the above is a fool-proof indication that an event is an explosion. Evidence of a small, short-lasting, shallow event with compression rather than dilatation as the first motion in all observable directions is not in itself sufficient to point unequivocally to an explosion. More sophisticated discriminants are needed, particularly at low yields (less than 20 kT). Almost all research is concentrated in this low yield range at the moment.

Propagation

Elastic energy from a seismic event radiates outward by four distinct types of waves. Two of these (body waves) exist in all solids displaying elastic behaviour, two (surface waves) exist only by coupling with the surface of the Earth. The wave types are:

24.2 An impression of the generation of seismic waves, and a definition of some seismological terms. The period range for the best observation of body waves is 0.5 s to 2.0 s, and for surface waves is 10 s to 60 s.

(i) P-waves (body). The small displacements in these waves are along the direction of propagation of the wave.

(ii) S-waves (body). The displacements are in a plane perpendicular to the direction of propagation of the wave.

(iii) Rayleigh waves (surface). Near the surface the ground exhibits a rolling motion, not unlike that of gravity waves in deep water but with a retrograde elliptical particle orbit. The amplitude of this disturbance drops off rapidly with depth by a factor of, typically, ten in 50 km.

(iv) Love waves (surface). They display the same general amplitude features as Rayleigh waves, but the particle motion is horizontal and perpendicular to the direction of propagation. They are similar to guided electro-magnetic waves, energy being constrained by total internal reflection to remain in the upper layers of the Earth.

Figure 24.2 shows schematically these wave types and defines certain seismological terms.

In theory, explosions should only generate P-waves and Rayleigh waves. In practice these wave types dominate, but not to the total exclusion of S-waves and Love waves.

Surface waves radiate from the epicentre with a velocity of about 4 km s^{-1}, and theoretically decay in amplitude in inverse proportion to the distance from the epicentre. They also spread out from being a sharp phenomenon (with a time span about that of the event itself near the epicentre) to become a long *dispersed* wave train in which, because of layering within the Earth, the long period waves travel faster than the short period ones. Such an observed Rayleigh wave is shown in Fig. 24.3. Clearly without a very detailed knowledge of the path over which the wave has travelled it would be impossible to reconstruct the exact character of the source from this wave train. At present however a rather empirical measure of the surface wave generating strength of seismic source—the surface wave magnitude M_s—is measured routinely. If at a distance Δ (in degrees), the amplitude in micrometres of the surface wave displacement at a period of 20 seconds is A_{20}, then

$$M_s = \log_{10} A_{20} + \tfrac{5}{3} \log_{10} \Delta + 2.0 .$$

The last term is present only to align this magnitude scale with other magnitude measurements. The middle term shows that the observed amplitude varies inversely as the cube root of the fifth power of the distance, as opposed to the

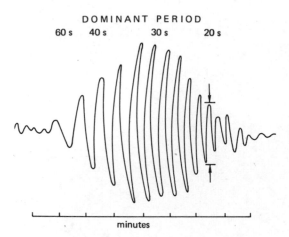

DOMINANT PERIOD

60 s   40 s   30 s   20 s

minutes

24.3 A Rayleigh wave from an earthquake in Colombia recorded in Montana. The amplitude A_{20} used for determining the magnitude is half the displacement between the two arrows—in this case 5 micrometres, corresponding to an M_s of 4.9. Note the dispersion leading to a steady change in the dominant period.

theoretical inverse first power. This arises through non-elastic absorption behaviour of the Earth's crust and upper mantle.

Recently efforts have been made to broaden the description of the source by quoting values for surface wave excitation at a range of periods·from 15 to 60 s, or ideally the spectrum of surface wave excitation.

P-waves from a seismic source travel deep into the mantle before emerging at *teleseismic* distances. They are usually recorded in the band around one second simply because the signal to background noise ratio is larger in this region than in others. Figure 24.4 shows the fundamentals of P-wave propagation. For the first ten degrees or so the waves first to arrive at a receiver have travelled in the Earth's crust or the topmost part of the mantle, where the P-wave velocity is 8.1 km s^{-1}. Just below the top of the mantle the P-wave velocity

generally diminishes owing to thermal effects (possibly slight partial melting). This decrease is not observed in some continental regions. Below this low velocity layer, about 100 km thick, in which the velocity may have fallen to 7.7 km s^{-1}, the P-wave velocity climbs steadily until the core is reached at a depth of 2900 km. The presence of the low velocity layer leads to variable and unpredictable signals in the range $10° < \Delta < 25°$, but beyond that distance, waves that have traversed the deeper mantle and been refracted up are clearly recorded as the first arriving energy. These rays obey a modified form of Snell's Law applicable to a spherical symmetric Earth:

$$\frac{v_1}{r_1 \sin i_1} = \frac{v_2}{r_2 \sin i_2}$$

which predicts the inclination of the ray to the vertical (i_2) at a radial distance r_2 where the velocity is v_2 provided the values v_1 and i_1 are known at another radial distance r_1 (for instance at the Earth's surface). As Figure 24.4 shows, a relatively narrow cone at the source can be seen teleseismically. This, amongst other things, makes discrimination based on a first motion criterion difficult to apply because of the restricted coverage.

The signal amplitude changes predictably and relatively little across the teleseismic zone from 25° to 100°. Such a signal is shown in Figure 24.5. It is very simple in comparison with a signal less than 25° away, and this zone is often called the 'teleseismic window', as it records the signal clearly and unaffected by diffraction, refraction and reflection which affect signals recorded at shorter distances. As we shall see, the effort in body-wave discrimination is now very largely concentrated in teleseismic methods. As with surface waves, a magnitude of excitation is defined for P-waves as

$$m_b = \log_{10}\frac{A}{T} + B(\Delta).$$

A and T (in micrometres and seconds) are defined by Figure 24.5. $B(\Delta)$ is an empirical term to counterbalance the decay of the signal with distance. Thus both m_b and M_s measurements should be independent of distance from the earthquake or explosion. m_b or body wave magnitude is often described as the *Richter Scale*, and ideally the magnitude of any seismic event is obtained as the average of as many individual station determinations as possible.

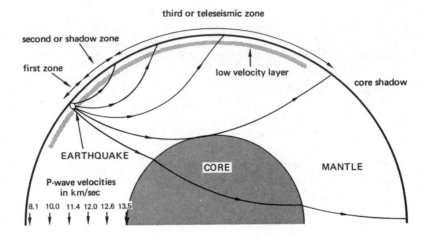

24. 4   The interior of the Earth, and some typical seismic rays.

third or teleseismic zone

second or shadow zone

first zone

low velocity layer

core shadow

EARTHQUAKE

P-wave velocities in km/sec

8.1 10.0 11.4 12.0 12.6 13.5

CORE

MANTLE

Receiver

The vast majority of seismometers in use to-day (and there are at least 600 stations in operation around the globe) work on the principle of continuous recording of the position of a spring-suspended mass relative to a frame which is firmly attached to the ground. Almost invariably this is done by attaching a coil to the mass and a magnet to the case or *vice versa*. The current induced by ground motions, in which the coil moves through the field of the magnet, is passed directly through a galvanometer which records either on moving photographic paper or on magnetic tape. By appropriate choice of spring constants, masses, galvanometer natural period and damping it is possible to make band pass instruments. Both short period and long period instruments are commonly deployed in threes—to measure vertical, N-S and E-W components of ground motion respectively.

What prevents us from using one broad band instrument to cover the whole seismic spectrum from 0.1 to 100 s or beyond? Such instruments exist, but an issue we always have to face is that of noise. The surface of the Earth is in continuous motion. Nearby disturbances (wind, cars, etc.) feed energy into the ground in the period range 0.1 to 1.0 s. At a typical seismographic station this noise will be on average 2 to 10×10^{-9} metres. Distant disturbances such as storms in ocean basins generate so-called microseisms, Rayleigh waves of period 4 to 8 s, with amplitudes of anything from 0.1 to 10×10^{-6} metres. It is desirable to run

instruments at magnifications at which the noise is visible but does not hinder visual observations. For this reason it is necessary to make instruments relatively unresponsive to the 4 to 8 s band. The record shown in Figure 24. 5 is a good example of an instrument running at the correct magnification.

Until 1958 instrumental developments had been haphazard, and it was clear that a more unified approach was needed. In the spirit of the report of the Geneva Conference, the United States designed and deployed over 100 so-called World Wide Standard Seismographic Network (WWSSN) stations. Each had identical instrumentation and recording facilities, 3 short period and 3 long period instruments, and the data were stored on microfilm and available to any who paid 30 c. per station per day. The benefit to seismology from this network has been inestimable. From the discrimination point of view it has answered most of the questions posed.

The U.K. started developing another approach —noise suppression—at the same time, and now has installed five high quality arrays in Scotland, Canada, India, Australia and Brazil. The principle behind the array stations is that signal from a teleseismic P-wave over an aperture of about 20 km should be coherent, whereas noise should be incoherent. Thus if N short-period seismometers are placed on the ground and the signal recorded appropriately, a theoretical suppression of $\sqrt{N}$ in noise is possible.

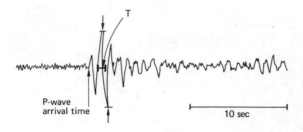

24. 5 The P-wave record of an explosion in the Aleutians recorded in Japan. The period T and twice the amplitude A are defined. The corresponding ground displacement is about 2×10^{-7} m.

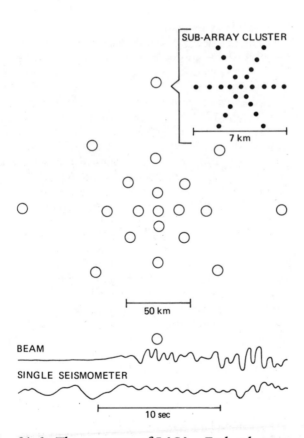

24. 6 The geometry of LASA. Each sub-array contains 25 seismometers. Also shown is the improvement in signal to noise ratio when 525 seismometers are beamed at the event.

This is achieved in practice by the following technique. Twenty or so seismometers are sited along the two arms of an L, the separation of each seismometer from its neighbour being about 2 km. All record on one central magnetic tape unit. If an event occurs 60° away along the direction of one of the arms of the L, the P-waves from it will approach the station with an inclination to the vertical of about 25° and will hit successive seismometers down the arm at intervals of about 0.15 s. The wave front will reach all seismometers in the other arm simultaneously. Thus, by appropriate shifting of time origins on the magnetic tape for different seismometers before adding the signals, it is possible to 'beam' the array to look at the appropriate azimuth and distance. It is a simple matter to compute the time shifts for any distance and azimuth.

In 1964–5 the United States built a Large Aperture Seismic Array (LASA) in Montana with an aperture of 200 km and 525 seismometers. Data from all sensors (63 long-period instruments are also in the array) are stored on digital tape, and there is a certain amount of automatic detection and location of events by continuous beamforming and searching for signal power with a large digital computer.

DISCRIMINATION

We have dwelt at length on the tools and concepts of seismology, and can now deal fairly rapidly with the problem of discriminating between explosions and earthquakes. The process of discrimination is not a cut and dried one in which one technique is greatly superior to all others. Rather does one go through a chain of elimination. We shall discuss the elements of the chain briefly.

Location and Depth of Focus

By the use of a global network of seismometers it is generally possible to locate the epicentre of an event to an accuracy of about 20 km unless the magnitude m_b is less than 4. 5. At this magnitude the teleseismic amplitude of ground motion from the event is about 10^{-8} metres, and so the signal to noise ratio of many stations is approaching a value at which it is impossible to say exactly when the P-wave starts. At $m_b = 4.0$ the probability is above 50% that an event will not be detected teleseismically by four conventional stations (the smallest number needed for location) but the

TABLE II

Some Approximate Figures for Earthquakes and Explosions

Body wave magnitude m_b	No. of Earthquakes yearly with magnitude greater than m_b	Yield of nuclear explosion fired in granite (in kt)	P-wave ground displacement at 60° from event (in micrometres)
4	5 000–10 000	1–3	0.002
5	500–1 000	10–30	0.02
6	50–100	100–200	0.2
7	5–10	600–2 000	2.0

probability is around 80% that LASA (and NORSAR, in Norway—the second large aperture array) will detect the event and be capable of giving a location accurate to five hundred kilometres.

The location process will clearly rule out the vast majority of the 5 000 to 10 000 events reported annually, and the next process is to find if the event is deep. The travel times from the focus to a network of stations will be depth-dependent, and it is possible to estimate the depth by finding the best-fitting travel times. This however is a somewhat insensitive technique and is unsatisfactory for events between depths of 0 and 50 km. A second method is more powerful when it can be applied. The P-wave that travels to the Earth's surface near an event, is reflected back with reversed polarity, and then travels along almost the same path as the initial P-wave is called 'pP', and the time interval between the arrival of P and pP is a powerful indicator of depth, providing pP can be seen. Figure 24. 7 shows such a case, but frequently the initial P-wave masks the pP-wave. If the event is shallower than 5 km (i.e. a possible explosion) pP will certainly not be seen as a separate phase. Such an example is also given.

The presence of a pP phase therefore marks a shallow focus disturbance as being of natural origin.

First Motion

The Geneva Conference saw great importance in the identification of the direction of first motion. There is now little support for its use, because at the low magnitudes (below 5.0) where discrimination is now an issue the signal to noise ratio may be insufficient to permit clear identification of the first half cycle of the signal as compressional. This is also shown in Figure 24. 7. Further, many earthquakes at this magnitude level may be recorded at only a few stations, all of which happen to be in compressional lobes of the radiation pattern.

Complexity

We expect explosions to register short signals, earthquakes more drawn out or complex ones. This is indeed true in a large majority of cases, as Figure 24. 7 also shows, but some unfortunate counter-examples occur. Small earthquakes are occasionally so compact in time and space that they look like explosions and, for reasons which we are beginning to appreciate now, explosions from certain areas do not have a dominant pulse to begin with. Complexity of signal has proved its value in a restricted number of cases and it may yet prove useful more widely.

Spectra of Signals

The advent of good digital recordings of seismic records allows us to perform routine operations on signals such as spectral analysis, i.e. the resolution of a wave train into the different oscillations of distinct frequency of which it is the sum total.

The sharpness of the source of an underground explosion leads to distinctly enhanced excitation of high frequency seismic energy. If a ratio is formed of the energy in a band from 1.4 to 1.8 Hz compared with that in the band from 0.5 to 0.9 Hz, explosions down to a magnitude of at least 5.0 clearly have more energy in the high frequency

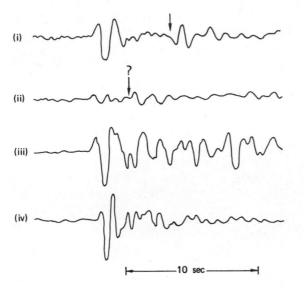

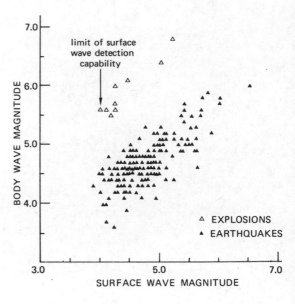

24. 7 Some typical seismograms showing P waves from

(i) an earthquake 15 km deep (pP shown by arrow);

(ii) a very shallow earthquake (pP difficult to find, queried);

(iii) a complex earthquake;

(iv) an underground explosion. Note higher frequency energy.

24. 8   Discrimination between explosions and earthquakes by the surface wave/body magnitude criterion. A study performed by the U.K. Atomic Energy Authority Seismic Group.

band. Figure 24. 7 shows this. The difficulties with discrimination at low magnitudes are compounded by lack of events for comparison, in addition to low signal to noise ratios.

The most powerful discriminant at present is the relative excitation of P- and Rayleigh waves expressed in terms of the m_b and M_s values for each event. Rayleigh waves are relatively poorly excited by explosions, as shown in Figure 24. 8. This diagram, based on a 1968 study of WWSSN data for one year, showed that the deployed capability of seismic instruments had a discrimination threshold at a magnitude M_s of 4 (below that value Rayleigh waves from explosions are frequently not detected by the WWSSN, although body waves are). Recent results from arrays and high quality single stations indicate that it is now possible to push this threshold down

to at least an M_s of 3.2. An M_s of 3.2 corresponds to an m_b for an explosion of about 5.0 and a yield of about 20 kT in hard rock. There are clear indications based on regional studies in the Western United States that these spectral excitation differences persist down to much lower levels, perhaps a few kilotons, but monitoring a test ban in a global framework (it would have to be assumed for instance that, as at present, the Soviet Union would not supply seismograms from its network for discrimination purposes) probably constrains us to an ultimate threshold for identification of 5–10 kT.   An investment to go to a lower threshold would be several hundred million dollars, and would be largely unproductive, as a potential violator would then use evasion techniques which for low yields can diminish his risk of detection very substantially, as we shall see.

Other Discriminants

We have not mentioned S-waves, Love waves, reflections from the core and a whole host of seismic signals that make up the total seismogram. A rich field still remains to be studied here. One of the attractions of the discrimination problem to the academic geophysicist is that it is a very diverse field requiring a wide understanding of the physics of the Earth's interior. In dealing with small signals, information has to be culled from all possible sources, and it is desirable to have as good an understanding, say, of plate tectonics as of explosion phenomenology.

Snags

We are fairly sure than an explosion in the range 1 to 3 kT in granite gives a signal recorded as magnitude 4, and likewise 10 to 30 kT is equivalent to a magnitude 5 event. But the same explosion fired in a softer, more porous material appears to have a lower magnitude. For instance, the volcanic tuffs in Nevada reduce the magnitude by 0.3 to 0.5 units, whilst dry alluvium (if it can be found) can lower the magnitude a whole unit. The less compacted the material, the more energy is mopped up in compacting it and the less is transmitted to the outside world as compressional energy. Surrounding an explosion with sponge-cake undoubtedly would help a determined violator even more, and it turns out that the most effective 'decoupling' is obtained simply by firing the explosion in a large hole. A 50 kT test could be conducted and give no more seismic disturbance than $\frac{1}{4}$ kT in granite. The economics however are sobering. Such a cavity would need to be 100 metres in radius—20 million tons of rocks would have to be removed! It is also possible to make identification much more difficult by firing explosions in the wake of very large earthquakes or by firing them in a ripple pattern to simulate an earthquake. On the other hand we may mention a small number of earthquakes which fail the $M_s : m_b$ discriminant. These are usually events of less than m_b 5.0, and come from fairly well defined regions at the rate of one or two a year. It is important that a total evaluation include a consideration of evasion and problem events.

EARTHQUAKES TRIGGERED BY EXPLOSIONS

Aftershocks are a phenomenon commonly associated with large shallow earthquakes. The readjustment of rocks in the vicinity of the fault is unlikely completely to release stored elastic energy with the first catastrophic break. However, once a new rupture has been opened, it is likely that successive earthquakes will complete the process. Generally, the first shock has to be of magnitude 6 or more for aftershocks to occur. The magnitude of the largest aftershock is at least one unit less than the main shock, and the activity dies away within weeks.

Rather similar rules seem to apply after very large underground nuclear explosions. About five tests with magnitudes greater than 6 have led to some activity, although no earthquake clearly associated with a test has yet been recorded further than 40 km from the test site in Nevada. A megaton test (Milrow) on the Amchitka test site in the Aleutians also triggered some relatively small activity. The energy released was probably local and stored by the explosion itself rather than tectonic on the large scale. More recently Cannikin (5 megatons) caused essentially no post-shot seismicity.

FURTHER READING:

C. F. RICHTER 1958. Elementary Seismology. *W.H. Freeman & Co.*

D. DAVIES (ed.) 1968. Seismic Methods for Monitoring Underground Explosions. *Gerald Duckworth & Co.*

U. K. ATOMIC ENERGY AUTHORITY 1965. The Detection and Recognition of Underground Explosions.

The next article in this book is quite different from those which have preceded it. The object of all the others was to describe and discuss the scientific results which are, after all, the end product of the scientific endeavour and the whole point of carrying out research. They tell us what results have been obtained; they discuss the consequences of the various data; in some cases they explain how the data were obtained; and a few even discuss how science affects Man in a practical way rather than by generally enriching his cultural heritage.

But, by the same token, they also omit a great deal. They tell us nothing, for example, about the organization of science, for that was not their point. But how is science organized? How does a scientist go about getting the financial support he requires? What problems does he encounter before he can even perform his first experiment? And who, for that matter, decides which people to support in the first place?

Nowadays the bulk of the money for research comes, either directly or indirectly, from Government. But money is not unlimited; and from all the potential projects which are intrinsically worthwhile, a selection must be made. The choice is difficult. The worth of a project in pure, as opposed to applied, science is not always obvious before the project is complete. Experiments which set out to prove one thing, frequently end up proving something quite different; many of the really significant advances in knowledge were unexpected.

In a difficult situation, Governments have chosen, as one might have expected, to give preferential support to the better scientists and, in so far as it is possible to judge in advance, the better ideas. By and large it is a policy which has paid off; but good scientists and good ideas do not guarantee success. These days scientific projects are sometimes so large and complex that purely scientific ability is no longer sufficient.

Such was the case with *Project Mohole*, the United States' plan to drill a hole through the Earth's crust to the mantle. Eminent scientists backed it; and from the scientific point of view the idea was eminently worthwhile. But, because the planning and organizational bases were not well-founded, Mohole floundered long before it was complete, and became perhaps the greatest fiasco in the history of science.

Daniel S. Greenberg, who is on the staff of the American journal *Science*, is well known for his investigations into the darker corners of United States science policy and has made a special study of the Mohole disaster. Here he recounts the history of the project—what it was, what its special problems were, and why, in the end, it was such a spectacular failure.

25 Mohole: Geopolitical Fiasco

by DANIEL S. GREENBERG

The project for drilling through the Earth's crust was one that required funds on a scale which only Government could provide. Scientists, administrators and politicians had to plan in concert—and on this occasion they failed signally.

IN 1957, A SMALL GROUP of scientists met in Washington, D.C., to evaluate applications for government support of research in the Earth Sciences. It was a routine sort of meeting, commonplace in the management of science in the U.S., Britain, and the many other countries where government organizations regularly call in scientists, usually from universities, to judge the worth of proposed research projects. The outcome of this particular meeting, however, was to be far from routine.

As the scientists proceeded with their review of the applications, they agreed that—as one of them later wrote—'none of these proposals was really fundamental to an understanding of the Earth'. Then one of them, Walter Munk, a distinguished oceanographer, suggested a project that would meet this criticism: the drilling of a hole through the rocky crust of the Earth, down to the next and so far unreachable layer, known as the mantle. Samples of the mantle could help answer many critical questions about the origin and history of the Earth, continental drift, heat flow, and mineral formation. And thus was born what was soon christened *Project Mohole*, derived from the name of a Yugoslav geologist, Andrija Mohorovičić, who had pioneered the use of seismic transmissions to detect the progression from crust to mantle.

Viewed as an exercise in science and technology, Mohole was an incredibly ambitious venture, for the crust, covering the Earth like a hard, bumpy skin, is anywhere from 4500 to 50000 metres thick—thinnest where the ocean is very deep, thickest on dry land—and existing drilling technology was nowhere near the capability of boring to such prodigious depths, either on land or from a vessel in deep water. Thus the technical aspects were indeed extraordinary. But as Mohole progressed from merely being a startling idea toward becoming an operational scientific research programme, another aspect of it became even more extraordinary, namely its politics. Eventually it was to become perhaps the greatest *cause célèbre* in the relationship between basic science and government in the United States—that is before it became so enveloped in rancour and controversy that the U.S. Congress took the unprecedented step of decreeing that Mohole was to be wholly terminated.

Mohole is not a pleasant story; in fact, American scientists still recoil at the mention of it, in large part, it may be, because the Mohole episode marked the beginning of the end of the post-war honeymoon between American science and government. But it is in many respects a uniquely instructive episode, simply because the battles that arose from it became so intense at times that they spilled from the closed committee rooms where science policy is normally formulated, out into public view. Mohole revealed little about the inner composition of the Earth, but it did reveal a great deal about the inner workings of science and government in the United States.

Before examining those workings in the light cast upon them by Mohole, it is necessary to consider, very briefly, some basic and distinctive features of the American science and government relationship. First of all, in contrast to European experience, that relationship is predominantly a post-World War II phenomenon. Prior to the war, the U.S. government showed little interest in supporting basic research. In contrast to the situation that prevails to-day, Western Europe, with its well-established traditions of government patronage for research, led the world in most fields of science. The emergence of radar and atomic weapons from wartime research persuaded the U.S. government that it could no longer afford to ignore the relationship between science and national security. As a result, with a few fits and starts, the post-war years saw the U.S. government becoming increasingly involved in the financing and administration of research, both in government research establishments and in independent universities, the latter being the principal locale for basic research in the U.S. Since national security was the prime motive for government's post-war entry into the support of research, military agencies figured large in these developments, as they do to this day. But the post-war creation most dear to the hearts of American scientists was a U.S. government agency, the National Science Foundation (NSF), established by Act of Congress in 1950—after five years of political battles—for the sole purpose of providing money for university scientists. Having felt neglected for many years, American scientists responded with gusto to the opportunity for generous financing of research. When government was laggard, as it was at times, some scientists were quick to warn that if the U.S. did not make rapid progress in this or that field—well, the Russians might.

In 1957, when the Mohole idea was conceived, NSF was a mere seven years old. It was growing rapidly, but its annual budget, all of it from government, had to stretch over all the scientific disciplines. Most of its funds were in small packages, and, it is critically important to note, NSF had accumulated virtually no experience in the management of big engineering programmes. Mohole, which was clearly destined to be a multi-million dollar venture, was, at the time, the biggest proposal ever formulated in the Earth sciences. It would be a vast engineering undertaking, involving construction of drilling equipment based on radically new concepts, as well as many years of uninterrupted drilling. From its director, the late Alan T. Waterman—a physicist who had been a research administrator since the beginning of World War II—down, NSF was a deskbound organization, staffed by academic scientists, assisted by others who periodically served as part-time consultants (like the group, mentioned at the beginning, that conceived the Mohole plan).

NSF did no research; it simply gave money, on the basis of expert opinion, to working scientists seeking government finance for their research. NSF was no more qualified to supervise the drilling of a hole through the Earth's crust than it was to operate an international airline. With that bit of background in mind, let us return to the scientists who were the parents of Mohole.

Having NSF's bankroll in mind, they realized that some sort of organization—'institutional base' is the formal term—would be needed to apply formally for a research grant and to administer the project. It occurred to them that there already existed an organization that might fill this role. It was a whimsical organization, really nothing but a very 'in' joke among geophysicists: the American Miscellaneous Society. AMSOC, as it was known, had existed for a number of years solely as a harmless caricature of conventional scientific organizations and their ponderous ways. It did nothing, held no meetings, had no membership rolls, bye-laws, officers, or publications. A get-together for drinks at a club in Washington now and then was the sole manifestation of its existence. But many prominent geophysicists had joined in the joke and proudly asserted their association with AMSOC. Let AMSOC at last do something useful, was the decision of Mohole's progenitors. And so, a number of AMSOC 'members,' meeting at a 'wine breakfast' in California, constituted themselves into a special group—the 'drilling committee'—and wrote out an application asking NSF for $30 500 to finance detailed planning of the project.

Under the conservative hand of Director Waterman, who had to account for his activities to an often critical Congress, NSF was not disposed to turn over 30 000 of its scarce dollars to such an organizational peculiarity. It advised the group to get itself moored to an established organization before seeking government funds; so AMSOC got itself attached to one of the oldest, stablest and organizationally impeccable institutions in the United States—the National Academy of Sciences. Established by Act of Congress in the time of President Lincoln to honour scientific merit and to advise the government on technical matters, the Academy was a drowsy, ineffectual organization —unknown to even most of the government that it was supposed to advise—but it was there in Washington, with a handsome headquarters building, a staff of hundreds, and a great deal of prestige, at least within the scientific community. It was actually unprecedented for the tradition-bound Academy to take an outside organization into its prestigious fold; but several AMSOC members were influential members of the Academy, and they easily prevailed upon the Academy's officers to make an exception. Nor was it difficult, once AMSOC was within the Academy, for it to get a grant from the National Science Foundation. Reports began to circulate —unfounded as it turned out—that the Soviets were planning to drill a hole through the Earth's crust. In these circumstances, the project acquired considerable momentum, plus $15 000 in NSF money as a down payment. But what was the aim of the project?

The question may appear naïve, but eventually Mohole was to fall apart administratively and politically on the very issue of what precisely it was intended to accomplish. So, let us carefully examine what AMSOC said it wanted to achieve when it applied to the U.S. government for research funds.

In its formal application to NSF, AMSOC solicited funds 'for support of the study of the feasibility of drilling *a hole to the Moho discontinuity*'. Reference was later made in the application to various geophysical studies that could be made in '*the hole*'. It was stated that sedimentary materials might be sampled at the upper levels, but on the face of it, it appeared that AMSOC's ambitions were directed toward a single, crust-piercing hole. Hindsight shows the absurdity of suggesting the construction of a multi-million dollar apparatus simply to drill one hole; not un-like constructing an immensely powerful micro-scope simply to make a single observation. But the creators of Mohole, eager for a project that would have dramatic as well as scientific appeal— Earth scientists were feeling a bit envious of the money and glamour then attaching to the space programme—formulated their proposal with an eye toward public relations impact. And NSF, with its tight budget and small, inexperienced staff, took up the proposal with the understanding that Mohole was a one-shot affair. The ultimate cost could not be reckoned until detailed studies had been completed. But in 1958, two key figures in AMSOC published an article in which the cost was estimated at $5 million, spread over several years, and it may be assumed that NSF was thinking of a figure in that vicinity when it de-

cided to commit its money and reputation to the support of Mohole.

By 1965, estimates had risen to at least $125 million. But, long before that, a remarkable figure came into the affairs of Mohole, a young oceanographic engineer named Willard Bascom. One of the few live wires on the staff of the venerable Academy, Bascom naturally gravitated to this exciting project, and, in short order, the university-based scientists who made up AMSOC, and who came to Washington only now and then to look into it among other matters, asked Bascom to become the first full-time employee of Mohole. Bascom enthusiastically accepted, be-coming executive secretary of the project, with instructions to turn the idea into a reality. As plans for Mohole crystallized in Bascom's mind, he published an article in which he rejected the one-shot concept. 'The ultimate objective is to reach the mantle,' he wrote, but 'an intermediate step is likely to yield equally valuable and interest-ing results. . . . No one site or hole,' concluded Bascom, 'will satisfy the requirements of the Mohole project.' Noting reports that the Russians had set up a committee to look into very deep drilling, Bascom speculated, 'Perhaps there will be a race to the mantle.' Bascom then pro-ceeded to propose a technical strategy for the project, one that involved a seaborne approach to the mantle. It would be difficult, of course, to drill through thousands of metres of rock from a vessel riding on thousands of metres of water. But the Earth's crust was thinnest under the deep sea, and Mohole's chief planner felt certain that when all the difficulties were calculated, it would be faster, cheaper and easier to drill a shorter dis-tance in deep water than a far greater distance on dry land. But since deep drilling at sea was then in its infancy, Bascom suggested that the necessary experience be acquired by starting out with a relatively modest waterborne project. For this purpose, he proposed that AMSOC hire an oil-drilling barge, modify and re-equip it for Mohole's specialized needs, and undertake a preliminary drilling exercise, preparatory to designing and building a special vessel for going all the way to the mantle. Meanwhile, NSF provided $80 700 to finance studies of promising sites for drilling. This was shortly followed by another $80 500 to pay the salaries of the growing AMSOC staff.

The time was now June 1959, and planning was sufficiently advanced for the AMSOC executive committee to give NSF some idea of the costs that

25.1 CUSS 1, the drilling ship used in Phase I of Project Mohole. 18-m lengths of drill pipe can be seen aft of the drilling tower. The starboard bow motor is correcting the attitude of the ship.

would ultimately be incurred. AMSOC had no special funds of its own; NSF was footing all the bills. The committee assembled at the site of the now-famous 'wine breakfast' where Mohole was hatched two years earlier, and came up with a figure of $14 million 'for the entire project'. Only a few months earlier, the cost had been placed at $5 million. Why had it risen nearly three-fold? The answer, quite simply, was that since the project had never been clearly defined, the costs could not possibly be estimated with any reasonable accuracy. Furthermore, AMSOC consisted mainly of scientists, not engineers, and they had little or no experience in the intricacies of estimating the costs of a unique engineering system. Finally, and perhaps most important, the Cold

War and the early days of the space age had virtually obliterated financial conservatism in scientists' dealings with government. Money in unprecedented quantities was flowing into many fields of science and technology. As a White House science adviser later explained, 'I took a quick look at the project and decided, "Why not? It's only going to run about the cost of one space shot"'. AMSOC did not require the $14 million all at once. It informed NSF that $1250000 would suffice for the first stage of the drilling scheme proposed by Bascom. (Shortly afterwards, a club of geologists in Washington put on its annual show. The title that year was *Mo-Ho-Ho and a Barrel of Funds*.)

Employing a barge equipped with engines and a navigation system designed to keep it nearly motionless at any point on the high sea, Mohole's Phase I was a stunning success. Brilliantly directed by Bascom, at a site off the California coast, it set a world record for deep drilling by boring 197 metres into the ocean bottom at a water depth of nearly three km. This remarkable feat drew wide public acclaim, and was even hailed in a message from President John F. Kennedy, who, to the delight of American scientists, had brought an enthusiastic appreciation of science into the White House. But Phase I, successful as it was, constituted no more than a mere scratch on the ocean bottom compared to the objective of drilling through the Earth's crust. That feat called for securing a vessel in nearly five km of water, and drilling about 5000 metres into the ocean floor, most of the way through rock so dense that two years of nearly continuous drilling would be required. High temperatures, corrosion, the crushing weight of mile upon mile of piping, weather uncertainties—all together posed engineering problems far beyond the state of the art. At that point, spirits were high. What was not realized was that though Mohole was so far a great technical success, its administrative and, ultimately, its political underpinnings were beginning to crumble.

With Phase I successfully completed, the Academy began to feel uncomfortable about having an operational research programme headquartered on its premises. As a statutory adviser to government, the Academy had always remained aloof from any matter on which its advice might be sought, and clearly Mohole, with all of its technical and financial complexities, was the very sort of matter on which the Academy's counsel might be solicited. So, the Academy wanted to get out; but then so did AMSOC, whose members found themselves increasingly burdened with the demands of the project, despite the fact that the energetic Bascom was in full-time charge.

The two organizations agreed that it would be desirable to establish a new administrative base, though, of course, they hoped that NSF would continue to foot the bills for the continuation of the project. (These were estimated at $4.2 million for the fiscal year beginning 1 July 1961). But what was to be done with Bascom and the staff that had so successfully carried out Phase I? It was concluded that they should go into the employ of whatever organization was selected for the next phase. At this point, various sub-committees of AMSOC split on the question of whether Phase I had produced sufficient knowledge for building a vessel that could drill through the crust, or whether, as Bascom proposed, it would be advisable to build an 'intermediate' ship and acquire still further knowledge, before going on to the ultimate ship. NSF, which was still the sole source of financial support for the venture, publicly agreed that it would rely on the advice of AMSOC. In response to the suggestion that it hire a new organization to run the project, NSF invited public bids from various academic and industrial organizations. But bids for what? '. . . the drilling of a series of holes in the deep ocean floor, *one of which* will completely penetrate the Earth's crust.' That is how the invitations for bidders read. In response to this and related specifications, 11 individual and combined organizations submitted bids. The initial evaluation, by a group of NSF officials, gave first place to a consortium headed by the Socony Mobil petroleum company. Fifth place was given to a late-comer that just barely met the deadline for bid, a vast engineering combine from Houston, Texas, Brown and Root, whose business fortunes, mainly in the form of government contracts, were alleged to have been closely linked to the Democratic Party, to which it regularly contributed campaign funds. Houston was also the home base of a publicity-shy but immensely powerful member of Congress, the late Albert Thomas, the chairman of the very sub-committee that annually decided how much money would be voted for NSF—the sole source of finance for Project Mohole. Brown and Root got the contract, touching off angry allegations of a political fix.

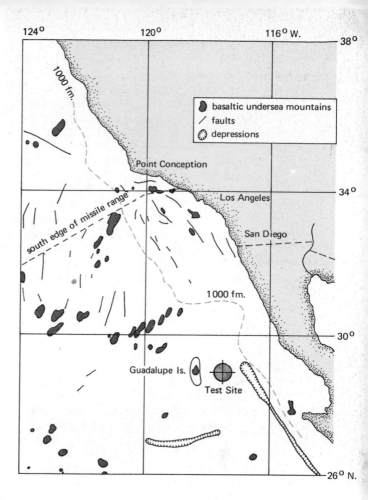

25.2 **The site of the holes drilled in Phase I of Project Mohole.**

Brown and Root was now responsible for Phase II of Project Mohole. In accord with the expressed desire of AMSOC and the Academy, it duly hired Bascom and his crew to help in the design of the vessel needed to carry out the project. Two months later, Bascom and his colleagues resigned, asserting that Brown and Root was unqualified for the task. Meanwhile, an important change occurred within AMSOC. A tough-minded geologist, Hollis Hedberg, a vice-president of Gulf Oil and a part-time professor of geology at Princeton University, became chairman of the committee. Hedberg decided that before the project proceeded any further, all uncertainties should be resolved about methods and objectives. The one-shot concept, he declared, was publicity-seeking nonsense. 'The project . . should in no way be considered merely a stunt in deep drilling; it is a most serious search for fundamental information concerning the rock character and the rock record of the Earth on which we live. And,' continued Hedberg, 'the scope of the project should be such as to take advantage of opportunities for contributions to this geological end wherever they may be found—water or land, deep or shallow.' Hedberg then came out for a cautious approach to Phase II, one that would involve construction of an 'intermediate' drilling vessel, as proposed by Bascom. However, Brown and Root, the Texas firm that had been awarded the contract for Phase II, was proceeding with the understanding—as written into the contract—that its task was to design a ship that could drill a hole through the crust. Much time had been lost through Brown and Root's inability to assemble a staff, and several members of the U.S. Congress, aroused by the political allegations surrounding the contract, had taken to sniping at NSF and its contractual choice. Finally, early in 1963, some 13 months after the award of the contract, Brown and Root unveiled its plans. These called for building a gigantic, self-propelled sea-going platform, 76 by 82 metres, resting on two submarine-shaped hulls, each 120 metres long. The cost of construction and drilling was estimated at $67 700 000.

When this figure was stated, the White House baulked and directed NSF to suspend the project pending clarification. NSF set up a special study committee; meanwhile, several congressional committees began to probe into the tangled affairs of Mohole. Hedberg went before one of these committees and declared that, though Mohole had unfortunately started out as a gimmick, there was time to salvage and build upon its true scientific merit. Whereupon the President of the National Academy of Sciences—AMSOC was still affiliated with the Academy—demanded that Hedberg clear all future public statements with him. Hedberg promptly resigned, and thus both he and Bascom, perhaps the two most effective men ever associated with Mohole, were out of the picture.

In the meantime, Waterman, then aged 71, had retired as director of NSF, and was succeeded by Leland J. Haworth, a veteran research administrator long associated with the U.S. Atomic Energy Commission. Working closely with the Academy, Haworth finally persuaded the White House to permit the project to proceed, though NSF officials now conceded that the costs might ultimately be as high as $125 million. For a time, the prospects looked good, but then in January 1966, Mohole's chief congressional guardian, Representative Albert Thomas of Texas, died. One of the first steps taken by his successor as chairman of the sub-committee that controlled Mohole's funds was to cut off the project. There was a brief chance of a reprieve by the Senate. Then it was revealed that a few days after the Senate had been asked to attempt to salvage the project, members of the Brown and Root family gave $23 000 to a Democratic Party fund-raising organization.

Politically that was the last straw. In August 1966, a little less than a decade after a group of scientists in Washington had proposed drilling a hole through the Earth's crust, the U.S. Congress voted to eliminate all further funds for *Project Mohole*.

FURTHER READING:

WILLARD BASCOM 1967. A Hole in the Bottom of the Sea. *Doubleday and Co., New York.*

DANIEL S. GREENBERG 1969. The Politics of American Science. *Pelican Books.*

WILLARD BASCOM, April 1959. The Mohole. *Scientific American.*

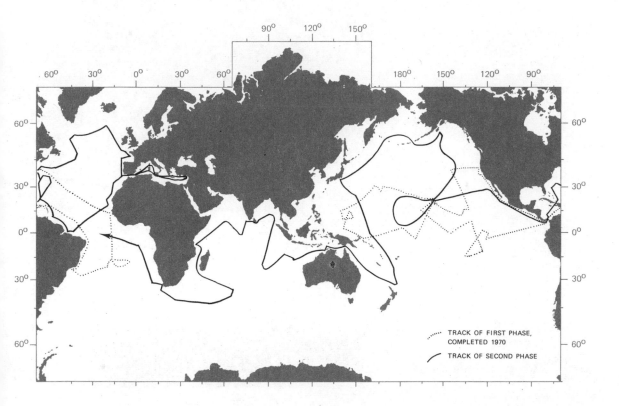

TRACK OF FIRST PHASE,
COMPLETED 1970

TRACK OF SECOND PHASE

25. 3   One positive outcome of the Mohole Project has been the development of the JOIDES Deep-Sea Drilling Project.   The map shows the track of the specially constructed vessel, the 'Glomar Challenger', whose work has been of outstanding scientific significance.

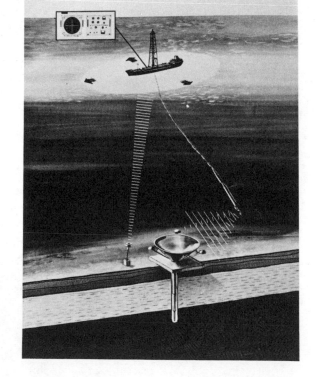

25. 4   Drilling deep into hard rock requires the ability to re-enter bore-holes after dulled drilling bits have been changed. 'Glomar Challenger' achieved this by placing a re-entry cone on the ocean floor before the start of drilling.   A system of sonar scanning, acoustic reflectors and electronic controls, combined with lateral water-jets located just above the drill-bit, makes it possible to guide bits into the same hole repeatedly.

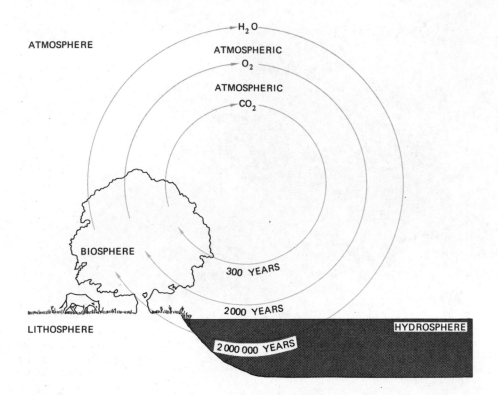

ATMOSPHERE

H$_2$O

ATMOSPHERIC
O$_2$

ATMOSPHERIC
CO$_2$

BIOSPHERE

300 YEARS

2000 YEARS

LITHOSPHERE

2 000 000 YEARS

HYDROSPHERE

26.1   The biosphere exchanges water vapour, oxygen and carbon dioxide with the atmosphere and hydrosphere in a continuing cycle, shown here in simplified form. All the Earth's water is split by plant cells and reconstituted by animal and plant cells about every two million years.   Oxygen generated in the process enters the atmosphere and is recycled in about 2 000 years.   Carbon dioxide respired by animal and plant cells enters the atmosphere and is fixed again by plant cells after an average atmospheric residence time of about 300 years.

In the next chapter we return to the theme of the interaction between geological processes, biological processes and the evolution of the Earth's atmosphere, to which Chapters 10, 11 and 12 have already contributed. Atmospheric oxygen takes part in a number of cyclic exchanges, whose complexity serves to emphasize the inter-dependence of all the 'working parts' of our global and essentially indivisible environment.   These cycles are discussed in the following article, which was written by Preston Cloud Jr. and Aharon Gibor, Professors of Bio-Geology and Biology respectively at the University of California.   It first appeared in *Scientific American* of September 1970.

26 The Oxygen Cycle

by Preston Cloud Jr. and Aharon Gibor

The oxygen in the atmosphere was originally put there by plants. Hence the early plants made possible the evolution of the higher plants and animals that require free oxygen for their metabolism.

The history of our planet, as recorded in its rocks and fossils, is reflected in the composition and the biochemical peculiarities of its present biosphere. With a little imagination one can reconstruct from that evidence the appearance and subsequent evolution of gaseous oxygen in the Earth's air and water, and the changing pathways of oxygen in the metabolism of living things.

Differentiated multicellular life (consisting of tissues and organs) evolved only after free oxygen appeared in the atmosphere. The cells of animals that are truly multicellular in this sense, the Metazoa, obtain their energy by breaking down fuel (produced originally by photosynthesis) in the presence of oxygen in the process called respiration. The evolution of advanced forms of animal life would probably not have been possible without the high levels of energy release that are characteristic of oxidative metabolism. At the same time free oxygen is potentially destructive to all forms of carbon-based life (and we know no other kind of life). Most organisms have therefore had to 'learn' to conduct their oxidations anaerobically, primarily by removing hydrogen from foodstuff rather than by adding oxygen. Indeed, the anaerobic process called fermentation is still the fundamental way of life, underlying other forms of metabolism.

Oxygen in the free state thus plays a role in the evolution and present functioning of the biosphere that is both pervasive and ambivalent. The origin of life and its subsequent evolution was contingent on the development of systems that shielded it from, or provided chemical defences against, ordinary molecular oxygen (O_2), ozone (O_3) and atomic oxygen (O). Yet the energy requirements of higher life forms can be met only by oxidative metabolism. The oxidation of the simple sugar glucose, for example, yields 686 kilocalories per mole; the fermentation of glucose yields only 50 kilocalories per mole.

Free oxygen not only supports life; it arises from life. The oxygen now in the atmosphere is probably mainly, if not wholly, of biological origin. Some of it is converted to ozone, causing certain high-energy wavelengths to be filtered out of the radiation that reaches the surface of the earth. Oxygen also combines with a wide range of other elements in the earth's crust. The result of these and other processes is an intimate evolutionary interaction among the biosphere, the atmosphere, the hydrosphere and the lithosphere.

Consider where the oxygen comes from to support the high rates of energy release observed in multicellular organisms and what happens to it and to the carbon dioxide that is respired (see Fig. 26.1). The oxygen, of course, comes from the air, of which it constitutes roughly 21 per cent. Ultimately, however, it originates with the decomposition of water molecules by light energy in photosynthesis. The 1 500 million cubic kilometres of water on the earth are split by photosynthesis and reconstituted by respiration once every two million years or so. Photosynthetically generated oxygen temporarily enters the atmospheric bank, when it is itself recycled once every 2 000 years or so (at current rates). The carbon dioxide that is respired joins the small amount (0.03 per cent) already in the atmosphere, which is in balance with the carbon dioxide in the oceans and other parts of the hydrosphere. Through other interactions it may be removed from circulation as a part of the carbonate ion (CO_3^-) in calcium carbonate precipitated from solution. Carbon dioxide thus sequestered may eventually be returned to the atmosphere when

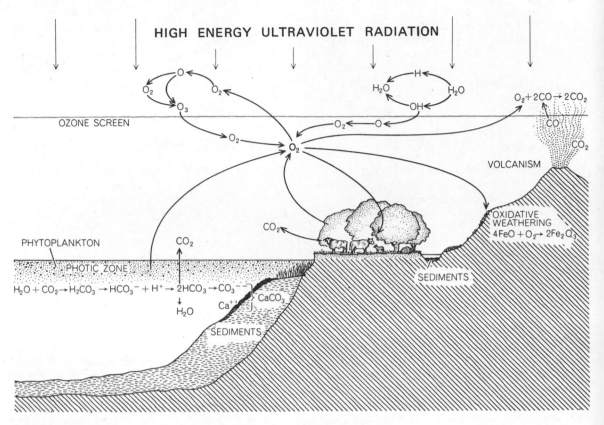

HIGH ENERGY ULTRAVIOLET RADIATION

OZONE SCREEN

$O_2 + 2CO \rightarrow 2CO_2$

VOLCANISM

PHYTOPLANKTON

PHOTIC ZONE

CO_2

CO_2

OXIDATIVE WEATHERING
$4FeO + O_2 \rightarrow 2Fe_2O_3$

$H_2O + CO_2 \rightarrow H_2CO_3 \rightarrow HCO_3^- + H^+ \rightarrow 2HCO_3 \rightarrow CO_3^{--}$

Ca^{++} $CaCO_3$

H_2O

SEDIMENTS

SEDIMENTS

26.2 The Oxygen Cycle is complicated because oxygen appears in so many chemical forms and combinations, primarily as molecular oxygen (O_2), in water and in organic and inorganic compounds. Some global pathways of oxygen are shown here in simplified form.

limestone, formed by the consolidation of calcium carbonate sediments, emerges from under the sea and is dissolved by some future rainfall.

Thus do sea, air, rock and life interact and exchange components. Before taking up these interactions in somewhat greater detail let us examine the function oxygen serves with individual organisms.

Oxygen plays a fundamental role as a building block of practically all vital molecules, accounting for about a fourth of the atoms in living matter. Practically all organic matter in the present biosphere originates in the process of photosynthesis, whereby plants utilize light energy to react carbon dioxide with water and synthesize organic substances. Since carbohydrates (such as sugar), with the general formula $(CH_2O)_n$, are the common fuels that are stored by plants, the essential reaction of photosynthesis can be written as $CO_2 + H_2O + \text{light energy} \rightarrow CH_2O + O_2$. It is not immediately obvious from this formulation which of the reactants serves as the source of oxygen atoms in the carbohydrates and which is the source of free molecular oxygen. In 1941 Samuel Ruben and Martin D. Kamen of the University of California at Berkeley used the heavy oxygen isotope oxygen 18 as a tracer to demonstrate that the molecular oxygen is derived from the splitting of the water molecule. This observation also suggested that carbon dioxide is the source of the oxygen atoms of the synthesized organic molecules.

The primary products of photosynthesis undergo a vast number of chemical transformations in plant cells and subsequently in the cells of the animals that feed on plants. During these processes changes of course take place in the atomic composition and energy content of the organic molecules. Such transformations can result in carbon compounds that are either more 'reduced' or more 'oxidized' than carbohydrates. The oxidation-reduction reactions between these compounds are the essence of biological energy supply and demand. A more reduced compound has more hydrogen atoms and fewer oxygen atoms per carbon atom; a more oxidized compound has fewer hydrogen atoms and more oxygen atoms per carbon atom. The combustion of a reduced compound liberates more energy than the combustion of a more oxidized one. An example of a molecule more reduced than a carbohydrate is the familiar alcohol ethanol (C_2H_6O); a more oxidized molecule is pyruvic acid ($C_3H_4O_3$).

Differences in the relative abundance of hydrogen and oxygen atoms in organic molecules result primarily from one of the following reactions: (1) the removal (dehydrogenation) or addition (hydrogenation) of hydrogen atoms; (2) the addition of water (hydration), followed by dehydrogenation; (3) the direct addition of oxygen (oxygenation). The second and third of these processes introduce into organic matter additional oxygen atoms either from water or from molecular oxygen. On decomposition the oxygen atoms of organic molecules are released as carbon dioxide and water. The biological oxidation of molecules such as carbohydrates can be written as the reverse of photosynthesis: $CH_2O + O_2 \rightarrow CO_2 + H_2O + energy$. The oxygen atom of the organic molecule appears in the carbon dioxide and the molecular oxygen acts as the acceptor for the hydrogen atoms.

The three major non-living sources of oxygen atoms are therefore carbon dioxide, water and molecular oxygen, and since these molecules exchange oxygen atoms, they can be considered as a common pool. Common mineral oxides such as nitrate ions and sulphate ions are also oxygen sources for living organisms, which reduce them to ammonia (NH_3) and hydrogen sulphide (H_2S). They are subsequently reoxidized, and so as the oxides circulate through the biosphere their oxygen atoms are exchanged with water.

The dynamic role of molecular oxygen is as an electron sink, or hydrogen acceptor, in biological oxidations. The biological oxidation of organic molecules proceeds primarily by dehydrogenation: enzymes remove hydrogen atoms from the substrate molecule and transfer them to specialized molecules that function as hydrogen carriers. If these carriers become saturated with hydrogen, no further oxidation can take place until some other acceptor becomes available. In the anaerobic process of fermentation organic molecules serve as the hydrogen acceptor. Fermentation therefore results in the oxidation of some organic compounds and the simultaneous reduction of others, as in the fermentation of glucose by yeast: part of the sugar molecule is oxidized to carbon dioxide and other parts are reduced to ethanol.

In aerobic respiration oxygen serves as the hydrogen acceptor and water is produced. The transfer of hydrogen atoms (which is to say of electrons and protons) to oxygen is channelled through an array of catalysts and cofactors. Prominent among the cofactors are the iron-containing pigmented molecules called cytochromes, of which there are several kinds that differ in their affinity for electrons. This affinity is expressed as the oxidation-reduction or *redox* potential of the molecule; the more positive the potential, the greater the affinity of the oxidized molecule for electrons. For example, the redox potential of cytochrome *b* is 0.12 volt, the potential of cytochrome *c* is 0.22 volt and the potential of cytochrome *a* is 0.29 volt. The redox potential for the reduction of oxygen to water is 0.8 volt. The passage of electrons from one cytochrome to another down a potential gradient, from cytochrome *b* to cytochrome *c* to the cytochrome *a* complex and on to oxygen, results in the alternate reduction and oxidation of these cofactors. Energy liberated in such oxidation-reduction reactions is coupled to the synthesis of high-energy phosphate compounds such as adenosine triphosphate (ATP). The special copper-containing enzyme cytochrome oxidase mediates the ultimate transfer of electrons from the cytochrome *a* complex to oxygen. This activation and binding of oxygen is seen as the fundamental step, and possibly the original primitive step, in the evolution of oxidative metabolism.

In cells of higher organisms the oxidative system of enzymes and electron carriers is located in the special organelles called mitochondria. These organelles can be regarded as efficient low-temperature furnaces where organic molecules are burned with oxygen. Most of the released energy is converted into the high-energy bonds of ATP.

Molecular oxygen reacts spontaneously with organic compounds and other reduced substances. This reactivity explains the toxic effects of oxygen above tolerable concentrations. Louis Pasteur discovered that very sensitive organisms such as obligate anaerobes cannot tolerate oxygen concentrations above about 1 per cent of the present atmospheric level. Recently the cells of higher organisms have been found to contain organelles called peroxisomes, whose major function is thought to be the protection of cells from oxygen. The peroxisomes contain enzymes that catalyze the direct reduction of oxygen molecules through the oxidation of metabolites such as amino acids and other organic acids. Hydrogen peroxide (H_2O_2) is one of the products of such oxidation. Another of the peroxisome enzymes, catalase, utilizes the hydrogen peroxide as a hydrogen acceptor in the oxidation of substrates such as ethanol or lactic acid. The rate of reduction of oxygen by the peroxisomes increases proportionately with an increase in oxygen concentration, so that an excessive amount of oxygen in the cell increases the rate of its reduction by peroxisomes.

Christian de Duve of Rockefeller University has suggested that the peroxisomes represent a primitive enzyme system that evolved to cope with oxygen when it first appeared in the atmosphere. The peroxisome enzymes enabled the first oxidatively metabolizing cells to use oxygen as a hydrogen acceptor and so reoxidize the reduced products of fermentation. In some respects this process is similar to the oxidative reactions of the mitochondria. Both make further dehydrogenation possible by liberating oxidized hydrogen carriers. The basic difference between the mitochondrial oxidation reactions and those of peroxisomes is that in peroxisomes the steps of oxidation are not coupled to the synthesis of ATP. The energy released in the peroxisomes is thus lost to the cell; the function of the organelle is primarily to protect against the destructive effects of free molecular oxygen.

Oxygen dissolved in water can diffuse across both the inner and the outer membranes of the cell, and the supply of oxygen by diffusion is adequate for single cells and for organisms consisting of small colonies of cells. Differentiated multicellular organisms, however, require more efficient modes of supplying oxygen to tissues and organs. Since all higher organisms depend primarily on mitochondrial aerobic oxidation to generate the energy that maintains their active mode of life, they have evolved elaborate systems to ensure their tissues an adequate supply of oxygen, the gas that once was lethal (and still is, in excess). Two basic devices serve this purpose: special chemical carriers that increase the oxygen capacity of body fluids, and anatomical structures that provide relatively large surfaces for the rapid exchange of gases. The typical properties of an oxygen carrier are exemplified by those of haemoglobin and of myoglobin, or muscle haemoglobin. Haemoglobin in blood readily absorbs oxygen to near-saturation at oxygen pressures such as those found in the lung. When the blood is exposed to lower oxygen pressures as it moves from the lungs to other tissues, the haemoglobin discharges most of its bound oxygen. Myoglobin, which acts as a reservoir to meet the sharp demand for oxygen in muscle contraction, gives up its oxygen more rapidly. Such reversible bonding of oxygen in response to changes in oxygen pressure is an essential property of biochemical oxygen carriers.

Lungs and gills are examples of anatomical structures in which large wet areas of thin membranous tissue come in contact with oxygen. Body fluids are pumped over one side of these membranes and air, or water containing oxygen, over the other side. This ensures a rapid gas exchange between large volumes of body fluid and the environment.

How did the relations between organisms and gaseous oxygen happen to evolve in such a curiously complicated manner? The atmosphere under which life arose on the Earth was almost certainly devoid of free oxygen. The low concentration of noble gases such as neon and krypton in the terrestrial atmosphere compared with their cosmic abundance, together with other geochemical evidence, indicates that the terrestrial atmosphere had a secondary origin in volcanic outgassing from the Earth's interior. Oxygen is not known among the gases so released, nor is it found as inclusions in igneous rocks. The chemistry of rocks older than about two thousand million years is also inconsistent with the presence of more than trivial quantities of free atmospheric oxygen before that time. Moreover, it would not have been possible for the essential chemical precursors of life—or life itself—to have originated and persisted in the presence of free oxygen before the evolution of suitable oxygen-mediating enzymes.

On such grounds we conclude that the first living organism must have depended on fermentation for its livelihood. Organic substances that orginated in non-vital reactions served as substrates for these primordial fermentations. The first organism, therefore, was not only an anaerobe; it was also a heterotroph, dependent on a pre-existing organic food supply and incapable of manufacturing its own food by photosynthesis or other autotrophic processes.

The emergence of an autotroph was an essential step in the onward march of biological evolution. This evolutionary step left its mark in the rocks as well as on all living forms. Some fated eobiont, as we may call these early life forms whose properties we can as yet only imagine, evolved and became an autotroph, an organism capable of manufacturing its own food. Biogeological evidence suggests that this critical event may have occurred more than three thousand million years ago.

If, as seems inescapable, the first autotrophic eobiont was also anaerobic, it would have encountered difficulty when it first learned to split water and release free oxygen. John M. Olson of the Brookhaven National Laboratory recently suggested biochemical arguments to support the idea that primitive photosynthesis may have obtained electrons from substances other than water. He argues that large-scale splitting of water and release of oxygen may have been delayed until the evolution of appropriate enzymes to detoxify this reactive substance.

We nevertheless find a long record of oxidized marine sediments of a peculiar type that precedes the first evidence of atmospheric oxygen in rocks about 1.8×10^9 years old; we do not find them in significant amounts in more recent strata. These oxidized marine sediments, known as banded iron formations, are alternately iron-rich and iron-poor chemical sediments that were laid down in open bodies of water. Much of the iron in them is ferric (the oxidized form, Fe^{+++}) rather than ferrous (the reduced form, Fe^{++}), implying that there was a source of oxygen in the column of water above them. Considering the problems that would face a water-splitting photosynthesizer before the evolution of advanced oxygen-mediating enzymes such as oxidases and catalases, one can visualize how the biological oxygen cycle may have interacted with ions in solution in bodies of water during that time. The first oxygen-releasing photoautotrophs may have used ferrous compounds in solution as oxygen acceptors—oxygen for them being merely a toxic waste product. This would have precipitated iron in the ferric form ($4FeO + O_2 \rightarrow 2Fe_2O_3$) or in the ferro-ferric form (Fe_3O_4). A recurrent imbalance of supply and demand might then account for the cyclic nature and differing types of the banded iron formations.

Once advanced oxygen-mediating enzymes arose, oxygen generated by increasing populations of photoautotrophs containing these enzymes would build up in the oceans and begin to escape into the atmosphere. There the ultraviolet component of the sun's radiation would dissociate some of the molecular oxygen into highly reactive atomic oxygen and also give rise to equally reactive ozone. Atmospheric oxygen and its reactive derivatives (even in small quantities) would lead to the oxidation of iron in sediments produced by the weathering of rocks, to the greatly reduced solubility of iron in surface waters (now oxygenated), to the termination of the banded iron formations as an important sedimentary type and to the extensive formation of continental red beds rich in ferric iron. The record of the rocks supports this succession of events: red beds are essentially restricted to rocks younger than about 1 800 million years, whereas banded iron formation is found only in older rocks.

So far we have assumed that oxygen accumulated in the atmosphere as a consequence of photosynthesis by green plants. How could this happen if the entire process of photosynthesis and respiration is cyclic, representable by the reversible equation $CO_2 + H_2O + energy \rightleftharpoons CH_2O + O_2$? Except to the extent that carbon or its compounds are somehow sequestered, carbohydrates produced by photosynthesis will be reoxidized back to carbon dioxide and water, and no significant quantity of free oxygen will accumulate. The carbon that is sequestered in the Earth as graphite in the oldest rocks and as coal, oil, gas and other carbonaceous compounds in the younger ones, and in the living and dead bodies of plants and animals, is the equivalent of the oxygen in oxidized sediments and in the Earth's atmosphere! In attempting to strike a carbon-oxygen balance we must find enough carbon to account not only for the oxygen in the present atmosphere but also for the 'fossil' oxygen that went into the conversion of ferrous oxides to ferric oxides, sulphides to sulphates, carbon monoxide to carbon dioxide and so on.

Interestingly, rough estimates made some years ago by William W. Rubey, now of the University of California at Los Angeles, do imply an approximate balance between the chemical combining equivalents of carbon and oxygen in sediments, the atmosphere, the hydrosphere and the biosphere (see Fig. 26.3). The relatively small excess of carbon in Rubey's estimates could be accounted for by the oxygen used in converting carbon monoxide to carbon dioxide. Or it might be due to an underestimate of the quantities of

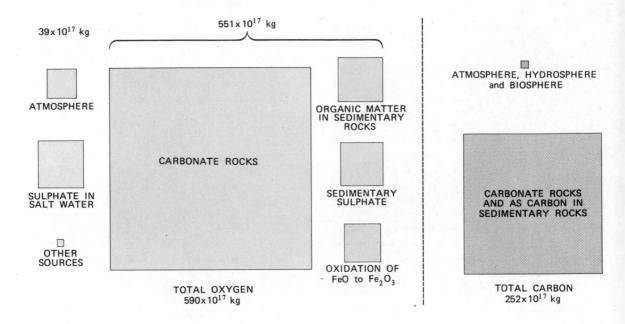

26.3 The Oxygen–Carbon Balance Sheet suggests that photosynthesis can account not only for all the oxygen in the atmosphere but also for the much larger amount of 'fossil' oxygen, mostly in compounds in sediments. The diagram, based on estimates made by William W. Rubey, indicates that the elements are present in about the proportion, 12/32, that would account for their derivation through photosynthesis from carbon dioxide (one atom of carbon, atomic weight 12, to two of oxygen, atomic weight 16).

sulphate ion or ferric oxide in sediments. (Rubey's estimates could not include large iron formations recently discovered in western Australia and elsewhere.) The carbon dioxide in carbonate rocks does not need to be accounted for, but the oxygen involved in converting it to carbonate ion does. The recycling of sediments through metamorphism, mountain-building and the movement of ocean-floor plates under the continents is a variable of unknown dimensions, but it probably does not affect the approximate balance observed in view of the fact that the overwhelmingly large pools to be balanced are all in the lithosphere and that carbon and oxygen losses would be roughly equivalent. The small amounts of oxygen dissolved in water are not included in this balance.

Nonetheless, water does enter the picture. Another possible source of oxygen in our atmosphere is photolysis, the ultraviolet dissociation of water vapour in the outer atmosphere followed by the escape of the hydrogen from the Earth's gravitational field. This has usually been regarded as a trivial source, however. Although R. T.

Brinkmann of the California Institute of Technology has recently argued that nonbiological photolysis may be a major source of atmospheric oxygen, the carbon–oxygen balance sheet does not support that belief, which also runs into other difficulties.

When free oxygen began to accumulate in the atmosphere some 1 800 million years ago, life was still restricted to sites shielded from destructive ultraviolet radiation by sufficient depths of water or by screens of sediment. In time, enough oxygen built up in the atmosphere for ozone, a strong absorber in the ultraviolet, to form a shield against incoming ultraviolet radiation. The late Lloyd V. Berkner and Lauriston C. Marshall of the Graduate Research Center of the Southwest in Dallas calculated that only 1 per cent of the present atmospheric level of oxygen would give rise to a sufficient level of ozone to screen out the most deleterious wavelengths of the ultraviolet radiation. This also happens to be the level of oxygen at which Pasteur found that certain micro-organisms switch over from a fermentative type of metabolism to an oxidative one. Berkner and

YEARS BEFORE PRESENT	LITHOSPHERE	BIOSPHERE	HYDROSPHERE	ATMOSPHERE
20 Million	GLACIATION	MAMMALS DIVERSIFY GRASSES APPEAR		OXYGEN APPROACHES PRESENT LEVEL
50 Million				
	COAL FORMATION VOLCANISM			
100 Million		SOCIAL INSECTS, FLOWERING PLANTS		ATMOSPHERIC OXYGEN INCREASES AT FLUCTUATING RATE
		MAMMALS		
200 Million	GREAT VOLCANISM		OCEANS CONTINUE TO INCREASE IN VOLUME	
	COAL FORMATION			
		INSECTS APPEAR LAND PLANTS APPEAR		
500 Million				
	GLACIATION SEDIMENTARY CALCIUM SULFATE	METAZOA APPEAR RAPID INCREASE IN PHYTOPLANKTON	SURFACE WATERS OPENED TO PHYTOPLANKTON	OXYGEN AT 3-10 PERCENT OF PRESENT ATMOSPHERIC LEVEL
1 000 Million	VOLCANISM			OXYGEN AT 1 PERCENT OF PRESENT ATMOSPHERIC LEVEL, OZONE SCREEN EFFECTIVE
		EUCARYOTES		OXYGEN INCREASING, CARBON DIOXIDE DECREASING
2 000 Million	RED BEDS	ADVANCED OXYGEN-MEDIATING ENZYMES	OXYGEN DIFFUSES INTO ATMOSPHERE	OXYGEN IN ATMOSPHERE
	GLACIATION	FIRST OXYGEN-GENERATING PHOTOSYNTHETIC CELLS PROCARYOTES ABIOGENIC EVOLUTION	START OF OXYGEN GENERATION WITH FERROUS IRON AS OXYGEN SINK	
	BANDED IRON FORMATIONS OLDEST SEDIMENTS OLDEST EARTH ROCKS			
5 000 Million	(ORIGIN OF SOLAR SYSTEM)			NO FREE OXYGEN

26.4 **The chronology that inter-relates the evolutions of atmosphere and biosphere is gradually being established from evidence in the geological record and in fossils. According to calculations by Lloyd V. Berkner and Lauriston C. Marshall, when oxygen in the atmosphere reached 1 per cent of the present atmospheric level, it provided enough ozone to filter out the most damaging high-energy ultraviolet radiation so that phytoplankton could survive everywhere in the upper, sunlit layers of the seas. The result may have been a geometric increase in the amount of photosynthesis in the oceans that, if accompanied by equivalent sequestration of carbon, might have resulted in a rapid build-up of atmospheric oxygen, leading in time to the evolution of differentiated multicelled animals.**

Marshall therefore jumped to the conclusion (reasonably enough on the evidence they considered) that this was the stage at which oxidative metabolism arose. They related this stage to the first appearance of metazoan life somewhat more than 600 million years ago. (See Chapter 10.)

The geological record has long made it plain, however, that free molecular oxygen existed in the atmosphere well before that relatively late date in geologic time. Moreover, recent evidence is consistent with the origin of oxidative metabolism at least twice as long ago. Eucaryotic cells (see Fig. 12. 2)—cells with organized nuclei and other organelles—have been identified in rocks in eastern California that are believed to be about 1.3×10^9 years old. Since all living eucaryotes depend on oxidative metabolism, it seems likely that these ancestral forms did too. The oxygen level may nonetheless have still been quite low at this stage. Simple diffusion would suffice to move enough oxygen across cell boundaries and within the cell, even at very low concentrations, to supply the early oxidative metabolizers. A higher order of organization and of atmospheric oxygen was required, however, for advanced oxidative metabolism. Perhaps that is why, although the eucaryotic cell existed at least 1 200 million years ago, we have no unequivocal fossils of metazoan organisms from rocks older than perhaps 640 million years.

In other words, perhaps Berkner and Marshall were mistaken only in trying to make the appearance of the Metazoa coincide with the onset of oxidative metabolism. Once the level of atmospheric oxygen was high enough to generate an effective ozone screen, photosynthetic organisms would have been able to spread throughout the surface waters of the sea, greatly accelerating the rate of oxygen production. The plausible episodes in geological history to correlate with this development are the secondary oxidation of the banded iron formations and the appearance of sedimentary calcium sulphate (gypsum and anhydrite) on a large scale. These events occurred just as or just before the Metazoa first appeared in early Palaeozoic time. The attainment of a suitable level of atmospheric oxygen may thus be correlated with the emergence of metazoan root stocks from pre-metazoan ancestors beginning about 640 million years ago. The fact that oxygen could accumulate no faster than carbon (or hydrogen) was removed argues against the likelihood of a rapid early build-up of oxygen.

That subsequent biospheric and atmospheric evolution were closely inter-linked can now be taken for granted. What is not known are the details. Did oxygen levels in the atmosphere increase steadily throughout geologic time, marking regular stages of biological evolution such as the emergence of land plants, of insects, of the various vertebrate groups and of flowering plants, as Berkner and Marshall suggested? Or were there wide swings in the oxygen level? Did oxygen decrease during great volcanic episodes, as a result of the oxidation of newly emitted carbon monoxide to carbon dioxide, or during times of sedimentary sulphate precipitation? Did oxygen increase when carbon was being sequestered during times of coal and petroleum formation? May there have been fluctuations in both directions as a result of plant and animal evolution, of phytoplankton eruptions and extinctions and of the extent and type of the terrestrial plant cover? Such processes and events are now being seriously studied, but the answers are as yet far from clear.

What one can say with confidence is that success in understanding the oxygen cycle of the biosphere in truly broad terms will depend on how good we are at weaving together the related strands of biospheric, atmospheric, hydrospheric and lithospheric evolution throughout geologic time. Whatever we may conjecture about any one of these processes must be consistent with what is known about the others. Whereas any one line of evidence may be weak in itself, a number of lines of evidence, taken together and found to be consistent, reinforce one another exponentially. This synergistic effect enhances our confidence in the proposed time scale linking the evolution of oxygen in the atmosphere and the management of the gaseous oxygen budget within the biosphere (see Fig. 26. 4).

The most recent factor affecting the oxygen cycle of the biosphere and the oxygen budget of the Earth is man himself. In addition to inhaling oxygen and exhaling carbon dioxide as a well-behaved animal does, man decreases the oxygen level and increases the carbon dioxide level by burning fossil fuels and paving formerly green land. He is also engaged in a vast but unplanned experiment to see what effects oil spills and an array of pesticides will have on the world's phytoplankton. The increase in the albedo, or reflectivity, of the Earth as a result of covering its waters with a molecule-thick film of oil could also affect plant growth by lowering the temperature and in other unforeseen ways. Reductions in the length of growing seasons and in green areas would limit terrestrial plant growth in the middle latitudes. (This might normally be counter-balanced by increased rainfall in the lower latitudes, but a film of oil would also reduce evaporation and therefore rainfall.) Counteracting such effects, man moves the Earth's fresh water around to increase plant growth and photosynthesis in arid and semi-arid regions. Some of this activity, however, involves the mining of ground water, thereby favouring processes that cause water to be returned to the sea at a faster rate than evaporation brings it to the land.

He who is willing to say what the final effects of such processes will be is wiser or braver than we are. Perhaps the effects will be self-limiting and self-correcting, although experience should warn us not to gamble on that. Oxygen in the atmosphere might be reduced several per cent below the present level without adverse effects. A modest increase in the carbon dioxide level might enhance plant growth and lead to a corresponding increase in the amount of oxygen. Will a further increase in carbon dioxide also have (or renew) a 'greenhouse effect', leading to an increase in temperature (and thus to a rising sea level)? Or will such effects be counterbalanced or swamped by the cooling effects of particulate matter in the air or by increased albedo due to oil films? It is anyone's guess. (Perhaps we should be more alarmed about a possible decrease of atmospheric carbon dioxide, on which all forms of life ultimately depend, but the sea contains such vast amounts that it can presumably keep carbon dioxide in the atmosphere balanced at about the present level for a long time to come.) The net effect of the burning of fossil fuels may in the long run be nothing more than a slight increase (or decrease?) in the amount of limestone deposited. In any event the recoverable fossil fuels whose combustion releases carbon dioxide are headed for depletion in a few more centuries, and then man will have other problems to contend with.

What we want to stress is the indivisibility and complexity of the environment. For example, the Earth's atmosphere is so thoroughly mixed and so rapidly recycled through the biosphere that the next breath you inhale will contain atoms exhaled by Jesus at Gethsemane and by Adolf Hitler at Munich. It will also contain atoms of radioactive strontium 90 and iodine 131 from atomic explosions and gases from the chimneys and exhaust pipes of the world. Present environmental problems stand as a grim monument to the cumulatively adverse effects of actions that in themselves were reasonable enough but that were taken without sufficient thought to their consequences. If we want to ensure that the biosphere continues to exist over the long term and to have an oxygen cycle, each new action must be matched with an effort to foresee its consequences throughout the ecosystem and to determine how they can be managed favourably or avoided. Understanding also is needed, and we are woefully short on that commodity. This means that we must continue to probe all aspects of the indivisible global ecosystem and its past, present and potential interactions. That is called basic research, and basic research at this critical point in history is gravely endangered by new crosscurrents of anti-intellectualism.

FURTHER READING:

H. C. UREY 1959. The Atmospheres of the Planets. From 'Handbuch der Physik', v. lii. *Springer-Verlag.*

L. V. BERKNER AND L. C. MARSHALL 1965. History of Major Atmospheric Components. *Proc. Nat. Acad. Sci.*, v. 53, No. 6.

R. T. BRINKMANN 1969. Dissociation of Water Vapor and Evolution of Oxygen in the Terrestrial Atmosphere. *J. Geophys. Research*, v. 74, No. 23.

J. M. OLSON 1970. The Evolution of Photosynthesis. *Science*, v. 168, No. 3930.

Several of the authors in this book have discussed some of the interactions between the Earth sciences and Earth scientists on the one hand, and the wider society outside the purely academic confines of the subject on the other. In this final chapter, Dr. Stephen Drury of the Open University introduces what may well be the most crucial of all problems facing both the Earth sciences and society: the maintenance of the supply of the natural physical resources without which modern civilization as we know it cannot survive. For as well as developing the revolutionary concepts embodied in the 'new global tectonics', Earth scientists have come to appreciate more and more that the Earth is finite, and that what it can provide in the way of useful materials can be used up much as a lake can be overfished.

Here Dr. Drury outlines what is meant by the term *resources*, and more particularly *physical resources*, and the implications of the continuing use of physical resources in ever increasing quantities. In the first part he discusses types of resources, how resources interact, how long they take to form, how they are distributed in the Earth and how they are being used up. He then goes on to examine the questions of how long physical resources will last and how two particular types of resource exploitation affect the environment. Finally, he discusses some of the social implications of the use of the Earth's physical resources.

27 Resources and Environment

by STEPHEN DRURY

An overall view indicates strongly that a crisis of physical resources is developing.   The forecasts may err on points of detail — but can we ignore the risk that they are right in principle?

EVERY PART OF THE EARTH that is put to use in the service of Man is a resource.   Resources not only include food, natural fibres and physical matter extracted from the atmosphere, hydrosphere and crust, such as water, metals and fuels, but also, more indirectly, the effect of the environment, and its alteration, on health and intellect.   Basically then a resource comes into existence when something is thought of as relevant and of use to Man.

Because of economic, cultural, scientific and technological changes, our notions of resources change.   For instance, until aluminium's properties became better known it was just a metallurgical curiosity sometimes used in ornaments. In the early days of the oil industry, the only uses were in lamps and as lubricants.   It needed the invention of engines which could burn petroleum before it became a major resource.   As it has become more obvious that high incidence of crime, suicide and divorce is connected with urban life and its associated stresses, the long recognised 'fresh air cure' and access to unspoilt open places is an increasing necessity.   Here we should only be concerned with the physical type of resource, those inanimate aggregations of atoms and molecules like water, coal, copper and brick clay, extracted from the Earth for the service of Man.   But as physical resources interact with the other, non-physical resources, the picture is by no means clear cut or precise. Obviously, inorganic fertilisers such as potash directly control the availability of agricultural produce, but taking this a stage further, without the need to supply food over and above the natural crop yields to feed expanding populations, fertilisers mined from rocks would never have be-

come resources.   Consider a tractor or a trawler; both require refined metals and petroleum fuels and represent direct interaction between physical and biological resources.   Fertile land has a direct influence on growing food, but what if it is more valuable because there are physical resources beneath its surface, or if it is covered by highways and urban sprawl?   Its value as an area of natural environment is lessened by *any* of its interactions with other types of resource.

Although there are interactions between different types of resource, we restrict our terms of reference to the physical resources.   These are the sources of energy, such as coal, oil and fissionable elements; metallic ores; raw materials for the chemical industry; building materials, water and oxygen.

In a closed dynamic system such as the Earth, every component is taking part in one or many different types of cycle.   So, resources are parts of cycles.   The periods of natural cycles are all different (Fig. 27.1) and they often proceed irregularly.   Also, the cycles are interrelated in a complex way.   For instance, biological resources go in life cycles and crop cycles interacted upon and interlocking with such cycles as those of nitrogen and phosphorus.   Environmental resources such as natural vegetation, landforms and soil go in a whole gamut of biological and geological cycles.   Physical resources are mainly controlled by geological cycles including those inherent in plate-tectonic processes, rock-forming processes, erosional cycles, hydrological cycles and in some cases parts of, say, the carbon cycle which has a geological component in the formation of coal and oil.   It is worth following this concept a little further.

Extracting physical resources and using them breaks the continuity of the relevant cycle. Certainly the cycle will continue, and may even, with time, naturally modify itself to the extractive processes. For instance, carbon dioxide from coal burning will end up in the carbon cycle as a matter of course, and a tin can on the ocean floor will be incorporated back into the bowels of the Earth when it has been subducted! The question is though, how long will it take a particular resource to renew itself after it or its waste products have been discarded by us and re-entered natural cycles? To be usefully *renewable* any resource must be replenished roughly within those time cycles relevant to Man—the life cycles—say at most 100 years.

Of the physical resources only water has a reasonably rapid cycle. This is because the hydrological cycle has a periodicity of about a year. As for the rest, they are all products of geological cycles—erosion, sedimentation, burial, metamorphism, igneous activity, the so-called rock cycle whose periodicities range up to hundreds of million years.

Figure 27.1 shows the time ranges involved in the formation of some common physical resources. Consider as an example that part of the carbon cycle involving burial of plant debris. There is a progressive compaction during burial to give the succession *peat—soft coal (lignite)—coal—anthracite*. Even peat may take 10 000 years to accumulate to a useful fuel, whilst coal in most world coalfields was formed 300 Ma ago. For sub-surface oil to form and accumulate involves the transformation of organic remains and migration of the resultant fluid. This takes at least 10 Ma, although most oil is much older. Most of the common metallic ores seem to have their origin in dynamic and thermal reworking of crustal rocks, or in emanations originating in the upper mantle; these processes probably took place over hundreds, or even thousands, of millions of years.

So, with the exception of water, physical resources do *not* renew themselves as far as Man is concerned. Indeed some may never be renewed, for the geological cycles and processes which formed them no longer function. Such is the case with iron, as a large proportion of iron ores seem to have formed some 2 000 to 3 000 Ma ago under peculiar sedimentary conditions by chemical or primitive biological action which can no longer take place because of changes in the atmosphere and the extinction of the responsible organism (see Chapters 10, 11 & 26). So there is a distinction between *renewable* and *non-renewable* resources.

Physical resources are usually found as localised concentrations of a material or element above its normal abundance. Because of this it depends on the level of technology at the time whether it is physically or economically possible to exploit them. For example, to be extracted to-day mercury must occur in concentrations 100 000 times greater than in common rocks, uranium 1 000 times, copper 100 times and iron about 10. Furthermore, there is a tendency for physical resources to be clustered in *provinces*, as Chapter 22 suggests. In these provinces, resources have been produced by particular combinations of geological processes. There are petroleum provinces where organic sediments have accumulated within thick piles of sediments, and various metallic provinces where elements have been concentrated by igneous or metamorphic processes from a more evenly dispersed state at low concentrations. Because different combinations of processes are often mutually exclusive in a geological sense, as with metamorphism and the formation of petroleum, some resources may be present in a certain geological province, and others excluded. Furthermore, what resources are present in a province may be inconveniently situated. They may be deeply buried or far from centres of industry. Clearly, there is wide disparity in the distribution of physical resources on the Earth, both in terms of quantity and diversity. It is these factors which determine whether an area can develop or not. There must be resource *diversity* in an area, as even with large amounts of metal ores or mineral fertilisers, fuels must be present to provide energy for smelting and for driving machines.

Early civilised Man's impact on physical resources was very small. This continued to be the case until the need for raw materials rocketed during the rise in technological expertise of the Industrial Revolution. Since then Man has used the Earth's physical resources far faster than geological processes can regenerate them. Indeed, many elements are now being extracted faster than they are being transported by erosion and suspension in flowing water. For instance, 25 million tons of iron are transported by the Earth's rivers in a year, while 319 million tons are mined. Similarly, the natural movement of 375 thousand tons of copper is matched by a Man-induced rate of 4.5 million tons per year.

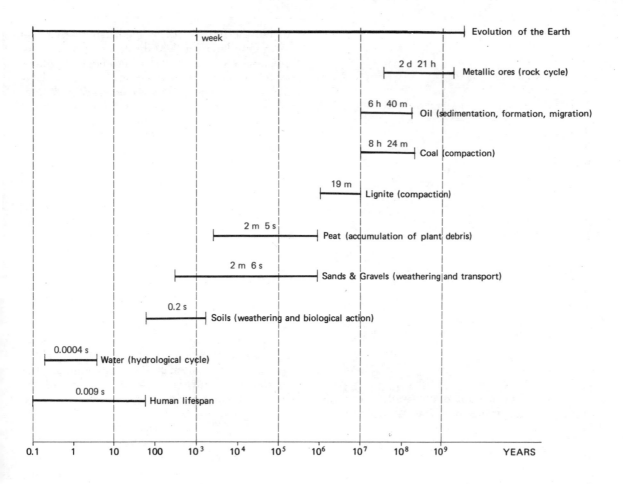

27. 1 Comparison of times of formation of various physical resources. Time is shown on a logarithmic scale; to facilitate comparisons, durations are also expressed (in colour) on a scale which compresses the evolution of the Earth into one week.

There are two basic implications of these excesses. The first is obvious—exploitation causes an irreversible drain on naturally occurring raw materials. Secondly, potentially hazardous materials are being brought into contact with the environment by extraction at rates that exceed those that occur in a *natural* process such as weathering. It is a law of ecology that plant and animal life exists in equilibrium with naturally occurring toxic materials in their environment. Artificial increases in the amounts of these materials break this equilibrium.

DEPLETION OF NON-RENEWABLE RESOURCES

Before any statements can be made about the drain imposed on non-renewable physical resources by their exploitation and use, the amount available to draw on has to be known. This depends on the recognition of something as a resource, and knowledge about its geological occurrence. There is a fundamental difference between *resources* and *reserves* imposed by two other factors: economics and technology. Reserves are quantities which can be extracted under the economic climate and technical knowledge that exist at any given time. Because of the incompleteness of geological surveys, different levels of reliability exist for reserves. Those which are being worked and about which much is known quantitatively are *measured* reserves. In cases where sufficient geological evidence encourages exploitation ventures, the postulated amounts of resources are termed *indicated* reserves. In some cases, because of geological similarities

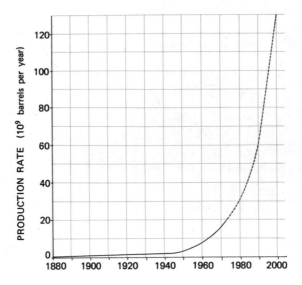

27. 2 World production of crude oil; past trends and exponential projection to 2000 A.D.

between producing and unexploited areas, knowledge about the modes of existence of particular physical resources and assumptions of various kinds, it is possible to estimate *inferred* reserves.

In addition to the reserves recoverable under current economic climates and with present machinery, the same materials are in many cases found in concentrations which are now uneconomic to extract or which cannot be exploited with existing tools. Depending on the necessary increases in extractive efficiency or financial returns for extraction, such occurrences of resources may be *marginal* or *submarginal*. However it is possible that some undiscovered deposits will be found and that these will eventually, after discovery, exploration and working give a similar hierarchy of measured, indicated and inferred reserves.

Adding up all these estimates of various levels of reliability for a physical resource, based upon geological, economic and technical assumptions and considerations, a figure can be arrived at for *potential reserves* (both known and undiscovered)

within the framework of predictable changes in method and market. At the moment estimates of potential reserves are more susceptible to philosophy of calculation than constrained by available data, so that there are always optimistic and pessimistic views of potential reserves. Taking oil as an example, an optimist suggests 6×10^{12} barrels (1 barrel $= 45$ gallons) for the world's potential reserves, a consistent pessimist favours 1.4×10^{12} barrels! The basic difference between the two estimates is the weight given to yet undiscovered oil.

To examine the future outlook for particular physical resources, one must know or be able to predict several things: these include past trends of exploitation, the cumulative amounts extracted in the past and models for the future features of the trends. If annual extraction is plotted against time it is possible to continue past trends into the future. The amount extracted over a particular time period is given by the area beneath the model curve between the time limits. Being able to calculate the progressively increasing amounts used enables the drain on reserves to be graphed by subtraction from the original amount of the particular physical resource, to give a curve expressing the change in potential reserves with time. In global terms the trend in exploitation has, up to now, been one of exponential increase with time for *every* type of physical resource. The increase has been due to a variety of reasons, the most important being exponential growth in world population and the staggering increase in per capita consumption of manufactured goods in the developed nations.

One important physical resource whose use exhibits the property of exponential growth particularly vividly is crude petroleum, for which a reasonable estimate of potential reserves is 2.1×10^{12} barrels. Since its recognition as a resource in the late nineteenth century, petroleum consumption has increased by 6.9 per cent per year in a remarkably regular fashion, so that consumption doubles roughly every ten years; the consumption in 1968 was 1.4×10^{10} barrels. The usage curve for petroleum is shown in Figure 27. 2 in its actual form up to 1968 and extrapolated as an exponential function beyond that, so that by 2 000 A.D. annual consumption could be 1.3×10^{11} barrels. From the area beneath the curve, in the century or so up to 1968, 2.2×10^{11} barrels had been used. The remainder of the potential reserves will be used in the next thirty years, as Figure 27. 3 shows. Thus the lifetime of what are now potential reserves of petroleum is only thirty years if the present exponential growth in consumption continues.

Knowing the rates of growth in consumption and estimates of the potential reserves of other physical resources one can draw a chart comparing their lifetimes with continued exponential growth (Fig. 27. 4). This might at first seem unreasonable, but it is worth bearing in mind that the growth in demand shows no sign of relenting, and if the aspirations of developing nations to approach the per capita consumption of the developed is only partly realised, the growth rate will be tremendously accelerated.

In many cases, as Figure 27. 4 shows, the damage seems to have been done already. Copper, zinc, lead, tin, molybdenum and mercury could be exhausted before the end of the century. To assume that consumption will suddenly level off at current rates is pleading a special case. On the other hand, to think that exponential growth can be maintained indefinitely is wholly unreasonable, for it would be physically impossible to maintain it, even if a deliberate attempt was made. Furthermore, it is becoming increasingly difficult to discover new sources of raw materials. Again this is well illustrated by the oil industry, where an index of successful exploration is the number of oil strikes per million feet of exploratory drilling. Considering the great advances in technique, including the whole range of geophysical methods of exploration, the record of oil discovery has shown a declining success rate for some decades. In the 1930's 3×10^8 barrels were discovered for each million feet of drilling, in the 70's this had decreased to 3×10^7 barrels, a tenth of the returns for the same effort. A time is bound to come, if demand continues to grow exponentially, when it will be impossible to supply fully the demand for oil no matter how intensive exploration is. The most obvious areas have been investigated, and little potentially rich ground remains. The result can easily be envisaged as a worsening situation in which the rate of supply grows more and more slowly until it goes over a 'hump' and begins to decline as difficulties of discovery and extraction increase. Such a situation has been suggested by M. K. Hubbert of the U.S. Geological Survey to hold for all non-renewable resources where the total available reserves are contained beneath a bell-shaped curve reflecting the history of development from its initiation through a period of exponential growth into a period of great difficulty of maintaining supplies and finally

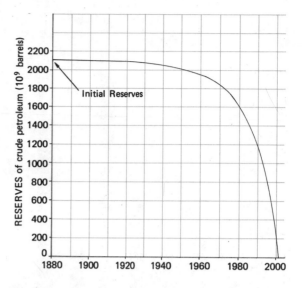

27. 3 Decline in world reserves of crude petroleum, assuming continued exponential growth of consumption.

one of exponential decline in supply (Fig. 27. 5). Using this more realistic model, and the same potential reserves as in Figure 27. 3, the lifetime of petroleum as a resource has 130 years to run, 80 per cent of the available petroleum being used in a 64-year golden age for the oil industry (Fig. 27. 5).

The real situation is that for all physical resources, the most obvious, the richest and the most easily exploited sources have been found and used. What remains of the original potential reserves are the now marginal, submarginal and poor, together with possibly a few rich deposits, as yet undiscovered, that will be increasingly more difficult to find and process. This situation is well illustrated by copper mining, where ever lower concentrations have to be worked to maintain supplies (Fig. 27. 6), yet it is more and more difficult to find even these meagre sources.

Obviously, more reserves of all physical resources will be found; they are being found all the time, and concepts of what constitutes a resource will change with technological advances. The

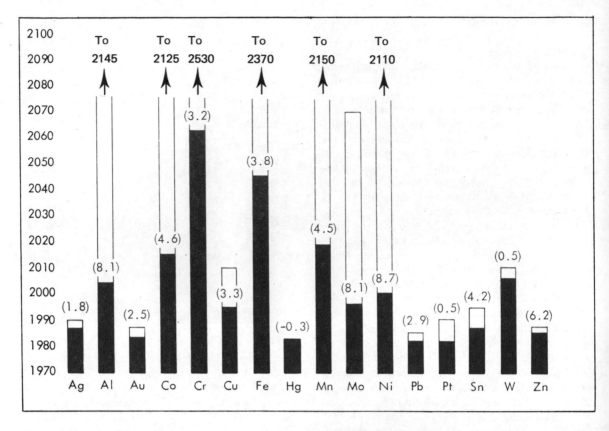

27. 4 Lifetimes of reserves, assuming current usage rate (coloured outline), and assuming continued exponential growth in consumption (solid colour). Figures in parentheses indicate growth each year as percentage of previous year.

question is, will these new concepts and discoveries markedly alter the situation if demands continue to grow exponentially? If production continues to keep up with demand and is exponential the answer is definitely 'No'. Even if reserves were doubled overnight, they would be gobbled up in a few doubling periods of the exponential trend. Large increases in reserves do not have a commensurate effect on the life histories of resources. By using the same model as in Figure 27. 5, an increase of 500 thousand million barrels of petroleum extends the lifetime of the resource only by 20 years or so.

It has been estimated that by the time they are fully operational the new Alaskan oil fields will supply the U.S.A.'s demands for a few years at the most. The oil in Britain's backyard—beneath the North Sea and possibly the Irish Sea—should be viewed in a realistic fashion. There is nowhere near as much there as in Alaska, and it is

proving tremendously difficult to find and extract it. Even if reserves of mercury were quadrupled, there would be no more to extract by this century's end.

This prognosis is based on past trends, which were influenced in the main by the consuming power of the *developed* nations. A few simple calculations show that to raise the standards and level of consumption of 3 400 million non-Americans to the level of the North American economy would place an insupportable burden on even the most abundant volumes of hidden mineral wealth.

What hope is there of maintaining the supplies needed for equalising standards on a world scale? The answer, it has been suggested, lies in new technologies. One approach is based on the notion that completely new methods of extraction will be able to tap vast tonnages of elements held at very low concentrations in common

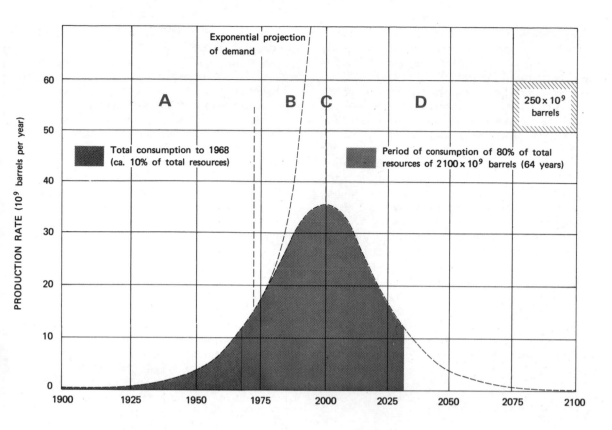

27. 5   Complete cycle of world crude-oil production.   A: exponentially growing demands are supplied. B: discovery and extraction become increasingly difficult; production is unable to meet demands. C: maximum attainable production rate.   D: declining production, ultimately decaying exponentially.

rocks such as granite and dissolved in the oceans. It goes without saying that such advances will have to arrive soon if they are to forestall world-wide shortages.

Such a technological solution depends on two fundamental assumptions.   The first is that nuclear power will provide a limitless source of cheap energy to drive the new extracting machines. But energy is already very cheap and amounts to only 4.6 per cent of industrial expenditure.   The second assumption is that technology progresses in jumps.   So it must, to move from exploiting what are now regarded as resources to what are concentrations several orders of magnitude lower in common rock and sea water.   Technological innovations have never arisen in such a fashion, and technology has always advanced on a front of gradually increasing efficiency and sophistication.

Several seemingly more reasonable ways of meeting the projected shortages have been sug-

gested.   One is that waste material, such as the tin layer on cans and the metals in wrecked cars can be refined again and put to new uses — that it can be *recycled*.   Another suggestion is that commonly occurring substitutes can be found for the rapidly dwindling *rare* resources, and recent rapid advances in ceramic and plastic technology might provide the necessary synthetic materials.

Recycling is the ideal in a *static* and stable economy; the pool of resources has only to make good the losses due to wear and tear.   Even then, it has its limitations; fuels cannot be recycled, they are burnt and that is that.   All that can be done is to burn them more efficiently.   In an economy of *growth*, recycling can only contribute a rotating stock of processed resources which continually has to be supplemented to supply growing demands, thereby only marginally easing the problem of depletion.   Furthermore, there are immense technological problems in devising methods of recycling. Refining waste metals, for instance, is not like smelting from an ore, which usually contains only one metallic element. Waste is usually a jumble of different metals and alloys, together with glass, plastic, rubber and paper. Everything has to be meticulously sorted—just melting it down will not do. Until now, recycling has always produced a material inferior to the bright 'new' article. For several resources we can envisage natural substitutes. For instance, copper can be replaced by aluminium for many purposes. But what of elements with unique properties, like the liquid metal mercury or the gas helium?

The most obvious synthetics are plastics, but even when they can serve as substitutes, can they in fact be considered providing a solution at all? They are usually manufactured from petroleum, but as this serves as a major fuel and in the present climate of growing use has been shown to be in danger of running out before long, there is not a great chance of much petroleum being used for making plastic substitutes.

ENVIRONMENTAL IMPACT

So far we have treated physical resources as technological building blocks, and have seen that the exploitation of one resource generally sets up significant demands on others. When the social and related environmental implications of their exploitation are considered, we find that the effects are transmitted to a highly complicated, delicate and *global* ecological web through a network of inter-related functions. Any manipulation of physical resources is bound to have a greater or lesser, good or bad effect at some points in this structure.

All newly evolved organisms modify previous natural cycles to some extent, but in the main the cycles adjust themselves to incorporate the newcomer harmoniously; that is, the ecological balances are restored or adjusted by actions or reactions within the particular eco-system. In most cases the changes are small or imposed gradually, so that they can be accommodated without marked alteration of the system.

Broadly, this remains true for Man in the case of primitive peoples, but highly evolved or populous civilisations have made breaks in natural regimes that are either too large or occurred too rapidly to be smoothed over. Civilised Man has outpaced or overstrained the restorative forces, and so scars have formed.

Environmental implications stemming from physical resource exploitation are of two types: *direct* or *indirect*. The direct type is inherent in the initial extraction of the commodity, and would include holes in the ground. A whole assortment of indirect effects ranges through many different degrees of removal from the source by processing, use and disposal as a used product or waste. It includes air pollution from burning of coal without any processing, air pollution by incomplete combustion of refined petroleum spiked with tetra-ethyl lead as an anti-knock agent, water pollution from metal-refining plants, and solid waste in the form of manufactured goods such as wrecked cars and plastic bottles. The more steps there are in the progress from raw material to ultimate potential waste, the more the chances of environmental impact. As increasingly sophisticated uses are found for a particular raw material, such as crude oil, the varied suite of processing steps expands, and the routes by which the environment can be affected multiply.

As well as breaking some natural cycles, exploitation of physical resources also modifies others, mainly affecting their rates. Pollution by this means can be defined as the exposure of naturally occurring materials, by excavation or use, so that they can enter natural cycles in larger volumes and at greater rates than would be the case if they were released by geological processes. (This leaves aside for the moment pollution by man-made and sometimes persistent compounds which do not occur naturally at all.) Two types of activity are discussed here, one with mainly direct implications, the second with more indirect implications.

Mining

Mining, the primary extraction of materials from the Earth, is the most obtrusive of man's modification of the environment, and it contributes a large volume of polluting materials. It is characterised by three main features, the physical removal of physical resources either from surface pits or underground mines, initial processing and dressing of the ore, and disposal of the unwanted proportion of the material extracted. The most common feature associated with mining is the production of derelict land; land lost as an amenity and a resource.

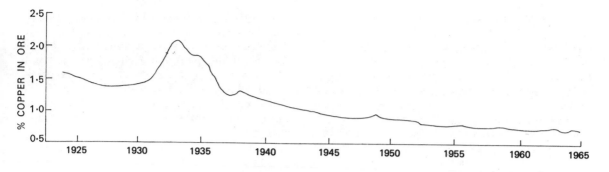

27.6 Decline in grade of mined copper ores since 1925.

In the first stages of surface mining, vegetation and topsoil are removed, haulage roads constructed and processing plant erected. Soon afterwards the holes appear, with corresponding large piles of unwanted rock. If the ore is milled at site, the fine-grained waste is settled out in large ponds, which by their very nature grow and occupy large acreages of valley bottom and flat land. Because of hole and heap construction, both natural and artificial slopes can become unstable and slumping is a threat. Subsurface mining, unless involving expensive underground repacking of waste, produces its share of spoil heaps. Furthermore, it upsets the surface stability, giving rise to subsidence. This changes directions of drainage and creates areas of flooding, buckles roads and causes building instability and deformation.

The polluting aspect of mining is manifold. In most cases the waste on spoil heaps contains certain minerals genetically associated with the geological history of the sought resource. With most metal ores and coal, a common, though almost valueless, mineral suite consists of the iron sulphides and arsenical sulphides. Exposure to air and rain water causes them to be oxidised to complex hydrated iron oxides or ochres and sulphuric acid, and to release arsenic to percolating rain water. By themselves, these three products can seriously affect downslope water-courses and vegetation, but when sulphuric acid dissolves other elements from the spoil, the impact is much greater. The on-site upgrading of ore by crushing and separating generates much dust, which by itself is a known cause of respiratory disease; if it contains the smallest volume of asbestos-like minerals it is exceptionally toxic. Though attempts are frequently made to settle out fine-

grained material from crushing in ponds, too often a load of fine suspended particles finds its way with ochres from spoil heaps into streams and rivers. The contents of some spoil heaps, notably those with high sulphide, coal or oil shale levels, are prone to spontaneous combustion in air. This produces uncontrollable smouldering with consequent air pollution by sulphur dioxide and smoke. Many coal mines, and a few mines of sulphide ores, have spontaneous smouldering underground with consequent pollution of the air through the mine's ventilating system.

Certainly in areas of high population density and high real and economic values of land, such as Britain and the east and west coasts of North America, popular demand has evoked much effort and legislation aimed at alleviating land dereliction and pollution. But although most of Britain's derelict land is a heritage from the past, some land is still being despoiled. The situation is somewhat different in exploited wilderness areas, where little impact falls on people. Sadly, it is the wild places where real ecological harm is caused by the destruction of habitats and modifying environments. The unspoilt areas of our planet, largely the little explored yet vast areas of Precambrian continental shields (Chapter 22) in the Americas, Africa and Australia, are increasingly attracting interest because of their prospects of undiscovered mineral wealth. Even those areas of Europe and North America that have been supposedly set aside as national parks and wilderness areas are beginning to bear the brunt of government-sponsored and approved mineral exploration and exploitation. Is it possible to reconcile the aims of nature conservation with the widespread despoliation of a frantic hunt for new minerals? Can mines be operated without affecting the areas around them?

Large-scale water projects

Water is a biological necessity, though less than a gallon per day is required for a human's survival. In a developed industrial nation, striving to manufacture goods and supply food for its growing population, and to provide water-consuming services (such as sewage disposal) to a standard of some efficiency, the water-cost per person rises to many orders of magnitude above that for mere survival. The result has been acute shortages in conurbations, even where there is high rainfall and abundant surface flow. The only means of alleviating this situation has so far been regional transfer. Projects of this type have steadily grown in size from impounding reservoirs and piping water to the area of need, as in the uplands of Britain, to recently conceived plans for reversals of flow on a continental scale. All involve modifying the natural course of surface water flow by means of dams, canals, pipelines and pumps.

The agricultural industries of the American Mid-West and the Russian steppes, and the manufacturing industries of the Atlantic states of America are short of water—so much so that demand will exceed supply in this decade. The unfortunate fact is that the greatest available untapped flows of water in North America and Russia are away from the areas of need—the great rivers of Canada and Russia flow northwards into the Arctic Ocean. Two unrelated but similar projects developed in recent years aim at diverting the flow of water from the Arctic wastes of the north to the thirsty south. These are respectively the plans of the North American Water and Power Alliance (NAWAPA) and the Davidoff Plan. They both necessitate impounding reservoirs to a level where the flow can be reversed through artificial channels, or by just diverting rivers into new channels cut through low continental divides. Apart from direct local effects of this sort of manipulation on the balances between landscape, ecology and water, more widespread modifications can easily be imagined. The most important of these is change in climatic patterns. At present the flow is such that the north is warmed by the river water derived from the south. This is reflected by northward extensions of the tree line and permafrost-free ground in broad swathes related to river courses. Reversal would have the opposite effect on the south, cold water flowing from the north having a cooling effect, whilst the climate of the north would get colder and drier. Vast lakes can create their own climate with respect to precipitation and winds because of evaporation. The Arctic Ocean itself has a profound effect on the climate of the Northern Hemisphere, and it is very sensitive to change because it is relatively closed off. To-day it is warmed by the rivers. What would happen if it became colder? What effect would large-scale transfers of river flow have on ocean currents, especially in the closed Arctic Ocean basin? It is slight physical changes such as these that many climatologists hold responsible for glacial and interglacial periods of the last $1\frac{1}{2}$ million years.

SOCIAL IMPLICATIONS

Physical resources impinge upon society in a wide and varied fashion, with long-term and short-term implications.

In the short term, exploitation of physical resources is mainly concerned with maintaining the status quo at the very minimum, whilst growth is expected to make some contribution to a rise in social standards. Thus every effort is made to maintain employment in extraction and processing of raw materials, and depressed areas are favoured in the setting up of new resource-based industries. Upon this primary source of employment and economic well-being of an area depend innumerable other industries such as transport, retailing and services. A decline in employment in resource-based industry causes a concomitant decline in dependent fields.

The use of physical resources in services, such as water supply, power supply and the provision of building materials, has deep social consequence. They are relied upon for a great diversity of uses connected with living standards, and the demand for them is bound to increase under the two-fold pressures of rising population and rising 'levels of expectation'—the standards to which people feel they can reasonably aspire. But, as more resources are used, as agricultural land is covered, as waste accumulates and pollution mounts, so the danger increases of environmental backlash. A vicious circle will be set up sooner or later when the environment has deteriorated due to a use of physical resources that, in the short term, seemed acceptable, but was in fact excessive.

The long-term view must concern itself with the social implications of a world which, quite apart from pollution and environmental changes, finds itself acutely short of raw materials. As there are wide disparities in the distribution of physical resources, some nation always holds a trump hand in one field or another. Wars have been fought in the past over these inequalities. To-day a network of international agreements attempts to regulate the situation. But, if social standards are threatened in a developed nation, because it cannot obtain the external supplies to feed its industrial demands, or in a developing country, because its need for, say, pesticides conflicts with the global interest, then pressures build up and ideological polarisations are intensified to a point that could cause world-wide conflict.

Energy resources are the key to the exploitation of all others. Without power, minerals cannot be extracted or processed to the extent that is necessary for the survival of a developed nation. Ever increasing amounts of energy are required to work lower-grade mineral deposits. Of the varied types of energy resource, only the fossil fuels, coal and oil, and nuclear fuels can currently meet the increasing world demands. Oil is already in short supply and could run out in a century. Coal pollutes. Nuclear power has not yet realised its potential, and if it continues to be based on 'burner' type reactors using ^{235}U exclusively, it will have a short life itself, as that isotope represents only a few per cent of naturally occurring uranium. A society with high energy consumption must depend for its future survival *exclusively* on the development of 'breeder' reactors, which use ^{238}U and ^{232}Th and manufacture their own artificial fuel to extend the life of fission power. But these isotopes can eventually run out too. So, in the more distant future, all depends on the essentially limitless source of nuclear *fusion* using deuterium from sea water—a method presently impossible to develop for common use. To make these two fundamental developments entails an increasingly well-instrumented and energy-consuming technology. If current energy supplies decline, the chances of reaching these goals diminish and a stage could be reached when they are unattainable, because technology itself is starved of energy.

So, should this energy/technology gap arrive, the world with its increasing population and diminishing energy supply would have less energy per person, and social standards *must* decay. This eventually would leave the human race without recourse to any source of power other than water power, solar radiation and its storage in wood by the carbon cycle, the wind and geothermal heat flow. These are basically the supplies that were available to early Man, and they are not particularly large. Such a society, without *extra* energy and with the only workable source of other raw materials being scrap, could not maintain the heights of technology and living standard enjoyed by the developed nations to-day. This dismal prospect lies in the very near future, and graphically explains why many thinking men are becoming ever more concerned with the planet Earth, its finite stock of resources and sensitive environment. Such new attitudes, as recorded in *Blueprint for Survival*, *The Limits to Growth* and *Resources and Man*, for all their well-publicised failings, are the writing on the wall. If they are not heeded, civilisation as we know it will inevitably deteriorate—possibly beyond the point of no return.

FURTHER READING:

EDWARD GOLDSMITH, ROBERT ALLEN, MICHAEL ALLABY, JOHN DAVOLL, SAM LAWRENCE 1972 (January). A Blueprint for Survival. *The Ecologist.*

DENNIS L. MEADOWS 1972. The Limits to Growth. *Earth Island Press.*

PRESTON CLOUD Jr. (Ed.) 1969. Resources and Man. *W. H. Freeman and Co.*

GLOSSARY

A short glossary such as this can be neither systematic nor complete. The entries listed below should be taken as rough guides to meaning, and not as formal definitions. The explanation of many additional terms can be found by means of the Index.

Abiogenic   not derived from living organisms.

Algae   large group of simple photosynthetic plants with unicellular organs of reproduction. Widely distributed in aquatic or damp surroundings.

Alluvial   pertaining to (the action of) rivers and floods.

Amphibole   a group of common rock-forming silicates, mainly of magnesium, iron, calcium and sodium, with added -OH groups and a prominent cleavage of 54° or 124°.

Anaerobic   not requiring the presence of free atmospheric or dissolved oxygen.

Andesite   an extrusive igneous rock, the surface equivalent of diorite.

Annelid   phylum of segmented worms including earthworms and leeches, with well-defined nervous and blood systems.

Anorthosite   plutonic rock, usually Precambrian, composed almost entirely of plagioclase.

Anticline   a fold that is (or was when it formed) convex upwards.

Archaeocyathea   group of Cambrian fossils resembling corals, but also showing affinity to sponges.

Arenite   consolidated rock with a sand-like texture.

Argillite   fine-grained rock derived from siltstone, shale or claystone, but considerably more hardened.

Arthropoda   the largest animal phylum, including insects, spiders, centipedes, crabs, etc., with a hard, articulated exoskeleton.

Asthenosphere   a weak, plastic layer of the globe beneath the Earth's crust, a few hundred kilometres in thickness.

Banded   consisting of narrow layers differing in colour, texture or composition.

Basalt   fine-grained, dark igneous rock; more specifically, rocks of this type containing principally calcic plagioclase and pyroxene, and variable amounts of olivine.

Batholith   large, intrusive igneous mass with steep sides extending to considerable, indeterminate depth.

Bathymetry   measurement of (especially oceanic) depths and contours; the topography of the sea-beds.

Bedding   stratification of sedimentary material parallel to the original surface of deposition (*true* bedding) or inclined to it (*false* or *current* bedding).

Bioherm   bed or mound consisting mainly or entirely of the fossil remains of colonial or gregarious marine organisms.

Breccia   a rock consisting of irregular, angular coarse fragments. The fragmentation may be due to friction, faulting or eruption.

Calcareous   containing calcium carbonate; usually applied to sedimentary rocks such as chalk or limestone.

Carbonaceous   predominantly made up of carbon and organic compounds.

Carbonatite   igneous carbonate rock closely associated with igneous intrusion; it is largely or wholly free of silicates.

Chert   a hard, flinty siliceous rock, often arising through chemical precipitation.

Chondrite   stony meteorite with chondrules, usually of olivine or pyroxene, embedded in a matrix.

Chondrule   small, rounded inclusion.

Clastic   a sedimentary rock largely made up of fragmental matter, e.g. sandstones and conglomerates.

Cleavage   the tendency of crystals or crystalline minerals to split easily along certain planes.

Coelenterata   phylum of aquatic animals including hydra, jelly-fish, sea-anemones, corals. They are among the most primitive Metazoans; the gut (*coelenteron*) has one opening only.

Cordierite $Mg_2Al_4Si_5O_{18}$, an orthorhombic brittle, dichroic mineral found in metamorphic rocks.

Craton   large, stable part of the Earth's crust relatively undisturbed by mountain-forming processes.

Cyanophyta   blue-green algae with procaryotic cell structure.

Diagenesis   gradual physical or chemical changes taking place in sediments during or after deposition.

Diatreme   volcanic vent or pipe blown by the explosive release of magmatic gases.

Dichroism   the property of displaying two colours.

Differentiation   separation into constituents according to some difference of mechanical or chemical property.

Diorite   plutonic rock, chiefly sodic plagioclase feldspar with hornblende, biotite and/or pyroxene.

Dipole   two equal and opposite, relatively closely spaced electric charges or magnetic poles, or the electric or magnetic fields equivalent to them.

Dolerite   a gabbroid or basaltic rock type occurring in sills or dykes.

Dyke   a large slab-like mass of igneous rock which cuts across neighbouring structures, usually steeply inclined.

Echinodermata   phylum of marine animals with radial symmetry and calcareous outer parts, including sea-urchins, starfish, sea-cucumber, etc.

Enstatite   an orthorhombic pyroxene, general composition $(Mg, Fe) O. SiO_2$.

Era   the largest type of division of geological time, of the order of 10^8 years or more, of which the last three are Palaeozoic, Mesozoic and Cenozoic. Eras are subdivided into Periods.

Eucaryote   organism with chromosomes embedded in a well-formed cell nucleus that is surrounded by a membrane.

Evaporite   the remains of a solution after most of the solvent (usually water) has evaporated.

Exponential Growth (or Decay)   continuous increase (or decrease) by a constant factor (greater or smaller than 1) for each unit of time. Exponential growth sooner or later becomes explosive, and then rapidly tends to infinity; exponential decay leads to ever-slower diminution but never attains 0.

Extrusive thrust out (at the Earth's surface). Extrusive rocks are usually fine-grained due to their rapid cooling.

Feldspar the most abundant group of rock-forming minerals in the Earth's crust, of which it provides nearly two-thirds. The chief constituents are alumino-silicates of the alkali metals.

Fluidization the effect whereby gas passing through powdery or finely-divided solid matter causes it to become as mobile as a liquid.

Foliation division into a laminated, leaf-like structure.

Foraminifera mainly marine protozoa with many-chambered calcareous or siliceous shells, mostly the size of a pin-head.

Fractionation separation of one or more constituents from a mixture by physical means or processes.

Gabbro any coarse-grained, dark plutonic rock, especially plagioclase feldspar with pyroxene.

Garnet group of minerals consisting mainly of double silicates of calcium or aluminium with other metals.

Gneiss metamorphic rock derived from igneous or sedimentary material, coarsely laminated and largely re-crystallized.

Granite a common, hard, massive plutonic rock composed essentially of alkalic feldspars and quartz. Plagioclase, hornblende, biotite and muscovite are frequently present. More generally, any light-coloured. coarse-grained igneous rock.

Greywacke dark-coloured, hard conglomerate of coarse sandstone and grit cemented together.

Guyot sea-mount with top that has been flattened and smoothed by erosion.

Haematite Fe_2O_3, the chief ore of iron.

Hornblende a complex calcium-magnesium-iron and sodium silicate with prismatic monoclinic crystals; an amphibole of several igneous and metamorphic rocks.

Hornfels a fine-grained, heavy and often speckled rock resulting from contact metamorphism in slates by granitic intrusions.

Hydrolysis mutual decomposition of water and a compound.

Igneous formed by the solidification of molten magma.

Ignimbrite a fine-grained rhyolitic rock resulting from the compaction of volcanic fragments.

Intrusion the penetration into older rock of a mass of igneous rock by means of magma flow or plastic deformation.

Iso- Greek prefix denoting 'equal'.

Isometric the crystal system with three equal, mutually perpendicular axes.

Isotope one of two or more forms of an element having the same number of protons but different numbers of neutrons in the atomic nucleus, and hence differing in atomic weight but being very similar chemically.

Joint a fracture in rock, often across stratifications or bedding planes, along which little or no movement has taken place.

Kimberlite a South African type of mica peridotite, sometimes bearing diamonds.

Limestone sedimentary rock composed mostly of calcium carbonate, often derived from consolidated shell fragments.

Lith-, Litho- Greek prefix, 'relating to stone'.

Lithification the gradual conversion of sediments into hard, rock-like substances.

Lithosphere the outer shell of the Earth (excluding the hydrosphere and atmosphere) formed by its crust.

Mafic relating to the dark-coloured minerals of igneous and metamorphic rocks.

Magma the molten mixture of complex silicates, oxides and volatile substances which forms igneous rocks on cooling.

Matrix a groundmass in which other substances are embedded.

Meiosis a sequence of two cell divisions involving only one division of the genes and chromosomes, so that each of the resulting cells has half the original complement.

Meso- Greek prefix meaning 'in the middle' or 'intermediate'.

Mesosphere thick strong layer within the Earth between the asthenosphere and the central core. (Also the layer of the Earth's atmosphere at 400–1000 km above the surface.)

Meta- Greek prefix relating to variation, change or transformation.

Metallogenic relating to the formation of ore deposits.

Metamorphism any change in the characteristics of consolidated rocks due to natural stresses such as heat or pressure.

Metaphyte one of a large group of plants — including most land plants — having multi-cellular sex organs.

Metazoon any animal other than a sponge whose body consists of many cells, as opposed to the non-cellular protozoa.

Mica a group of silicates with very finely spaced cleavage planes that allow easy splitting into thin transparent sheets.

Migmatite rock in which granitic matter has been injected between the foliation of schist.

Mineral a natural, inorganic substance of definite chemical composition and molecular structure; the term is sometimes also applied to substances of organic origin, e.g. mineral oil.

Mitosis the process whereby a cell nucleus divides into two nuclei, each with a complete set of chromosomes, during normal cell division.

Monoclinic System the crystal system with three axes of unequal length, precisely two of them at right angles to each other.

Morphology study of the form of animals and plants.

Nemertea phylum of marine worms with a flattened body, similar to flatworms but having distinct mouth and anus.

Nucleic Acids long, chain-like molecules which, in the various combinations of constituent groups, embody the genetic code (DNA) and assist with its transmission (RNA).

Olivine a mineral series of solid solutions of iron silicate with magnesium silicate in varying proportions. An important constituent of mafic and ultramafic rocks.

Oolith individual mineral spherule or ellipsoid, often formed by accretion round a nucleus and with radial or concentric structure.

Orogen belt of rocks deformed during mountain-building.

Orogeny, Orogenesis mountain building, especially when caused by lateral compression of a belt of the Earth's crust; one of the periods when this process was prominent in a region.

Ortho- Greek prefix meaning 'straight', 'correct' or 'at right angles'.

Orthorhombic System the crystal system with three axes of unequal length all at right angles to each other.

Palaeo- Greek prefix denoting 'ancient'.

Pennatulid a colonial form of coelenterate with polyps arranged in a feather-like pattern.

Period a subdivision of a geological era, e.g. the Tertiary and Quaternary of the Cenozoic era; the time interval during which a regular cyclic change (e.g. a wave motion) completes one cycle.

Plagioclase a very common group of rock-forming feldspar minerals with triclinic crystals, ranging from albite (almost wholly $NaAlSi_3O_8$) to anorthite (almost wholly $CaAl_2Si_2O_8$).

Plug the solidified mass of igneous rock filling a volcanic vent or pipe.

Plutonic Rock coarse-grained igneous rock which has cooled slowly at depth and in large masses.

Procaryote organism with the genetic material of DNA filaments not enclosed within a cell nucleus.

Pyro- Greek prefix denoting 'fire'.

Pyroclast rock fragment that was forcibly blown from a volcanic vent.

Pyroxene group of minerals consisting chiefly of the silicates of calcium, magnesium and iron; a common constituent of igneous rocks, with a cleavage of 87° and 93°.

Quartz natural crystalline silica, SiO_2. The clear varieties include amethyst and rock crystal.

Radioactivity the spontaneous emission from the nuclei of the atoms of certain elements of charged particles and/or strong radiation. The remaining nucleus is generally an isotope of another element. The half-life for a given radioactive change is the time during which half of the unchanged atoms of an element or isotope decay by radioactive emission.

Rhyolite granite so fine-grained that individual constituents cannot be seen with the naked eye.

Richter Scale a measure, based on the body wave magnitude, for comparing the total energy released in earthquakes.

Rift the intersection of a fault plane with the surface.

Rift Valley depressions resulting from the subsidence of a strip between two parallel rifts.

Sandstone rock formed from sand or quartz particles cemented together with clay, iron oxide, calcium carbonate or other binding material.

Schist a group of foliated rocks which split easily into thin laminae parallel to the layers of constituent platy minerals.

Sedimentary formed by the accumulated deposition of water-borne fragments or organic debris, or by precipitation from solution.

Shale fissile or laminated claystone or siltstone, hardened but softer and lighter than slate.

Siliceous pertaining to silica, SiO_2; rich in quartz.

Sill sheet-like minor intrusion of igneous rock lying roughly parallel to the stratification of surrounding rock.

Spicule small, hard, pointed structure, common in sponges.

Stromatolite laminated calcareous aggregation of fossil remains of algal-like organisms.

Syncline a fold in rocks that is (or was when it formed) convex downwards.

Tectonic relating to the results of deformation in the Earth's crust.

Trench long, narrow depression of ocean floor with relatively steep sides.

Triclinic the crystal system with three unequal axes, none of which are at right angles to each other. It is the system of least symmetry.

Ultrabasic relating to heavy, dark igneous rocks or stony meteorites containing less than 50% silica.

Xeno- Greek prefix denoting 'strange', 'alien'.

Xenolith rock fragment foreign to the body of igneous rock in which it occurs.

A **bold** figure indicates the number of a chapter which deals largely with the corresponding subject. An *italic* figure indicates the beginning of a titled section (either UPPER CASE or *italic* heading) which deals in detail with the corresponding subject. G indicates a term in the glossary.

distribution, 301
Iceland, *304*, 308
island arcs, *305*
islands, *302*
magma composition, *303*
mountain chains, *305*
ocean floors, *301*
ocean ridges, *301*
plateau basalts, *308*
rift valleys, *309*
sub-volcanic activity, *310*

Water
    climatic changes, 370
    large scale projects, *370*

Weathering
    chemical, *27*
    mechanical, *27*
Wiechert–Gutenberg Discontinuity, 61
World–Wide Standard(ized) Seismograph
    Network (WWSSN), 246, 337

Xeno-, (prefix), G
Xenoliths, 59, G

Yoredale cycles, *188*

Zinc, 315, 317, 319, 324
Zircon, 14, 19, 50